T0192879

Applied Statistical Inference
with MINITAB®

Applied Statistical Inference with MINITAB®

Second Edition

Sally A. Lesik
Central Connecticut State University

CRC Press
Taylor & Francis Group
Boca Raton London New York

CRC Press is an imprint of the
Taylor & Francis Group, an **informa** business

CRC Press
Taylor & Francis Group
6000 Broken Sound Parkway NW, Suite 300
Boca Raton, FL 33487-2742

© 2019 by Taylor & Francis Group, LLC
CRC Press is an imprint of Taylor & Francis Group, an Informa business

No claim to original U.S. Government works

First issued in paperback 2021

International Standard Book Number-13: 978-1-4987-7998-2 (hbk)
International Standard Book Number-13: 978-0-3677-8057-9 (pbk)

This book contains information obtained from authentic and highly regarded sources. Reasonable efforts have been made to publish reliable data and information, but the author and publisher cannot assume responsibility for the validity of all materials or the consequences of their use. The authors and publishers have attempted to trace the copyright holders of all material reproduced in this publication and apologize to copyright holders if permission to publish in this form has not been obtained. If any copyright material has not been acknowledged, please write and let us know so we may rectify in any future reprint.

Except as permitted under U.S. Copyright Law, no part of this book may be reprinted, reproduced, transmitted, or utilized in any form by any electronic, mechanical, or other means, now known or hereafter invented, including photocopying, microfilming, and recording, or in any information storage or retrieval system, without written permission from the publishers.

For permission to photocopy or use material electronically from this work, please access www.copyright.com (http://www.copyright.com/) or contact the Copyright Clearance Center, Inc. (CCC), 222 Rosewood Drive, Danvers, MA 01923, 978-750-8400. CCC is a not-for-profit organization that provides licenses and registration for a variety of users. For organizations that have been granted a photocopy license by the CCC, a separate system of payment has been arranged.

Trademark Notice: Product or corporate names may be trademarks or registered trademarks, and are used only for identification and explanation without intent to infringe.

Visit the Taylor & Francis Web site at
http://www.taylorandfrancis.com

and the CRC Press Web site at
http://www.crcpress.com

To the memory of my mother, Irene Lesik, and in honor of my father, John Lesik

To DHK, more than anything, forever and ever, ... totally unrehearsed.

Contents

Preface..xiii
Acknowledgments xvii

1. Introduction 1
    1.1   What is Statistics?...1
    1.2   What This Book Is About...2
    1.3   Summary Tables and Graphical Displays 2
    1.4   Descriptive Representations of Data 3
    1.5   Inferential Statistics 4
    1.6   Populations 5
    1.7   Different Ways of Collecting Data..5
    1.8   Types of Variables 6
    1.9   Scales of Variables...7
    1.10 Types of Analyses 9
    1.11 Entering Data into Minitab...10
    1.12 Best Practices 11
    Exercises 12

2. Graphs and Charts 15
    2.1   Introduction 15
    2.2   Frequency Distributions and Histograms.......................................15
    2.3   Using Minitab to Create Histograms 17
    2.4   Stem-and-Leaf Plots...21
    2.5   Using Minitab to Create Stem-and-Leaf Plots 22
    2.6   Bar Charts 24
    2.7   Using Minitab to Create a Bar Chart..24
    2.8   Boxplots 27
    2.9   Using Minitab to Create Boxplots..31
    2.10 Scatterplots...32
    2.11 Using Minitab to Create Scatterplots 33
    2.12 Marginal Plots 33
    2.13 Using Minitab to Create Marginal Plots..35
    2.14 Matrix Plots..36
    2.15 Using Minitab to Create a Matrix Plot 38
    2.16 Best Practices 38
    Exercises 41
    Extending the Ideas 44

3. Descriptive Representations of Data and Random Variables.............47
    3.1   Introduction 47
    3.2   Descriptive Statistics..47
    3.3   Measures of Central Tendency 48
    3.4   Measures of Variability 52
    3.5   Using Minitab to Calculate Descriptive Statistics.........................55

3.6   More on Statistical Inference 56
3.7   Discrete Random Variables 58
3.8   Sampling Distributions 61
3.9   Continuous Random Variables 64
3.10   Standard Normal Distribution 65
3.11   Non-Standard Normal Distributions 69
3.12   Other Discrete and Continuous Probability Distributions 73
3.13   The Binomial Distribution 74
3.14   The Poisson Distribution 75
3.15   The *t*-Distribution 77
3.16   The Chi-Square Distribution 78
3.17   The *F*-Distribution 79
3.18   Using Minitab to Graph Probability Distributions 79
       Exercises 85

4. Statistical Inference for One Sample 93
4.1   Introduction 93
4.2   Confidence Intervals 93
4.3   Using Minitab to Calculate Confidence Intervals for a Population Mean 99
4.4   Hypothesis Testing: A One-Sample *t*-Test for a Population Mean 100
4.5   Using Minitab for a One-Sample *t*-Test 106
4.6   Power Analysis for a One-Sample *t*-Test 115
4.7   Using Minitab for a Power Analysis for a One-Sample *t*-Test 116
4.8   Confidence Intervals and Hypothesis Tests for One Proportion 120
4.9   Using Minitab for a One-Sample Proportion 124
4.10   Power Analysis for a One-Sample Proportion 127
4.11   Confidence Intervals and Hypothesis Tests for One-Sample Variance 129
4.12   Confidence Intervals for One-Sample Variance 130
4.13   Hypothesis Tests for One-Sample Variance 132
4.14   Using Minitab for One-Sample Variance 134
4.15   Power Analysis for One-Sample Variance 136
4.16   Confidence Intervals for One-Sample Count Data 140
4.17   Using Minitab to Calculate Confidence Intervals for a One-Sample
       Count Variable 142
4.18   Hypothesis Test for a One-Sample Count Variable 144
4.19   Using Minitab to Conduct a Hypothesis Test for a One-Sample Count
       Variable 146
4.20   Using Minitab for a Power Analysis for a One-Sample Poisson 147
4.21   A Note About One- and Two-Tailed Hypothesis Tests 149
       Exercises 151
       References 155

5. Statistical Inference for Two-Sample Data 157
5.1   Introduction 157
5.2   Confidence Interval for the Difference Between Two Means 157
5.3   Using Minitab to Calculate a Confidence Interval for the Difference
       Between Two Means 160
5.4   Hypothesis Tests for the Difference Between Two Means 162
5.5   Using Minitab to Test the Difference Between Two Means 166

5.6   Using Minitab to Create an Interval Plot..167
5.7   Using Minitab for a Power Analysis for a Two-Sample *t*-Test.........................170
5.8   Paired Confidence Interval and *t*-Test..172
5.9   Using Minitab for a Paired Confidence Interval and *t*-Test..............................176
5.10   Differences Between Two Proportions...178
5.11   Using Minitab for Two-Sample Proportion Confidence Intervals and
       Hypothesis Tests..182
5.12   Power Analysis for a Two-Sample Proportion..184
5.13   Confidence Intervals and Hypothesis Tests for Two Variances..........................184
5.14   Using Minitab for Testing Two Sample Variances...191
5.15   Power Analysis for Two-Sample Variances..193
5.16   Confidence Intervals and Hypothesis Tests for Two-Count Variables 195
5.17   Using Minitab for a Two-Sample Poisson...198
5.18   Power Analysis for a Two-Sample Poisson Rate..199
5.19   Best Practices...201
Exercises..203

6. Simple Linear Regression...213
6.1   Introduction...213
6.2   The Simple Linear Regression Model..214
6.3   Model Assumptions for Simple Linear Regression...220
6.4   Finding the Equation of the Line of Best Fit...221
6.5   Using Minitab for Simple Linear Regression...224
6.6   Standard Errors for Estimated Regression Parameters.................................227
6.7   Inferences about the Population Regression Parameters.................................227
6.8   Using Minitab to Test the Population Slope Parameter.................................230
6.9   Confidence Intervals for the Mean Response for a Specific Value of the
       Predictor Variable...232
6.10   Prediction Intervals for a Response for a Specific Value of the
       Predictor Variable...233
6.11   Using Minitab to Find Confidence and Prediction Intervals 235
Exercises..242

7. More on Simple Linear Regression..247
7.1   Introduction...247
7.2   The Coefficient of Determination...247
7.3   Using Minitab to Find the Coefficient of Determination.................................249
7.4   The Coefficient of Correlation...250
7.5   Correlation Inference..254
7.6   Using Minitab for Correlation Analysis..257
7.7   Assessing Linear Regression Model Assumptions...259
7.8   Using Minitab to Create Exploratory Plots of Residuals................................259
7.9   A Formal Test of the Normality Assumption...264
7.10   Using Minitab for the Ryan–Joiner Test...266
7.11   Assessing Outliers...268
7.12   Assessing Outliers: Leverage Values..269
7.13   Using Minitab to Calculate Leverage Values..269
7.14   Assessing Outliers: Standardized Residuals 272
7.15   Using Minitab to Calculate Standardized Residuals.....................................273

7.16   Assessing Outliers: Cook's Distances 274
7.17   Using Minitab to Find Cook's Distances 275
7.18   How to Deal with Outliers 276
Exercises 277
References 283

8. Multiple Regression Analysis 285
8.1   Introduction 285
8.2   Basics of Multiple Regression Analysis 285
8.3   Using Minitab to Create Matrix Plots 287
8.4   Using Minitab for Multiple Regression 289
8.5   The Coefficient of Determination for Multiple Regression 290
8.6   The Analysis of Variance Table 292
8.7   Testing Individual Population Regression Parameters 296
8.8   Using Minitab to Test Individual Regression Parameters 299
8.9   Multicollinearity 300
8.10 Variance Inflation Factors 302
8.11 Using Minitab to Calculate Variance Inflation Factors 303
8.12 Multiple Regression Model Assumptions 304
8.13 Using Minitab to Check Multiple Regression Model Assumptions ... 305
Exercises 306

9. More on Multiple Regression 313
9.1   Introduction 313
9.2   Using Categorical Predictor Variables 313
9.3   Using Minitab for Categorical Predictor Variables 315
9.4   Adjusted R^2 321
9.5   Best Subsets Regression 324
9.6   Using Minitab for Best Subsets Regression 329
9.7   Confidence and Prediction Intervals for Multiple Regression ... 331
9.8   Using Minitab to Calculate Confidence and Prediction Intervals
      for a Multiple Regression Analysis 331
9.9   Assessing Outliers 333
Exercises 334

10. Analysis of Variance (ANOVA) 341
10.1   Introduction 341
10.2   Basic Experimental Design 341
10.3   One-Way ANOVA 342
10.4   One-Way ANOVA Model Assumptions 349
10.5   Assumption of Constant Variance 350
10.6   Normality Assumption 355
10.7   Using Minitab for One-Way ANOVAs 357
10.8   Multiple Comparison Techniques 370
10.9   Using Minitab for Multiple Comparisons 373
10.10 Power Analysis and One-Way ANOVA 374
Exercises 378
References 383

11. Nonparametric Statistics..385
    11.1   Introduction...385
    11.2   Wilcoxon Signed-Rank Test...385
    11.3   Using Minitab for the Wilcoxon Signed-Rank Test...........389
    11.4   The Mann–Whitney Test..395
    11.5   Using Minitab for the Mann–Whitney Test.....................400
    11.6   Kruskal–Wallis Test..400
    11.7   Using Minitab for the Kruskal–Wallis Test.....................405
    Exercises...411

12. Two-Way Analysis of Variance and Basic Time Series............417
    12.1   Two-Way Analysis of Variance....................................417
    12.2   Using Minitab for a Two-Way ANOVA..........................424
    12.3   Basic Time Series Analysis..440
    Exercises...449

Appendix..453
Index...461

Preface

What a difference a decade makes! It's hard to believe that it has been more than 10 years since I began writing the first edition of *Applied Statistical Inference with Minitab*. And while the core motivations for writing the first edition still remain, there have been a significant number of changes incorporated throughout the second edition. Some of the most significant changes were guided by my experience teaching statistics to students who are not mathematics or statistics majors, but who apply statistics in their discipline. My primary objective was and still remains to write a book that does not lack mathematical rigor, but provides readers with the tools and techniques to apply statistics in their given fields. Not only does the second edition expand on the core motivations of the first edition but it is also a reflection of my own growth as a teacher.

Many statistics books are written for readers learning statistics for the first time. But what distinguishes this book from others is the focus on the applications of statistics without compromising on mathematical rigor. The intent is to present the material in a seamless step-by-step approach so that readers are first introduced to a topic, given the details of the underlying mathematical foundations along with a detailed description of how to interpret the findings, and are shown an illustration of how to use the statistical software program Minitab to perform the same analysis.

Throughout my many years of teaching, I have found that readers often find it easier to learn statistics by being exposed to the underlying distributions and calculations and then understanding how the software essentially performs all of the same calculations. Presenting the details on sampling distributions in addition to how the calculations are done gives readers a solid foundation in applying the many different statistical methods and techniques on their own while also increasing their confidence in interpreting the output that is generated by a statistical software program such as Minitab. I deliberately avoided including any computational formulas because I want readers to be able to see what the formulas are actually calculating. For instance, in calculating the sample variance, while computational formulas are often perceived as "easier" to calculate, the more formal calculations do give readers the opportunity to see the underlying details of such a calculation, as the "average" squared difference between the actual data values and the sample mean. In my opinion, computational formulas only distract readers from gaining a deeper understanding of what the statistic is calculating only for the sake of simplicity.

This book is written to be user-friendly for readers and practitioners who are not experts in statistics, but who want to gain a solid understanding of basic statistical inference. While the presentation does not lack in the mathematics that underlies statistics, this book is oriented toward the practical use of statistics. The audience for this book may come from diverse disciplines, the examples, discussions, and exercises are based on data and scenarios that are common to readers in their everyday lives.

Most of the exercises in the second edition are new, where much of the focus is to give the reader some exposure to the many different types of questions that can be answered using statistics. However, the use of "big data" is not considered in this book because it is assumed that most of the methods described herein are focused on data obtained through empirical and/or experimental studies, not exhaustive data mining or data science techniques. However, many of the topics covered in this book can be extended to big data sets.

The second edition includes many new topics such as one- and two-sample variances, one- and two-sample Poisson rates, and more nonparametric statistics. There is also the addition of *Best Practices* sections that describe some common pitfalls and provide practical advice on statistical inference.

One of the challenges that I faced as a student learning statistics came from the many different perspectives that various disciplines have about statistics. And although such differences are often a source of tension, such disconnection motivated me to think more deeply about learning statistics and how to best present applied statistical inference in a general and understandable way so that readers can build on what is covered in this book and apply their knowledge to more advanced courses. It is for this reason that the underlying theory of random variables and sampling distributions remain as the primary foundation for the different inferential techniques that are emphasized throughout.

I have always been a big advocate for learning one single statistical software package at a time. Not only does focusing on a single statistical software package allow the reader to obtain a much more in-depth understanding of the given package, but I also believe that gaining a strong foundation with one statistical software package allows the reader to be able to easily adapt to other statistics programs with only a minimal amount of effort. Minitab was and still remains a natural choice for a first course in applied inference. Minitab has very intuitive menus and informative dialog boxes, as well as very clear and in-depth help menus. These help menus provide everything from the formulas used, to specific examples, to how to interpret the findings in a clear and meaningful way. In addition to being user-friendly for the beginner, Minitab also has some very advanced statistical methods and techniques.

Chapter 1 provides a basic introduction to some of the more common terminologies that one may likely encounter when learning statistics for the first time. And while conventions and definitions do differ across disciplines, I tried to stay consistent with the terminologies and notations used in Minitab. In the second edition, I added a discussion about physical and conceptual populations.

Chapter 2 gives a basic description of some of the more common graphs and charts that are used in applied inference such as histograms, stem-and-leaf plots, bar charts, boxplots, scatterplots, and marginal plots. I have tried to keep the presentation aligned with the conventions that are used in Minitab.

Chapter 3 presents basic descriptive statistics as well as a discussion of random variables and sampling distributions. The calculations for descriptive statistics have been carried out in great detail using the notations and symbols of a more traditional statistics course. The second edition also presents some of the more common discrete distributions such as the binomial and Poisson, as well as continuous distributions such as the t, Chi-Square, and F. I have also added a detailed discussion about the use of Minitab to graph probability distributions as well as finding probabilities.

Chapter 4 provides a first look at basic statistical inference. This chapter describes inference for one sample, such as the mean, proportion, variance, and count. Confidence intervals are described in addition to providing a detailed discussion of their interpretation. Hypothesis tests are introduced for a single mean, proportion, variance, and count. I have elaborated on how inferences can be made with both confidence intervals and hypothesis tests by continually referring back to the sampling distribution. There is also an initial discussion and conceptual introduction to conducting a power analysis for all the different one-sample tests in addition to detailed descriptions on how to use Minitab throughout.

Chapter 5 provides a more in-depth discussion of confidence intervals and hypothesis tests by using sample data collected from two populations such as two means, two proportions, two variances, and two counts. The material presented in Chapter 5 builds on that presented in Chapter 4 regarding the interpretation of the findings as well as how to conduct a power analysis for all of the different two-sample tests.

Chapter 6 gives an introduction to simple linear regression analysis. In order to understand simple linear regression, one needs to have a good intuitive feel for why "the line of best fit" is called so. I have elaborated on the importance of the line of best fit by providing a description of why this line is "better' when compared to a different line that connects two random points. This chapter also provides an introduction to confidence and prediction intervals by making inferences using the line of best fit.

Chapter 7 provides more detail on simple linear regression by describing statistics for model fit such as the coefficient of determination and the coefficient of correlation. Further details are provided regarding the assumptions underlying a simple linear regression analysis as well as using the Ryan–Joiner test as a formal test of the normality assumption. Further discussion about outliers and how Minitab identifies outliers has been presented.

Chapter 8 gives an introduction to multiple regression analysis. Details of the ANOVA table and the issue of multicollinearity are described.

Chapter 9 provides more detail on conducting a multiple regression analysis by introducing categorical predictor variables, how to use Best Subsets regression to find an optimal model fit, and how to assess the impact of outliers.

Chapter 10 begins by providing a conceptual introduction to basic experimental designs and randomized block designs. The basics of a one-way analysis of variance (ANOVA) are described in addition to learning the use of Bartlett's and Levene's tests as formal tests of the assumption of constant variance. Multiple comparison techniques such as the least significant difference (LSD) are introduced as a way of identifying the magnitude and direction for any significant differences found in an ANOVA.

Chapter 11 provides a basic discussion of nonparametric statistics. Calculations of the test statistics for nonparametric tests such as the Wilcoxon signed-rank test, Mann–Whitney test, and the Kruskall–Wallis test are described in detail, and the calculations are aligned with those that are used by Minitab.

Chapter 12 gives a discussion of two-way ANOVAs and a brief introduction to time series analysis.

This book can be used for a one-semester course in basic statistical inference for students who have some familiarity with introductory statistics. And while some exposure to introductory statistics is assumed, the first three chapters provide a concise review of the topics that this book builds on. A typical three-credit course would cover most of the material in Chapter 4 through Chapter 10. A four-credit course could include the material in the three-credit course in addition to some of the material presented in Chapters 11 and 12. The crux of the topics in any applied inference course are those that are presented in Chapters 4 and 5, and the coverage of these topics is needed in any course, whether beginning or advanced.

For a trial request page that allows access to a free 30-day trial of Minitab Statistical Software, Release 18, please see http://www.minitab.com/products/minitab/free-trial/.

Acknowledgments

Over the course of the last decade, I am very fortunate to have met so many extremely talented and dedicated students who helped guide most of the revisions in the revised book.

I am grateful to my friends and colleagues, Frank Bensics and Zbigniew Prusak, who provided suggestions and comments on the first edition and their input still remains a core component of the second edition. I am also grateful to my friend and colleague Daniel Miller, who has always been an advocate for being true to the discipline of statistics. I will always cherish his words of wisdom and admire his dedication to the profession.

I am fortunate to have wonderful colleagues who have always been supportive: Roger Bilisoly, Robert Crouse, Darius Dziuda, Chun Jin, Daniel Larose, Krishna Saha, and Gurbakhshash Singh. They are a remarkable group of people and a joy to work with.

David Grubbs, Sherry Thomas, Emeline Jarvie, and the entire staff at Taylor and Francis have always been very helpful and supportive throughout the process of this revision. Special thanks to Jay Margolis and Kevin Craig from CRC, Christine Bayley and Denise Macafee from Minitab®, and to the previous adopters who provided feedback. I will always be grateful that I was given the opportunity to work on this project that was based solely on my experience as a teacher.

Correspondence

Although a large amount of effort has gone into making this text clear and accurate, if you have any suggestions regarding errors or content, or feel that some clarification is needed, please contact me at lesiks@ccsu.edu. I would be interested in hearing your feedback and comments.

Portions of the input and output contained in this publication/book are printed with permission of Minitab Inc. All such material remains the exclusive property and copyright of Minitab Inc. All rights reserved.

Minitab® and all other trademarks and logos for the Company's products and services are the exclusive property of Minitab Inc. All other marks referenced remain the property of their respective owners. See minitab.com for more information.

1

Introduction

1.1 What is Statistics?

Statistics is a branch of mathematics that deals with collecting, analyzing, presenting, and interpreting data. Many different fields use statistics as a way to understand complex relationships by collecting and analyzing data and presenting the results in a meaningful way. For instance, a researcher in education may want to know if using computers in an algebra classroom can be effective in helping students build their mathematical skills. To answer this question, the researcher could collect data about computer usage and mathematical skills and try to extract information in a meaningful way. A marketing manager for a luxury car manufacturer may want to know whether or not customers are satisfied with their luxury car purchase. The manager may collect data for a sample of their customers and then use this data to make a generalization to the larger group of all their luxury car buyers. In other fields, such as environmental science, researchers may want to figure out the factors that may contribute to global warming by collecting data about the makes and models of automobiles emitting larger amounts of greenhouse gas.

Since statistics deals with collecting, analyzing, presenting, and interpreting data, we first need to develop an idea about what data is. Generally speaking, *data* is information, characteristics, or attributes about some observation of interest or a set of observations of interest. A *set of data* represents a collection of variables that represent information about characteristics or attributes for a number of different observations. Data sets usually consist of a collection of observations and each observation can have measures of a variable or variables of interest. For instance, Table 1.1 gives an example of a small data set that describes the number of credits attempted, the number of hours worked each week, major, and gender for a random sample of five university freshmen. The *rows* of the data set represent each observation consisting of an individual student. The *columns* of the data set represent the different variables, and each student has measures for each of the four different variables that represent the number of credits, number of hours worked each week, major, and gender.

There are two basic types of data that can be collected, *quantitative data* and *qualitative data*. *Quantitative data* is data that is numeric in form. The main purpose of collecting quantitative data is to describe some characteristic or attribute using numbers. For example, quantitative data that represent the number of credits taken by students in a semester or the number of hours that students work each week can be collected. On the other hand, *qualitative data* is data that is categorical in nature and describes some characteristic or attribute with words or descriptions. For instance, qualitative data can be used to describe a student's major or their gender.

TABLE 1.1

Data Set Showing Different Characteristics for Five University Freshmen

Observation Number	Number of Credits	Number of Hours Worked Each Week	Major	Gender
1	15	22	Business	Male
2	12	20	Engineering	Male
3	9	0	Education	Female
4	18	18	Business	Female
5	15	9	Science	Male

The data set in Table 1.1 contains both quantitative and qualitative variables. The number of credits and the number of hours worked each week are examples of quantitative data since these variables are numeric in form, whereas major and gender are examples of qualitative data since these variables describe different categories or characteristics.

Determining whether to collect quantitative or qualitative data is driven by the characteristic or relationship that is being studied and the type of data that is available.

1.2 What This Book Is About

The purpose of this book is to introduce some of the many different statistical methods and techniques that can be used to analyze data along with effective ways to present the information extracted from data in a meaningful way. There are three broad areas of statistics that will be covered in this book: *summary tables and graphical displays*, *descriptive representations of data*, and *inferential statistics*.

1.3 Summary Tables and Graphical Displays

Graphical displays of data visually present some of the properties or characteristics of data by using different types of charts and graphs. The advantage of using charts and graphs to display data is that large amounts of information can be displayed in a concise manner.

For example, suppose you are interested in comparing the fuel efficiency of four different types of vehicles— Minivans, SUVs, Compact Cars, and Mid-Size Cars. You could obtain the average miles per gallon (MPG) for both city and highway driving for each of the types of vehicles and then create a summary table and/or a graph to illustrate the comparison between the different types of vehicles. A summary table is given in Table 1.2 and a graphical display is illustrated in Figure 1.1.

Note that Figure 1.1 graphically displays the city and highway MPG for each of the four different types of vehicles, and the graph allows you to compare and contrast mileages between the four different types of vehicles.

TABLE 1.2

Tabular Summary of the Average City and Highway MPG for
Minivans, SUVs, Compact Cars, and Mid-Size Cars

	Average City MPG	Average Highway MPG
Minivans	17.5	24.1
SUVs	17.4	22.4
Compact Cars	26.0	33.5
Mid-Size Cars	22.4	30.2

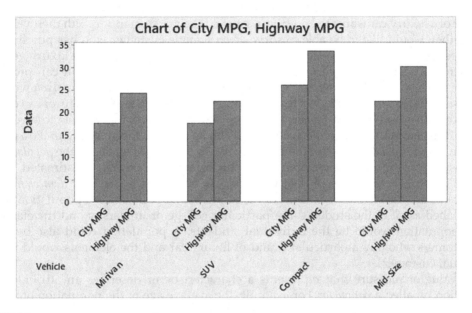

FIGURE 1.1
Bar chart comparing the average city MPG and the average highway MPG based on the four different types of vehicles.

1.4 Descriptive Representations of Data

Descriptive representations of data consist of methods and techniques that can be used to describe and summarize data. For instance, if you have ever shopped for a new car, you may have noticed that the sticker on the window of the car provides you with an estimate of the average miles per gallon (MPG) that you can expect for both city and highway driving. This number describes, on average, the gas mileage that you can expect from the vehicle. For example, if the sticker on the window of your new car is 25 MPG in the city and 36 MPG on the highway, then you can expect that your new vehicle will get approximately 25 MPG driving in the city and approximately 36 MPG driving on the highway. This does not necessarily suggest that you are guaranteed to get these mileages because these numbers are averages and they will vary based on your driving style and how you maintain your vehicle, along with a host of other factors. These numbers represent summarized data that you can use to estimate how efficient (or inefficient) the vehicle is. If you want to learn more about the information provided on the window labels of new cars, visit the website of the U.S Department of Energy (https://www.fueleconomy.gov/feg/label/learn-more-gasoline-label.shtml).

1.5 Inferential Statistics

Inferential statistics are methods and techniques that can be used to estimate or predict an unknown characteristic of interest by using available data. For example, suppose the manufacturer of a luxury car wants to know how satisfied their customers are with their vehicles. Ideally, they would like to be able to survey every single customer who has purchased a vehicle (in other words, the *population* of *all* customers who purchased the luxury vehicle) and then see what percentage respond that they are satisfied. But as you can probably imagine, it is usually either impossible or impractical to survey *every single person* who has purchased this luxury car and get a response from them. But if the manufacturer can obtain a representative portion or a *sample* of their customers and ask this sample to give their opinion, the manufacturer can then use the data obtained from this *sample* to make a generalization to the *population* of *all* of their customers. Thus, the percentage of the *population* of their customers who are satisfied with their vehicle purchase can then be estimated.

A *population* is a collection of data that consists of *every single possible item of interest*. These items are referred to as *elements* of the population. For instance, a population could be described as *all* of the students at a particular college or university and the elements of the population would be the individual students. A population could also be *all* of the customers who buy a particular brand of luxury car and the elements would be the individual customers.

Any value or measure that represents a characteristic or describes an attribute of a population is called a *parameter*. For example, the average age of the population of all the students at a particular college or university is considered a population parameter because it is a value or measure that represents a characteristic or attribute of the entire population. The percentage of the population of all customers who are satisfied with a particular brand of luxury car is also a parameter.

A *sample* is a collection of observations that consists of a subset, or portion, of a population of interest. For instance, a sample from the population of all the students at a particular college or university could be the seniors at the college or university. A sample could also be a subset that is selected at random from a given population. A *random sample* is a sample taken from a population such that each sample of a given size from the population has the exact same chance of being selected as does any other sample of the same size. A random sample is one of the best ways of obtaining a sample that is representative of the underlying population.

Any value or measure that describes a characteristic or attribute of a sample is called a *statistic*. For example, the average age of the seniors at a particular college or university would be a statistic because it represents a characteristic or attribute of a given sample. A statistic could also be the average age for a sample of students who were selected at random from the population of all the students at a particular college or university. Another example of a statistic is the percentage of a random sample of customers who are satisfied with a particular brand of luxury car. This is a statistic because it represents a value or measure that describes a characteristic or attribute of the given sample.

1.6 Populations

Throughout this book, we will be describing many different methods and techniques that can be used to make an inference, or prediction, about an unknown population parameter of interest based on information that is contained in a sample statistic. These different methods and techniques rely on using statistics collected from a representative sample to make generalizations about an unknown population parameter or parameters of interest.

There are two different descriptions of populations that you may encounter, and these are described as *physical populations* and *conceptual populations*.

Physical populations are populations whose elements physically exist. For example, the population of all the students currently enrolled at a college or university can be viewed as a physical population because the elements (i.e., the students) physically exist. A physical population can also be defined as a population that has a fixed or finite number of elements.

Conceptual populations are populations whose elements exist only in theory. A conceptual population can also be described as a population that has an infinite number of elements. For example, the population of all drivers who may drive through a particular intersection can be described as a conceptual population because there could be (at least theoretically) any number of drivers who could decide to drive through the intersection at any given time.

1.7 Different Ways of Collecting Data

Studies that generate data generally fall into two broad categories—*experimental studies* or *observational studies*. In an *experimental study*, subjects are assigned to participate in a treatment program based on a random assignment process. For example, suppose you want to conduct an experiment to determine if students who use computers in their statistics class perform better on the final examination as compared to those who do not use computers in their statistics class. In order to conduct a true experimental study, students would have to be randomly assigned to either the computer statistics class (this is called the *treatment group*) or the noncomputer statistics class (this is called the *control group*). To create treatment and control groups using a random assignment, we could flip a fair coin for each prospective student. Those who receive heads could be assigned to the treatment group (the computer statistics class) and those who receive tails could be assigned to the control group (the noncomputer statistics class). A comparison of how these two groups perform on the final examination could be used as an indication of which group outperforms the other.

The benefit of conducting a true experimental study is that a random assignment process creates treatment and control groups that have the greatest chance of being equivalent in all respects, except for the group assignment. By determining the group assignment based only on chance, this makes it less likely that any factors other than the group assignment are having an impact on the outcome of interest (which for our example is the score received on the final examination). However, one major problem with conducting an experimental study is that there could be instances where it may be difficult or even impossible to assign participants to treatment and control groups by using only a random assignment process. Random assignments are not always feasible, practical, or ethical.

For *observational studies*, data is collected by observation and subjects are not assigned by a random assignment process or in any other prescribed manner. For example, we could conduct an observational study to see if students from the computer statistics class performed better on the final examination as compared to those from the noncomputer statistics class by simply collecting data from two existing groups of students—one group who self-selected themselves to enroll in the computer statistics course and another group who self-selected themselves to enroll in the noncomputer statistics course.

Although it is usually much easier and often less expensive to collect observational data as compared to experimental data, observational studies are often plagued with *selection bias*. In other words, because students self-selected into either of these two types of courses, any findings regarding the effectiveness of the use of computers on student performance may be biased because of the possible differences between the two groups.

Now that we have discussed some of the typical ways that data can be collected, we will describe in more detail some of the different types and scales of data that we can expect to find.

1.8 Types of Variables

Before we can start to analyze data in a meaningful way, it is important to understand that there are many different types of variables that we are likely to encounter. One way to distinguish between the different types of variables is to consider whether additional observations can exist between any two values of a variable. We will begin by first describing two general types of variables—*discrete* and *continuous*.

We say that a variable is *discrete* if, between *any* two successive values of the variable, other values cannot exist within the context they are being used. Discrete variables can often be used to represent different categories.

Example 1.1

Suppose we code students at a university as either Part-Time = 0 or Full-Time = 1. This variable is a discrete variable because between the values of 0 and 1 there cannot exist any other values that have meaning within the context they are being used. Often discrete variables are coded using numbers, but it is important to keep in mind that these numbers do not have any numerical properties associated with them because you cannot add, subtract, multiply, or divide such numbers within the context that they are being used.

Example 1.2

Consider three different categories of temperature, coded as follows:

$$Cold = 0 \quad Warm = 1 \quad Hot = 2$$

This variable would be a discrete variable even though another possible observation could exist (namely Warm = 1) between the categories of 0 and 2. In order to be a discrete variable, there cannot be observations that exist between *any* two successive values of the variable, such as between the categories coded as 1 and 2.

Example 1.3

A variable that is quantitative can also be discrete. For example, the number of students in a classroom is discrete because between any two successive values of the variable, other values cannot exist. In other words, we cannot have a fraction of a student.

On the other hand, a variable is *continuous* if between *any* pair of observations, other values can theoretically exist.

Example 1.4

Consider two observations of the heights for two different individuals. If one individual is 5'11" and the other individual is 5'6", then we could observe another individual whose height lies somewhere between these two values, such as 5'8". Similarly, if we observe one individual with a height of 5'6" and another individual with a height of 5'7", we could (at least theoretically) observe an individual whose height lies somewhere in between these two values.

1.9 Scales of Variables

There are four different scales of variables that can be used to describe the type of information that a variable contains. These scales can also be used to further describe the mathematical properties of the variable. The different scales of variables are described as *nominal, ordinal, interval,* or *ratio.*

The weakest scale represents variables that have no mathematical properties at all. *Nominal scale* or *nominal variables* are variables that are mathematically the weakest because the values of a nominal variable only represent different categories, and therefore, they cannot be manipulated using basic arithmetic.

Example 1.5

Suppose we consider three different categories of temperature as described in Example 1.2:

$$\text{Cold} = 0 \quad \text{Warm} = 1 \quad \text{Hot} = 2$$

This variable is a nominal variable because it describes the different categories of temperature and there are no mathematical properties that can be associated with the values of this variable. For instance, if we take any two observations, such as Cold = 0 and Warm = 1 and if we add them together, $0+1=1$, this sum does not make sense within the context that these variables are being used. Furthermore, neither multiplication nor division makes any sense within the context that these variables are being used.

The second weakest scale of a variable is called an *ordinal scale* or an *ordinal variable.* An ordinal variable can be grouped into separate categories, but ordinal variables are different from nominal variables in that there is a natural *ordering* of the categories. Other than the ordering of the categories, ordinal variables are similar to nominal variables in that they are categorical variables with no numerical properties.

Example 1.6

Suppose we are looking at the four different ways in which undergraduates are typically classified:

$$1 = \text{freshman} \quad 2 = \text{sophomore} \quad 3 = \text{junior} \quad 4 = \text{senior}$$

Note that there is a natural ordering among the different categories since 1 = freshman can be seen as "less than" 2 = sophomore. Thus, the inequality 1 < 2 makes sense within the context that this variable is being used. However, there are no mathematical properties that are associated with these categories. If we add two values that represent any two of these categories, their sum or difference does not have any meaning within the context they are being used, and this also holds true for multiplication and division.

Another scale for variables is called an *interval scale* or an *interval variable*. With interval variables, equal distances between observations represent equal-sized intervals.

Example 1.7

Consider the variable that describes the elevation of where you live. In other words, how far you live with respect to sea level. This variable would be an interval variable because the measure of distance between two individuals who live 70 and 60 feet above sea level would represent the same-sized interval for two individuals who live 5 and 15 feet above sea level. In other words, a 10-foot difference has the same meaning, no matter where you live. Another example is the temperature in degrees Fahrenheit. If it is 40° on one day and 60° on another day, this 20° difference is the same no matter what temperatures we are measuring, such as the difference between 30° and 10°. We can add and subtract interval data, and we can measure and quantify such sums or differences within the context that the data are being used.

Another scale of a variable is where the value of "0" indicates the absence of the quantity of interest. This type of variable is called a *ratio scale* or a *ratio variable*. In other words, with ratio variables, quotients or ratios can be formed that have meaning within the context that they are being used because the value of "0" represents the *absence* of the quantity of interest.

Example 1.8

Consider the amount of money that two people have in their pockets. If one person has $100 and the other person has $200, then the ratio of the two amounts of money has meaning because $200 is twice as much as $100. Also, the value of $0 indicates having no money and so $0 represents the *absence* of the quantity of interest.

All ratio variables have the properties of interval variables, but not all interval variables will have the properties of ratio variables. For example, consider

TABLE 1.3

Mathematical Properties of Nominal, Ordinal, Interval, and Ratio Variables

Scale of Variable	Mathematical Properties
Nominal	None
Ordinal	Inequalities
Interval	Addition and subtraction
Ratio	Addition, subtraction, multiplication, division

temperature in degrees Fahrenheit. If it is 84° Fahrenheit on one day and 42° Fahrenheit on another day, it does not make sense to say that 84° is twice as hot as 42° even though we can form the quotient 84/42 = 2. Furthermore, the value of 0° on the Fahrenheit scale does not represent the absence of temperature because it is still a temperature, just a very cold one!

Table 1.3 illustrates the four different scales of variables based on the mathematical properties that were described.

1.10 Types of Analyses

There are many types of analyses that we can use to make inferences to a larger population of interest by using available data obtained from a sample. Each of these analyses can be used in their own unique way by using sample data to describe some characteristic of a population or to assess the relationships amongst a set of variables in a population. When to use which method or technique often depends on the type of study that is being done and the type of data that can be collected.

We will now describe the general idea behind *basic statistical inference, regression analysis*, and *analysis of variance* (ANOVA). These analyses use sample data to make inferences or predictions to a larger population.

Basic statistical inference consists of various statistical methods and techniques that can be used to make an inference or prediction about some population parameter or parameters of interest by using information obtained from a sample. For instance, suppose we want to determine whether a weight loss program is effective in helping people to lose weight. We could get a sample of program participants and measure their weight before and after participating in the program and use this sample data to see if the average weight loss is significant for the entire population of program participants. Chapters 4 and 5 describe many different statistical methods that are often used to make inferences to population averages, population proportions, population variation, and population counts.

Regression analysis is an inferential technique that consists of using sample data to develop and validate a population model by describing how one variable is related to another variable or to a collection of different variables. For instance, the asking price of a house is often determined by many different factors, such as the square footage, lot size, number of bedrooms, number of bathrooms, and age. A regression analysis would allow you to develop a population model that describes how the price of a house is related to these factors using data collected from only a sample of houses. Regression analysis uses sample data to develop a model to predict the asking price for the population of *all* houses based on the different factors as mentioned earlier. A regression analysis will also allow you to use such a model for estimation and prediction and there are techniques that can assess how well the model fits the data. We will be covering regression analysis in Chapters 6–9.

Analysis of variance (often abbreviated as ANOVA) is an inferential technique that can be used to estimate whether there are differences in averages between more than two groups based on some characteristic. For example, suppose you are interested in determining whether there is a difference in the number of pages you can print with four different brands of printer toner. One way to assess this could be to set up an experiment where you have four identical brands of printers and a total of 16 printer cartridges (4 printer cartridges for each brand). Then you could put the printer cartridges in each of the different printers and count up the total number of pages that each printer printed with the given cartridge. An ANOVA could then be used to see if there is a difference in the average number of pages printed based on the different brands of printer cartridges. The sample would be the 16 printer cartridges (4 of each brand), and the inference would be made to the population of *all* printer cartridges of the four different brands. Thus, by using a sample of 4 printer cartridges from each brand, we could generalize whether there is a difference in the number of pages printed between the four different brands.

1.11 Entering Data into Minitab

Minitab is a statistical software program that can simplify the analysis of a set of data. One very nice feature of Minitab is that data can easily be entered in the form of a worksheet. Minitab is easy to use and has some very powerful features that will be described in detail throughout this book.

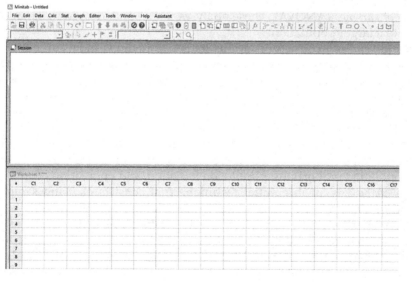

FIGURE 1.2
Minitab session and worksheet window.

⊞ Worksheet 1 ***			
↓	C1-D	C2-T	C3
	Date	Salesperson	Yearly Sales
1	March 20, 2018	Rena	56320
2	February 13, 2018	Susan	54290
3	April 7, 2018	Jackie	53800
4	May 12, 2018	Michael	54000
5	November 7, 2017	Rebecca	52780
6	July 17, 2018	Matthew	58500
7			

FIGURE 1.3
Minitab worksheet illustrating date, text, and numeric forms of data.

Figure 1.2 illustrates what a basic Minitab session and worksheet window looks like. The worksheet section on the bottom portion of Figure 1.2 is used to display data and the session window on the top portion displays the results of any calculations or analyses that you may run. The top bar of the Minitab screen provides all the different types of pull-down menus that are available.

Data are typically entered into a Minitab worksheet in row and column forms. Each column represents a variable and a collection of variables can be represented by different columns, namely C1, C2, ..., and so on. The rows of a Minitab worksheet consist of the individual observations for each of the variables and these are labeled in the left-most column with the numbers 1, 2, 3, ..., and so on.

Data can be entered into Minitab by typing manually, but this is usually only appropriate if you have a very small amount of data. If you have a large amount of data, you can also cut and paste data from other programs or import data from other files.

Minitab only allows three types of data to be entered and they are *numeric, text,* or *date*. For example, in Figure 1.3, a collection of sales data is entered that represents the date, name, and amount of sales in dollars for a sample of six salespeople. Notice in Figure 1.3 that the date column is C1-D (column one, date), the column that contains the names of the salespeople is a text column and is represented as C2-T (column 2, text), and the column that consists of yearly sales data is C3 (column 3). When there is no letter associated with a column label, it is considered to be numeric data.

Once your data is entered into a Minitab worksheet, you can save your project data from the session window and the data entered in the worksheet window. This can be done by selecting **File** on the top menu bar followed by **Save Project As**. You can also save the current worksheet as well by selecting **File** on the top menu bar and then **Save Current Worksheet As**.

1.12 Best Practices

Throughout this book, there are sections listed as Best Practices that are located at the end of some of the chapters. These sections describe some of the best practices in statistics as well as some potential pitfalls that can occur when collecting and analyzing data.

- Know the population you are generalizing to!

  Since statistical inference relies on making a generalization to a larger popula-tion, it is important to understand what that population is. Is it a conceptual or physical population? What are the characteristics of the population that you are interested in? How can you describe your inferences to the larger population?

- Know your data!

  Before you decide to conduct a study where you are interested in presenting data or using data to make inferences to a larger population of interest, it is impor-tant to keep in mind the scale of the data that you plan to collect. As you have already seen and will see throughout this book, data can take on many different forms such as being discrete or continuous, nominal, ordinal, interval, or ratio, and the type of analyses that you do largely depends on the type of the data that is being used. It is important to always keep in mind the scale of the variables and how they are used in your field.

- Ethical practice in statistics.

  There are ethical considerations to be kept in mind when collecting and analyzing data. One resource for ethical practice in statistics is the American Statistical Association. They offer a set of detailed guidelines that can help you understand what good statistical practice is. These guidelines are available on their website at http://www.amstat.org/about/ethicalguidelines.cfm.

Many other fields of study in the physical and social sciences may also have adopted ethical guidelines for collecting, analyzing, and presenting data. It is usually a good practice to familiarize yourself with the expectations of ethical practice for collecting and analyzing data in your discipline before beginning a study.

Exercises

1. A survey asked 50 consumers to rate their level of satisfaction with a certain brand of laundry detergent. The responses to the survey were either very satisfied with the detergent, satisfied with the detergent, not satisfied with the detergent, or have not used the detergent. Would you classify this variable as nominal, ordinal, inter-val, or ratio? Justify your answer.

2. Describe each of the following variables as quantitative or qualitative, discrete or continuous, and classify the scale as nominal, ordinal, interval, or ratio:

   a. Social security numbers
   b. IQ score
   c. College class (freshman, sophomore, junior, and senior)
   d. The different models of cars on the road
   e. The length of a fiber-optic cable
   f. The number of shoppers in a checkout line

3. The partial data set given in Table 1.4 consists of a selection of variables for a ferry company that describe the revenue, time of day the ferry left the dock, number

TABLE 1.4

Partial Data Set of Revenue (in Dollars), Time of Day, Number of Passengers, Number of Large Objects, Weather Conditions, and Number of Crew Members

Revenue (in Dollars)	Time of Day	Number of Passengers	Number of Large Objects	Weather Conditions	Number of Crew Members
7,812	Morning	380	6	Calm	4
6,856	Noon	284	4	Calm	4
9,568	Evening	348	10	Calm	3
10,856	Morning	257	16	Rough	5
8,565	Morning	212	12	Rough	4
8,734	Noon	387	6	Calm	3

of passengers on the ferry, number of large objects on the ferry, the weather conditions, and the number of crew for a random sample of ferry runs over the course of a 1-month period. Describe each of the variables as quantitative or qualitative, discrete or continuous, and also classify the scale as nominal, ordinal, interval, or ratio.

4. Can your diet make you happier? A researcher wants to know if the food you eat has an effect on your happiness. A random sample of 41 adult participants completed a survey on happiness and a questionnaire on their diet (vegetarian, vegan, low carb, low fat, or no diet). The happiness survey is scored on a scale of 50–100, where higher scores are associated with a higher degree of happiness and satisfaction in the participants' life. Would you describe this as an experimental or an observational study? Describe the population that you believe the researcher can generalize to?

5. Explain why a random sample is one of the best ways to obtain a sample that is representative of the population.

6. A *stratified random sample* can be obtained by first partitioning a population into disjoint groups (or strata) and then taking random samples from each of the strata. For instance, a stratified random sample can be obtained by first partitioning all the undergraduate students at a university by their class (freshman, sophomore, junior, and senior) and then taking a random sample from each of these strata.

   a. Can you think of a situation where a stratified sample may improve on obtaining a representative sample as compared to drawing a simple random sample? Explain.

   b. Describe a population and a stratified sample of the population.

2

Graphs and Charts

2.1 Introduction

Often we are interested in creating a graphical display or a summary chart of the data that we have collected. One advantage to graphically displaying or summarizing data in a chart is the presentation of large amounts of information in a concise manner and this can provide a summary of the underlying characteristics of the variable of interest. Of the many different types of graphs and charts that can be used to display data, we will be discussing the following most common types in this chapter:

Histograms
Stem-and-leaf plots
Bar charts
Boxplots
Scatterplots
Marginal plots
Matrix plots

In addition to describing these different types of graphs, we will also be providing a few examples of how they can be used in practice and how Minitab can be used to generate them.

2.2 Frequency Distributions and Histograms

The first type of graph that we will be considering is called a *histogram*. A histogram is a graph that shows the *distribution* of a single continuous variable. The *distribution* of a variable can be used to illustrate all the different values of the variable along with how often or how likely those values are to occur. Distributions are often described or visualized using either a table or a graph. To create a histogram, the *x*-axis consists of data grouped into equal-sized intervals or bins that do not overlap each other, and the *y*-axis represents the number or frequency of the observations that fall into each of these bins. Perhaps the easiest way to create a histogram is from a table called a *frequency distribution*. A *frequency distribution* is a table that illustrates the number or frequency of the observations in the data set that fall within a given range of values. The purpose of drawing a histogram is to see the pattern of how the data is distributed over different ranges of values.

Example 2.1

The data set provided in Table 2.1 gives the round-trip commuting distance (in miles) for a random sample of 25 executives.

To construct a histogram for the data in Table 2.1, we first group the data into equal-sized intervals or bins that do not overlap each other and then count the number of observations that fall into each interval or bin. Table 2.2 illustrates what a frequency distribution with eight equal-sized bins would look like for the executive round-trip commuting data. Note that each bin is the same width and these bins do not overlap each other. Each bin has an upper and lower limit. For example, the first bin has a lower limit of 10 and an upper limit of 29.

A histogram can then be created by displaying these intervals on the *x*-axis along with the number of observations or frequencies that fall into each interval on the *y*-axis, as is illustrated in Figure 2.1.

The pattern of how often (or how likely) the different commuting distances are to occur can be seen in Figure 2.1. For instance, we can see that most executives commute less than 90 miles round-trip per day.

So, how many bins of a histogram are needed to get a good sense of the shape of the distribution of the data? Although there are no generally accepted rules

TABLE 2.1

Round-Trip Commuting Distance (in miles) for a Random Sample of 25 Executives

26	18	56	102	110
74	44	68	10	110
50	66	144	50	36
32	88	58	154	38
48	42	62	72	70

TABLE 2.2

Frequency Distribution for the Daily Round-Trip Commuting Distance (in miles) Using Eight Equal-Sized bins

Commuting Distance (in miles)	Frequency
10–29	3
30–49	6
50–69	7
70–89	4
90–109	1
110–129	2
130–149	1
150–169	1
Total	**25**

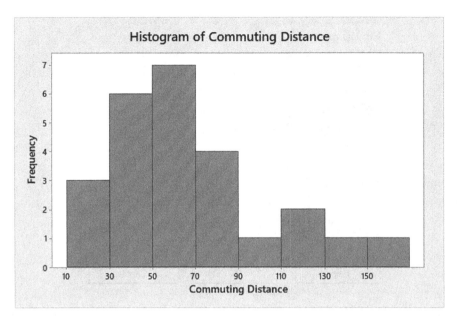

FIGURE 2.1
Histogram of the daily commuting distance data from the frequency distribution given in Table 2.2.

for the number of bins that a histogram must have, one rule of thumb is that seven or eight bins are usually an appropriate amount that will allow you to see the shape of the distribution of the data, but fewer or more bins can be used if necessary. However, keep in mind that a histogram with too many bins may only show a random pattern, and a histogram with too few bins may not reveal anything about the shape of the distribution of the data (see Exercise 1). Another rule of thumb for determining the number of bins for a histogram is to use the square root of the sample size.

Although a histogram can be created by hand, it is much easier and more efficient to use statistical software programs such as Minitab to create histograms.

2.3 Using Minitab to Create Histograms

The histogram for the commuting distance data can easily be created using Minitab by first entering the data in a single column in a Minitab worksheet. To draw the histogram, select **Graph** from the top menu bar, and then **Histogram**.

We can then select either a simple histogram, a histogram with a normal fit superimposed, a histogram with outline and groups, or a group of histograms with a normal curve superimposed, as illustrated in Figure 2.2.

After selecting a simple histogram, we then need to specify the variable that we wish to draw the histogram for. This requires highlighting the variable of interest and hitting **Select**, as is illustrated in Figure 2.3.

After selecting **OK**, the histogram for the commuting distance data is presented in Figure 2.4.

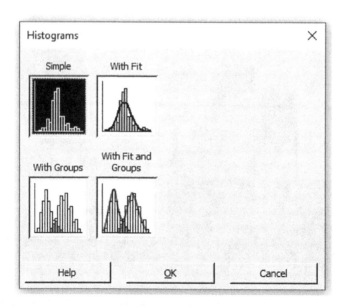

FIGURE 2.2
Minitab commands for drawing a histogram.

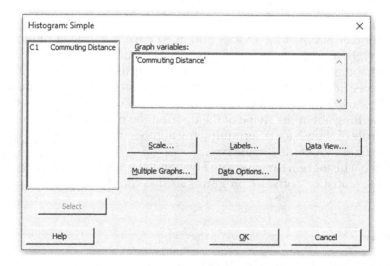

FIGURE 2.3
Minitab dialog box to select the variable for the histogram.

The **Automatic** setting in Minitab will give you the optimal number of bins based on an algorithm. But if necessary, you can easily change the shape of a histogram by specifying more or fewer bins. This is done by right-clicking on the bars of the histogram to obtain the **Edit Bars** dialog box, which is presented in Figure 2.5.

To change the number of bins on a histogram, select the **Binning** tab in the **Edit Bars** dialog box and specify the number of intervals (say 10), as is illustrated in Figure 2.6. You can also select to draw a histogram based on the midpoint or the cut-point positions.

This gives a histogram for the round-trip commuting distance data from Table 2.1 with ten bins, as is illustrated in Figure 2.7.

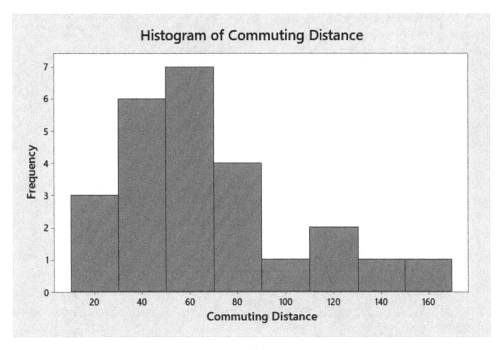

FIGURE 2.4
Histogram of the commuting data from Table 2.1 using Minitab.

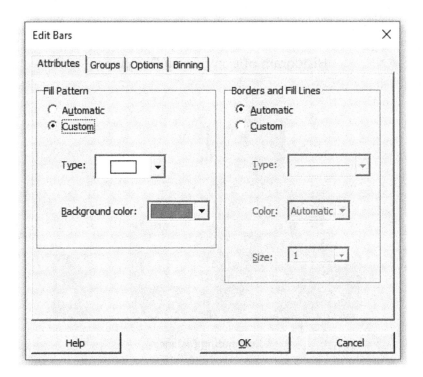

FIGURE 2.5
Minitab dialog box to edit the bars of the histogram.

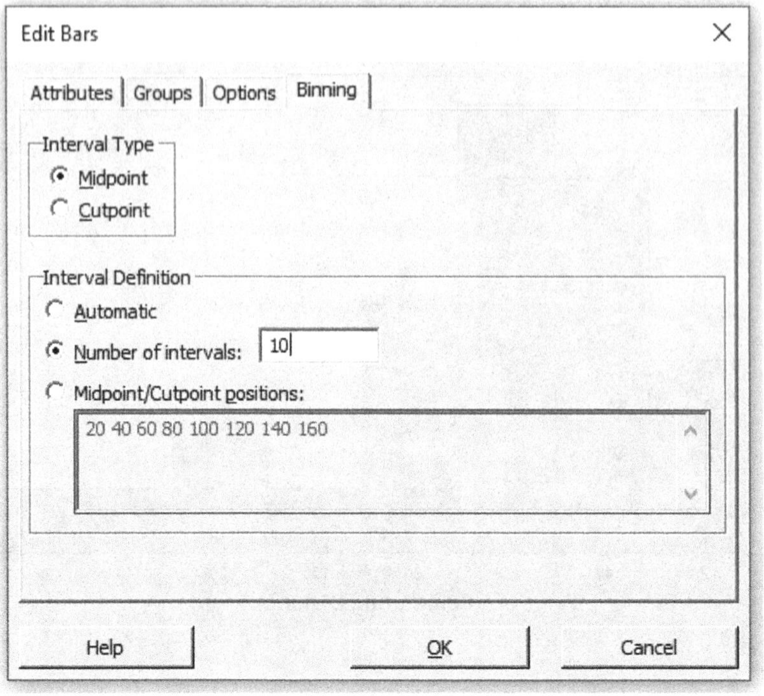

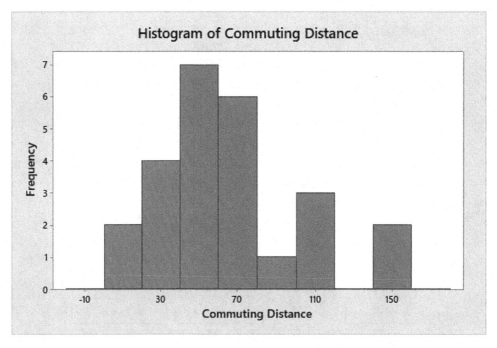

FIGURE 2.6
Minitab dialog box to specify the number of intervals as 10.

FIGURE 2.7
Minitab histogram of the round-trip commuting distance using ten bins.

Note that in Figure 2.7 the histogram has only seven bars showing, but there are ten bins with respect to the *x*-axis. This is because there are no observations contained in the bins that have a center point of –10 or in the bins that have center points of 130 or 170.

2.4 Stem-and-Leaf Plots

Stem-and-leaf plots are graphs that can be used to simultaneously show both the rank order and the shape of the distribution for a single continuous variable.

Example 2.2

To create a stem-and leaf plot for the commuting distance data given in Table 2.1, we need to create both a stem portion and a leaf portion. To create the stem portion, we need to list all of the possible leading digits for the values of the data set to the left of a vertical line. For the data provided in Table 2.1, we can see that the leading digits go from 1 to 15. This is because the range of the data is from 10 to 154. The smallest leading digit will be 1, corresponding to the value 10, and the largest leading digit will be 15, corresponding to the data value of 154.

The stem portion of the commuting distance data given in Table 2.1 is illustrated in Figure 2.8.

To create the leaf portion of the stem-and-leaf plot, begin by listing each observation to the right of the vertical line for its respective leading digit. For example, the observation 26 would have the value 2 in the stem portion (since it is the leading digit) and 6 in the leaf portion of the stem-and-leaf plot, as illustrated in Figure 2.9.

Entering the rest of the leaves for the commuting distance data and then putting all of the leaves in ascending order gives the finished stem-and-leaf plot that is illustrated in Figure 2.10.

Note that in Figure 2.10, the stem-and-leaf plot shows the distribution of the data along with the rank ordering of the data. The data is rank ordered on each

```
 1 |
 2 |
 3 |
 4 |
 5 |
 6 |
 7 |
 8 |
 9 |
10 |
11 |
12 |
13 |
14 |
15 |
```

FIGURE 2.8
Stem portion of the commuting distance data consisting of all possible leading digits to the left of a vertical line.

```
 1 |
 2 | 6
 3 |
 4 |
 5 |
 6 |
 7 |
 8 |
 9 |
10 |
11 |
12 |
13 |
14 |
15 |
```

FIGURE 2.9
Portion of a stem-and-leaf plot for observation 26.

```
 1 | 0  8
 2 | 6
 3 | 2  6  8
 4 | 2  4  8
 5 | 0  0  6  8
 6 | 2  6  8
 7 | 0  2  4
 8 | 8
 9 |
10 | 2
11 | 0  0
12 |
13 |
14 | 4
15 | 4
```

FIGURE 2.10
Stem-and-leaf plot for the round-trip commuting distance data given in Table 2.1.

row, and if you rotate the stem-and-leaf plot 90° to the left, you will be able to see the distribution of the data.

2.5 Using Minitab to Create Stem-and-Leaf Plots

To create a stem-and-leaf plot with Minitab, select **Stem-and-Leaf** under the **Graphs** tab. This brings up the stem-and-leaf plot dialog box that is presented in Figure 2.11. Notice that the value 10 is specified in the increment box. The increment represents the difference between the smallest values on adjacent rows. For this example, the smallest value on adjacent rows would be 10 because we want the stem-and-leaf plot to have adjacent rows separated by 10.

This gives the Minitab output shown in Figure 2.12.

You may have also noticed that the fifth row of the stem-and-leaf plot in Figure 2.12 has a count value of (4). This represents the row whose leading digit would be the leading digit

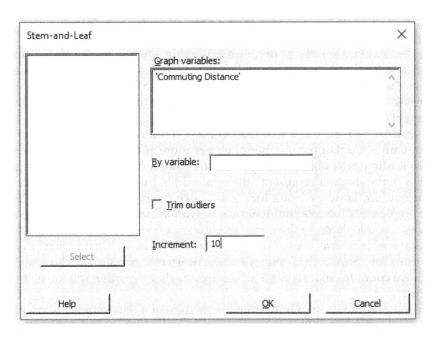

FIGURE 2.11
Minitab dialog box for a stem-and-leaf plot.

Stem-and-Leaf Display: Commuting Distance

Stem-and-leaf of Commuting Distance   N = 25

```
 2     1    08
 3     2    6
 6     3    268
 9     4    248
(4)    5    0068
12     6    268
 9     7    024
 6     8    8
 5     9
 5    10    2
 4    11    00
 2    12
 2    13
 2    14    4
 1    15    4
```

Leaf Unit = 1

FIGURE 2.12
Minitab stem-and-leaf plot for the round-trip commuting distance given in Table 2.1.

of the *median* of the data set. The *median* of a data set is defined as the value such that if the data is rank ordered, then 50% of the observations will lie at or below this value and 50% of the observations will lie at or above this value. For the commuting distance data, the median is actually 58, and the fifth row has a count value of 4, since there are four observations that fall in the same row as the median. Also notice that this value 4 is given in parentheses. This means that the median for the sample would be in the row that has a leading digit of 5.

You can also see in Figure 2.12 that the left-most column of the stem-and-leaf plot contains a cumulative count of the number of observations that falls below or above the row that has the leading digit of the median. The count for a row below the row that has the leading digit of the median represents the total number of observations in that row and the rows below. Similarly, the count for a row above the row that has the leading digit of the median represents the total number of observations in that row and the rows above. For instance, if you look at the third row of Figure 2.12, you will see that the cumulative count for that row is 6. This means that there are a total of six observations that lie in or below the third row. Similarly, if you look at the ninth row of Figure 2.12 you will see that the cumulative count for that row is 5, which means that five observations lie in or above the ninth row.

It is also important to keep in mind that the median of a data set may or may not be an actual observation from the data set.

2.6 Bar Charts

Bar charts can be used to illustrate the distribution of a single categorical variable or a set of categorical variables. Recall that categorical variables are variables whose observations represent distinct categories. However, there may be times when continuous variables need to be treated as categorical variables and this can be done by partitioning the continuous data into nonoverlapping categories. For example, the quantitative data for the round-trip commuting distance for the sample of 25 executives from Table 2.1 can be categorized as "near" if the distance is less than 50 miles or "far" if the distance is greater than or equal to 50 miles. Because discrete data is categorical in nature, we use a bar chart instead of a histogram to illustrate how the data is distributed within these two distinct and nonoverlapping categories.

2.7 Using Minitab to Create a Bar Chart

In order to create a bar chart to categorize the round-trip commuting distance as "near" or "far" using Minitab, we first need to create a categorical variable to represent whether the commuting distance is categorized as "near" or "far." This can be done by selecting **Code** and then **To Text** under the **Data** tab. This brings up the dialog box that is illustrated in Figure 2.13.

Notice in Figure 2.13 that we need to specify the data we wish to recode, the column of the worksheet where we want to store the recoded data, and the range of values that will

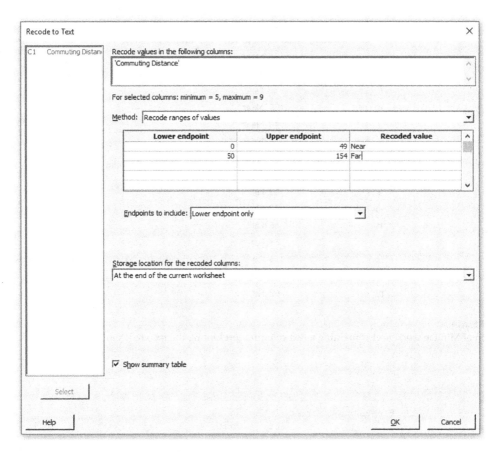

FIGURE 2.13
Minitab dialog box to code the continuous commuting distance variable as a categorical variable.

be assigned to each of the different categories. For our example, since we want to classify round-trip commuting distances with less than 50 miles as "near," we would specify that the values from 0 to 49 are to be recoded as "near." Similarly, since round-trip commuting distances of 50 miles or more are to be recoded as "far," we can assign the values of 50–154 as "far."

This gives a new categorical variable that is stored in column 2 (C2-T) of Figure 2.14.

Notice that the second column heading in Figure 2.14 is C2-T Recoded Commuting Distance, which means that this is a text column because the recoded data now represents the two distinct categories of "near" and "far."

To draw a bar chart using Minitab, select **Graph** from the top menu bar and then **Bar Chart**. We then need to select the type of bar chart that we want to draw. For instance, if we want to draw a simple bar chart, select **Simple**, as shown in Figure 2.15.

After hitting **OK**, we can then choose to graph the categorical variable labeled as Recoded Commuting Distance, which describes if the round-trip commuting distance is "near" or "far," as is shown in Figure 2.16.

Selecting **OK** gives the bar chart for the round-trip commuting distance that is presented in Figure 2.17.

↓	C1	C2-T
	Commuting Distance	Recoded Commuting Distance
1	26	Near
2	74	Far
3	50	Far
4	32	Near
5	48	Near
6	18	Near
7	44	Near
8	66	Far
9	88	Far
10	42	Near
11	56	Far
12	68	Far
13	144	Far
14	58	Far
15	62	Far

FIGURE 2.14
Portion of Minitab worksheet containing a text column representing the recoded commuting distance as near or far.

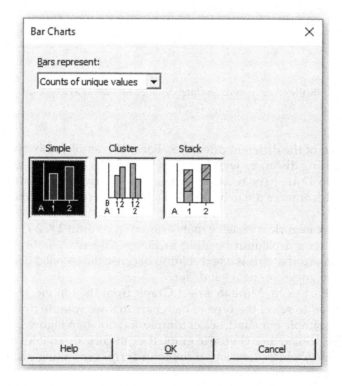

FIGURE 2.15
Minitab dialog box to select a simple bar chart.

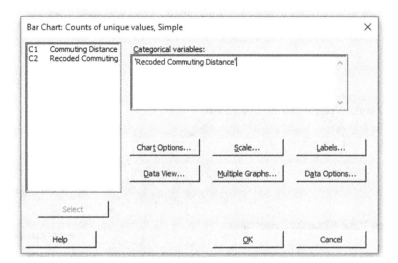

FIGURE 2.16
Bar chart dialog box.

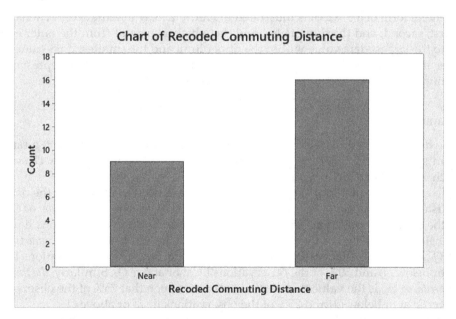

FIGURE 2.17
Bar chart illustrating the number of executives categorized as commuting less than 50 miles (near) or 50 miles or more (far).

2.8 Boxplots

In addition to histograms, *boxplots* (also known as box and whisker plots) can be created to illustrate the distribution of a single continuous variable. A boxplot can be constructed by first breaking the data into quarters, or quartiles, and then calculating upper and lower whiskers and identifying any outliers.

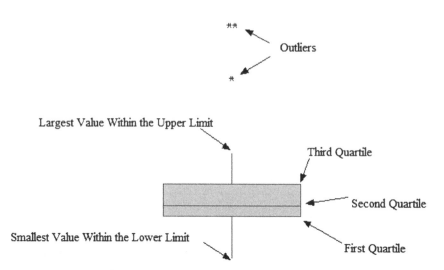

FIGURE 2.18
General form of a boxplot.

The general form of a boxplot is illustrated in Figure 2.18, where the box portion consists of the first, second, and third quartiles, and the whiskers extend from the outer edges of the box to the largest data value within the upper limit and the smallest data value within the lower limit. Any outliers are identified with an asterisk (*) and they represent those observations that lie outside of the upper and lower limits.

Example 2.3

The data presented in Table 2.3 gives the average number of hours that a random sample of 12 full-time college students spend studying for their classes each week (already in ascending order).

To create a boxplot for this set of data, we always need to make sure the data is listed in ascending order and then find the quartiles. The *first quartile*, or Q_1, is the value that partitions the data set such that 25% of the observations lie at or below Q_1 and 75% of the observations lie at or above Q_1. The *second quartile*, or Q_2, is the value that partitions the data such that 50% of the observations lie at or below Q_2 and 50% of the observations lie at or above Q_2. Similarly, *the third quartile*, or Q_3, is the value that partitions the data such that 75% of the observations lie at or below Q_3 and 25% of the observations lie at or above Q_3.

The value of Q_1 is the value that is found in position $(n + 1)/4$, where n is the sample size. If this value is not an integer, then interpolation is used. Similarly, the value of Q_3 is the value that is found in position $3(n + 1)/4$, and if this value is not an integer, then interpolation is used.

TABLE 2.3

Number of Hours a Random Sample of 12 Full-Time College Students Spend Studying for Their Classes Each Week and the Corresponding Position Number

Position	1	2	3	4	5	6	7	8	9	10	11	12
Data	5	6	7	8	8	9	10	12	15	18	19	40

For the data in Table 2.3, in order to find Q_1, we would need to locate the data value that is in the $(12 + 1)/4 = 3.25$th position (since $n = 12$), but since this position is not an integer, we would interpolate as follows:

$$Q_1 = x_3 + 0.25(x_4 - x_3) = 7 + 0.25(8 - 7) = 7.25$$

where x_3 is the value of the observation in the 3rd position, x_4 is the value of the observation in the 4th position, and 0.25 is the decimal portion of the 3.25th position.

To see what the first quartile is showing, first take 25% of the sample size:

$$(0.25)(12) = 3$$

The first quartile shows that 25% of the observations (i.e., there are three observations, namely 5, 6, and 7) lie at or below $Q_1 = 7.25$, as is illustrated in Table 2.4.

Similarly, for Q_3, we would locate the value in the $3(12 + 1)/4 = 9.75$th position, but since this position is not an integer, we would interpolate as follows:

$$Q_3 = x_9 + 0.75(x_{10} - x_9) = 15 + 0.75(18 - 15) = 17.25$$

where x_9 is the value of the observation in the 9th position, x_{10} is the value of the observation in the 10th position, and 0.75 is the decimal portion of the 9.75th position.

To see what the third quartile is showing, first take 75% of the sample size:

$$(0.75)(12) = 9$$

There are nine observations, namely 5, 6, 7, 8, 8, 9, 10, 12, and 15, that lie at or below $Q_3 = 17.25$, as is illustrated in Table 2.5.

The median position, or Q_2, is found by taking the value that partitions a rank-ordered set of data in half. For a sample of size n, if there are an odd number of observations, the median is the observation in position $(n + 1)/2$, and if there are

TABLE 2.4

Number of Hours a Random Sample of 12 Full-Time College Students Spend Studying for Their Classes Each Week and the Corresponding Position Number Showing the First Quartile (Q_1)

Position	1	2	3	4	5	6	7	8	9	10	11	12
Data	5	6	7	8	8	9	10	12	15	18	19	40

$Q_1 = 7.25$

TABLE 2.5

Number of Hours a Random Sample of 12 Full-Time College Students Spend Studying for Their Classes Each Week and the Corresponding Position Number Showing the Third Quartile (Q_3)

Position	1	2	3	4	5	6	7	8	9	10	11	12
Data	5	6	7	8	8	9	10	12	15	18	19	40

$Q_3 = 17.25$

an even number of observations, the median is the average of the observations that are in positions $n/2$ and $(n + 2)/2$.

Since there is an even number of observations for the data in Table 2.3, the median would be the average of the observations that fall in the $n/2 = 12/2 = $ 6th and $(n + 2)/2 = (12 + 2)/2 = $ 7th positions, as follows:

$$Q_2 = \frac{9 + 10}{2} = 9.50$$

To see what the second quartile is showing, first take 50% of the sample size:

$$(0.50)(12) = 6$$

There are six observations, namely 5, 6, 7, 8, 8, and 9, that lie at or below $Q_2 = 9.50$, as is shown in Table 2.6.

Once we have found the quartiles, the upper and lower limits of the boxplot can then be calculated using the following formulas:

$$\text{Upper Limit} = Q_3 + 1.5(Q_3 - Q_1)$$

$$\text{Lower Limit} = Q_1 - 1.5(Q_3 - Q_1)$$

The whiskers of the boxplot extend out to the *largest and smallest observations that lie within these upper and lower limits.* Outliers are then identified with an asterisk as those observations that lie outside of the whiskers.

For our example, the upper and lower limits can be calculated as follows:

$$\text{Upper Limit} = Q_3 + 1.5(Q_3 - Q_1) = 17.25 + 1.5(17.25 - 7.25) = 32.25$$

$$\text{Lower Limit} = Q_1 - 1.5(Q_3 - Q_1) = 7.25 - 1.5(17.25 - 7.25) = -7.75$$

The largest data value that lies within the upper limit is 19 and the smallest data value that lies within the lower limit is 5, so the whiskers extend to these values that lie within the upper and lower limits.

Any observations that fall outside of either the upper or the lower limits are outliers and they are identified with an asterisk. For this example, the only value that lies outside of the upper or lower limits is the observation 40. Putting this all together gives the boxplot that is presented in Figure 2.19.

TABLE 2.6

Number of Hours a Random Sample of 12 Full-Time College Students Spend Studying for Their Classes Each Week and the Corresponding Position Number Showing the Second Quartile (Q_2)

Position	1	2	3	4	5	6	7	8	9	10	11	12
Data	5	6	7	8	8	9	10	12	15	18	19	40

$$Q_2 = 9.50$$

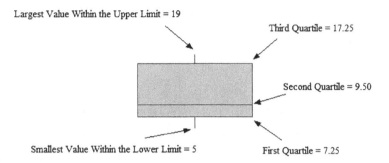

FIGURE 2.19
Boxplot of the data from Table 2.3.

2.9 Using Minitab to Create Boxplots

To create a boxplot with Minitab, select **Boxplot** under the **Graphs** bar. Notice that you can select to plot a single simple boxplot or you can select to plot multiple boxplots with groups. The dialog box for creating a simple boxplot is presented in Figure 2.20.

After selecting the appropriate variable, Minitab will create the boxplot that is presented in Figure 2.21.

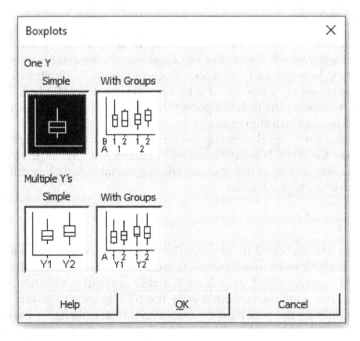

FIGURE 2.20
Minitab dialog box for creating a boxplot.

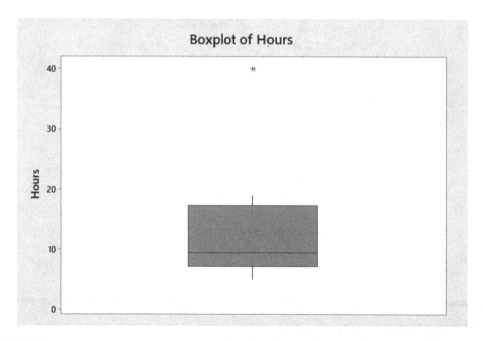

FIGURE 2.21
Boxplot of the number of hours a random sample of 12 full-time college students spends studying for their classes each week.

2.10 Scatterplots

Although histograms, bar charts, and boxplots can be used to graphically summarize the properties of a single variable, there may be occasions when we are interested in illustrating the relationship between *two* variables. We can graphically illustrate the relationship between two *continuous* variables x and y by creating what is called a *scatterplot*. A *scatterplot* is simply a graph of the ordered pairs (x, y) of the observations for two continuous variables plotted on the Cartesian plane.

Typically, we describe the relationship between two continuous variables by how one variable influences the other. We call the *response* or *dependent variable* (the y-variable) as the variable that is influenced by or depends on another variable, which is called the *predictor* or *independent variable* (the x-variable).

Example 2.4

Suppose we are interested in whether the score obtained on the mathematics portion of the SAT examination taken in high school is related to the first-year grade point average (GPA) in college. For this particular situation, the score received on the mathematics portion of the SAT examination (on a scale of 200–800) would be the predictor or independent variable (the x-variable) and the first-year GPA (on a scale of 0.00–4.00) would be the response or dependent variable (the y-variable).

TABLE 2.7

SAT Mathematics Scores (x) and First-Year GPA (y) for a Random Sample of Ten College
Students

Observation Number	SAT Math Score (SATM) x	Grade Point Average (GPA) y
1	750	3.67
2	460	1.28
3	580	2.65
4	600	3.25
5	500	3.14
6	430	2.82
7	590	2.75
8	480	2.00
9	380	1.87
10	620	3.46

Table 2.7 gives a data set that consists of the score received on the mathematics portion of the SAT examination (SATM) and the associated first-year GPA for a random sample of ten first-year college students.

We could either draw the scatterplot by hand by simply plotting each of the ordered pairs (x, y) on the Cartesian plane or have Minitab create the scatterplot for us.

2.11 Using Minitab to Create Scatterplots

Using Minitab to create scatterplots is very easy. The data is required to be entered in two separate columns in a Minitab worksheet—one column that consists of the independent variable (or x-variable) and the other column that consists of the dependent variable (or y-variable). Under the **Graph** tab, select **Scatterplot** to bring up the dialog box, as presented in Figure 2.22.

Selecting a simple scatterplot gives the dialog box in Figure 2.23, where the dependent variable (the y-variable) and the independent variable (x-variable) need to be specified in the appropriate order to generate the graph of the scatterplot as in Figure 2.24.

2.12 Marginal Plots

Another interesting plot that can be used to assess the relationship between two variables is called a *marginal plot*. A *marginal plot* consists of a scatterplot to assess the relationship between two variables along with a graph that shows the distribution for each of the two variables in the margins. A marginal plot consists of either a histogram for each variable along with a scatterplot, a boxplot for each variable along with a scatterplot, or a dotplot for each variable along with a scatterplot.

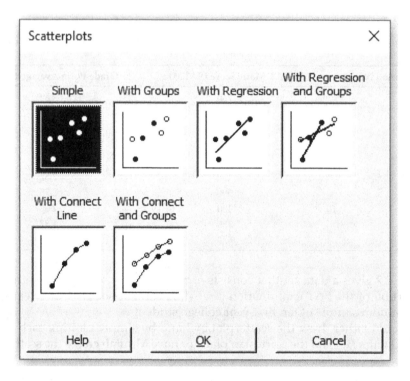

FIGURE 2.22
Minitab dialog box for selecting the type of scatterplot.

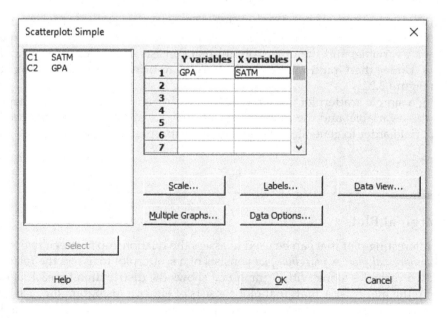

FIGURE 2.23
Minitab dialog box to select the independent and dependent variables.

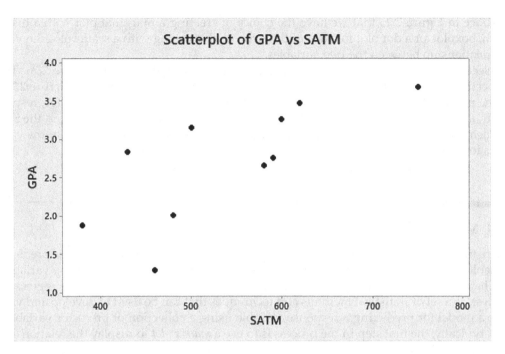

FIGURE 2.24
Scatterplot of first-year GPA versus SATM.

2.13 Using Minitab to Create Marginal Plots

Suppose that we want to create a marginal plot for the data given in Table 2.7. The Minitab commands require selecting **Marginal Plot** under the **Graph** tab. This would give the marginal plots dialog box that is provided in Figure 2.25.

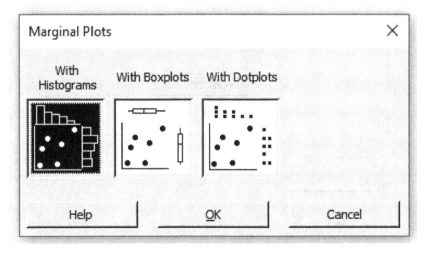

FIGURE 2.25
Marginal plot dialog box.

Notice in Figure 2.25 that we have the choice of creating a marginal plot with a histogram, boxplot, or a dot plot for each of the two variables along with a scatterplot showing the relationship between the two variables.

Selecting the marginal plot with histograms gives the dialog box in Figure 2.26. The respective graph formed on specifying the *y*- and *x*- variables is presented in Figure 2.27.

The marginal plot in Figure 2.27 provides the histograms for SATM and GPA, respectively, along with a scatterplot of the relationship between the first-year GPA and the SAT mathematics score. Figure 2.28 gives the marginal plot with boxplots for each of the variables for the data given in Table 2.4.

2.14 Matrix Plots

There may be situations where we want to consider whether a set of continuous predictor variables (i.e., a set of *x*-variables) is related to a single continuous variable (the *y*-variable). In Chapter 8, multiple regression analysis will be described in detail. Multiple regression analysis is a set of statistical methods and techniques that can be used to develop and validate a model for predicting a response variable using a collection of predictor variables. And typically, the first step in the process is to use a *matrix plot* to display the relationship of all the individual predictor variables to the given response variable. A *matrix plot* is simply a collection of scatterplots of the response variable versus each individual predictor variable.

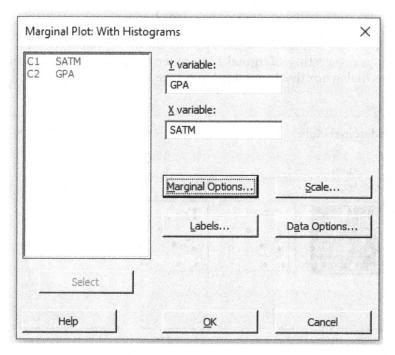

FIGURE 2.26
Dialog box for a marginal plot with histograms.

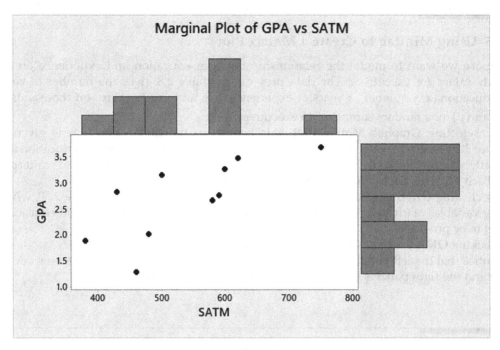

FIGURE 2.27
Marginal plot with histograms of first-year GPA versus SATM.

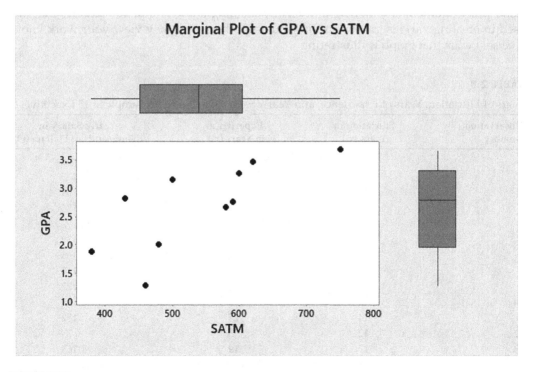

FIGURE 2.28
Marginal plot with boxplots of first-year GPA versus SATM.

2.15 Using Minitab to Create a Matrix Plot

Suppose we want to model the relationship between education and experience on the yearly salary for executives. The data provided in Table 2.8 gives the number of years of education (x_1), number of years of experience (x_2), and yearly salary on thousands of dollars (y) for a random sample of 15 executives.

By selecting **Graphs > Matrix Plot**, this brings up the dialog box that is given in Figure 2.29. Since we want to generete a scatterplot for yearly salary versus education and a scatterplot for the yearly salary versus experience, we would select a simple scatterplot for **Each Y versus Each X** as depicted in Figure 2.29.

By clicking **OK**, this would bring up the dialog box that is given in Figure 2.30, where the y-variable (yearly salary) and all the individual x-variables (experience and education) need to be provided.

Clicking **OK** gives the matrix plot that is depicted in Figure 2.31.

Notice that the left panel in Figure 2.31 is the scatterplot of yearly salary versus education and the right panel is the scatterplot of yearly salary versus experience.

2.16 Best Practices

- How to identify key descriptive measures on graphs.

Good practice entails identifying and clearly labeling any descriptive statistics that are used in creating various graphs so that you and others who may view your work know precisely what that graph is illustrating.

TABLE 2.8

Years of Education, Years of Experience, and Yearly Salary for a Random Sample of 15 Executives

Observation Number	Education in Years (x_1)	Experience in Years (x_2)	Yearly Salary in Thousands of Dollars (y)
1	12	12	95
2	13	19	145
3	16	20	164
4	14	24	186
5	15	30	197
6	12	16	139
7	13	19	163
8	15	15	125
9	16	25	122
10	13	26	173
11	14	25	152
12	12	5	75
13	13	19	165
14	13	22	167
15	14	28	187

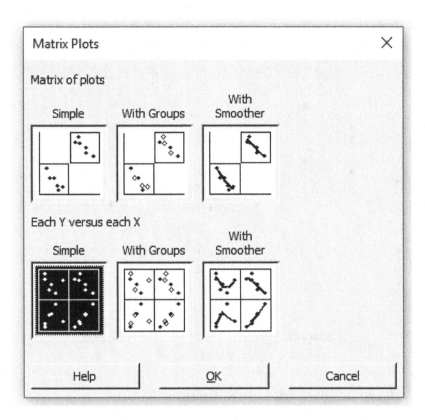

FIGURE 2.29
Minitab dialog box to generate a simple matrix plot.

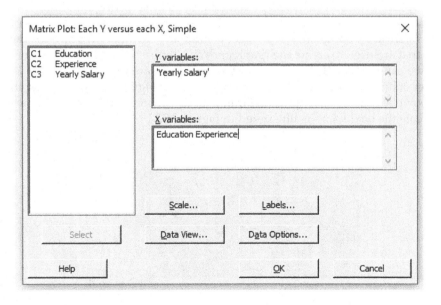

FIGURE 2.30
Minitab dialog box to specify the predictor variable (yearly salary) and the individual response variables (education and experience).

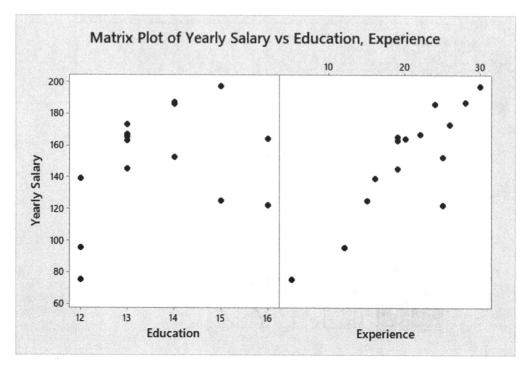

FIGURE 2.31
Matrix plot of yearly salary versus education and experience.

You can use Minitab to help you identify points and generate descriptive measures on graphs.

One interesting feature of Minitab is that when you are creating graphs, you can easily identify the descriptive measures on the graph by simply holding the mouse near the point or area of interest. By doing so, you will get a summary box that provides the descriptive measures that are used to create the graph and/or identify any specific points of data. For example, we can identify the descriptive statistics that were used to create the boxplot of the number of hours of studying data that is provided in Figure 2.21. By holding the mouse near the box portion of the boxplot, a summary box is obtained that lists all the statistics used to create the boxplot, as is illustrated in Figure 2.32.

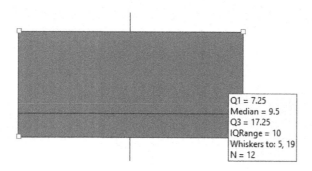

FIGURE 2.32
Portion of the boxplot and descriptive statistics of the hours studying data from Table 2.3.

Notice in Figure 2.32 that the quartiles, the interquartile range (IQR), and the data values of 5 and 19 to where the whiskers extend are provided in the summary box. And although Minitab can identify some of the descriptive measures used to create a graph, such as the quartiles and the *IQR*, it may not give you every descriptive measure that you are interested in.

Exercises

1. The data set in Table 2.9 consists of the profits/losses (in millions of dollars) for a random sample of 20 companies.
   a. Draw a histogram of the company profits/losses for this sample of companies using the default settings in Minitab.
   b. Using Minitab, draw a histogram with 20 bins of the company profits/losses for this sample of companies.
   c. Using Minitab, draw a histogram with 10 bins of the company profits/losses for this sample of companies.
   d. Which of these three histograms is more revealing as to the distribution of the company profits/losses?
2. Using Minitab and the profit/loss data provided in Table 2.9, create a boxplot. Find the values for Q_1, Q_2, and Q_3, the upper and lower limits, the upper and lower whiskers, and identify any outliers.
3. You may have heard that most car accidents happen within 25 miles of home. The data provided in Table 2.10 is the mileage away from home for a random sample of 36 car accidents that occurred over the course of a 2-week period.
   a. Using Minitab, draw a stem-and-leaf plot.
   b. Using Minitab, draw a boxplot.
   c. Using the stem-and-leaf plot and the boxplot, describe whether or not you believe that most accidents happen within 25 miles of home.

TABLE 2.9

Profits/Losses (in Millions of Dollars) for a Random Sample of 20 Companies

24.1	105.8	−8.5	10.5	12.0
37.2	−27.8	26.4	226.7	−5.2
55.7	−67.3	19.5	−17.2	18.6
−37.1	104.5	1.3	38.7	−5.7

TABLE 2.10

Random Sample of the Mileage Away from Home for 36 Car Accidents

2	18	27	47	29	24
18	46	38	36	12	8
26	38	15	57	26	21
37	18	5	9	12	28
28	17	4	7	26	29
46	20	15	37	29	19

4. The data set in Table 2.11 provides the yearly salaries (in thousands) for a sample of 15 executives by education, experience, gender, and type of management training (no training, management training, or financial training).

   a. Create a bar chart that shows the frequency of the sample based on gender.

   b. Using the **Cluster option,** draw a bar chart that shows the frequency of the sample based on type of management training and gender. This can be done by entering the categorical variables of gender and type of training in the categorical variables box.

5. The data set in Table 2.12 consists of a selection of variables for a ferry company that describe the revenue, time of day the ferry left the dock, number of passengers on the ferry, number of large objects on the ferry, the weather conditions, and the number of crew for a random sample of 27 ferry runs over the course of a month.

   a. Create a stem-and-leaf plot for the number of passengers.

   b. Create a bar graph that illustrates the number of ferry trips based on the time of day.

   c. Create a bar graph that illustrates the number of ferry runs based on the weather conditions.

   d. Create a boxplot for the number of large objects carried on the ferry.

   e. Create a matrix plot of revenue versus number of passengers and the number of large objects.

6. Can exercise improve your mood? A researcher was interested in whether there was a relationship between the number of hours exercised and the mood for a sample of six individuals. These individuals were asked to report the number of hours they exercised each week and they were also given a survey on their mood. The survey scores ranged from 0 to 150, where higher scores were reflective of a happier and more positive mood. The sample data is given in Table 2.13.

TABLE 2.11

Education, Experience, Gender, Training, and Yearly Salary for a Random Sample of 15 Executives

Observation Number	Education in Years (x_1)	Experience in Years (x_2)	Female (x_3)	Training (x_4)	Yearly Salary in Thousands of Dollars (y)
1	12	12	F	None	95
2	13	19	M	Management	145
3	16	20	M	Management	164
4	14	24	M	Financial	186
5	15	30	M	Financial	197
6	12	16	F	Management	139
7	13	19	M	Management	163
8	15	15	M	None	125
9	16	25	F	None	122
10	13	26	M	Financial	173
11	14	25	F	Management	152
12	12	5	F	None	75
13	13	19	M	Financial	165
14	13	22	M	Financial	167
15	14	28	F	Financial	187

TABLE 2.12

Revenue, Time of Day, Number of Passengers, Number of Large Objects, Weather
Conditions, and Number of Crew Members for a Ferry Company

Revenue	Time of Day	Number of Passengers	Number of Large Objects	Weather Conditions	Number of Crew Members
7,812	Morning	380	6	Calm	4
6,856	Noon	284	4	Calm	4
9,568	Evening	348	10	Calm	3
10,856	Morning	257	16	Rough	5
8,565	Morning	212	12	Rough	4
8,734	Noon	387	6	Calm	3
9,106	Evening	269	13	Calm	5
8,269	Evening	407	8	Rough	3
6,373	Noon	385	7	Calm	4
5,126	Noon	347	4	Calm	3
6,967	Noon	319	9	Rough	4
8,518	Morning	297	7	Rough	3
7,229	Evening	345	6	Rough	4
6,564	Morning	287	8	Rough	4
8,168	Morning	348	7	Calm	4
9,879	Evening	189	20	Calm	6
10,288	Evening	215	22	Calm	5
11,509	Morning	247	21	Rough	5
5,254	Noon	345	9	Rough	3
9,895	Morning	348	4	Calm	3
8,434	Evening	451	6	Calm	3
7,528	Noon	378	8	Rough	4
7,667	Morning	346	10	Calm	3
9,591	Evening	287	18	Calm	6
8,543	Evening	245	12	Rough	5
9,862	Morning	274	10	Calm	5
10,579	Morning	245	20	Rough	7

TABLE 2.13

Mood Score (1–150) and the Number of Hours Spent Exercising for a
Random Sample of Six Individuals

Mood Score	Number of Hours Exercising (Per Week)
105	15
75	10
90	10
60	8
35	3
20	6

a. Which variable is the response (or dependent variable) and which variable is the predictor (or independent variable)?

b. Using Minitab, draw a scatterplot of the mood score versus the number of hours exercised per week.

c. On the basis of the scatterplot, do you think exercise can improve your mood? How would you describe the trend between exercise and mood score?

Extending the Ideas

1. Table 2.14 gives the yearly starting salaries for males and females with bachelor's degrees who majored in the social sciences.

    Using Minitab, create two separate histograms in separate panels on the same graph and comment on whether you think there is a difference in yearly starting salaries for males and females with bachelor's degrees who majored in the social sciences. To put two histograms on the same graph, go to **Graph < Histogram < With Groups < Multiple Graphs** and **Show Graph Variable < In separate panels of the same graph**.

2. Suppose you want to compare the distributions for two different variables for a specific brand of coffee from ten different stores in New York City, New York, and

TABLE 2.14

Yearly Salaries for a Random Sample of 18 Males and Females

Yearly Salary for Males	Yearly Salary for Females
46,500	52,260
53,240	58,390
49,685	61,480
53,555	53,280
58,675	49,350
42,500	48,200
53,550	56,570
54,650	62,150
48,290	33,860
50,500	58,250
55,650	50,150
51,250	42,700
52,650	61,640
53,000	53,640
51,220	51,600
56,690	49,500
47,230	53,850
45,700	55,100

Kansas City, Missouri. Table 2.15 gives the coffee prices for a random sample of ten different coffee shops in New York City and Kansas City.

Using Minitab, draw a normal or bell-shaped curve with a histogram for the coffee prices from each of these different cities on the same graph by using the **With Fit** and **Groups** option in the histogram dialog box. You first need to put the data in two columns, one column that represents the coffee prices and the other column that represents the city.

3. The data set in Table 2.16 gives the enrollments for the four Connecticut State University institutions between 1993 and 2000. This data was obtained from the CSCU website: http://www.ct.edu/orse/research#facts

   a. Using Minitab, draw a boxplot that compares the enrollments for Central and Southern on a single graph. To do this with Minitab, select **Graph > Boxplot > Simple** under **Multiple Ys** as in Figure 2.33.

TABLE 2.15

Price of Coffee from Ten Different Coffee Shops in New York City, New York and Kansas City, Missouri

Price of Coffee (in Dollars)	
New York City	**Kansas City**
6.59	6.80
8.47	6.75
7.24	6.99
9.81	7.25
7.87	6.90
9.25	5.99
8.50	7.81
7.90	6.55
7.79	6.20
10.25	6.59

TABLE 2.16

Enrollment Trends for the Four Connecticut State University Institutions Between 1993 and 2000

Year	Central	Eastern	Southern	Western
1993	12,665	4,576	12,144	5,726
1994	11,959	4,523	11,652	5,583
1995	11,752	4,590	11,591	5,607
1996	11,646	4,527	11,412	5,397
1997	11,625	4,527	11,395	5,421
1998	11,686	4,724	11,264	5,372
1999	11,903	4,987	11,551	5,589
2000	12,252	5,145	12,127	5,806

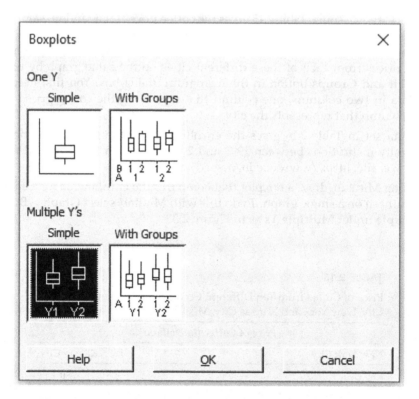

FIGURE 2.33
Minitab dialog box to draw multiple boxplots on the same graph.

   b. What does the boxplot tell you about the differences in enrollments between Central and Southern? (*Hint*: Think of the different measures of central tendency and spread).

   c. Using Minitab, draw a boxplot comparing Eastern and Western on a single graph.

   d. What does the boxplot tell you about the differences in enrollments between Eastern and Western?

3

Descriptive Representations of Data and Random Variables

3.1 Introduction

In this chapter, we describe descriptive statistics that provide us with summary measures of a variable. We begin by presenting some typical measures of central tendency such as the mean, median, and mode, and describe how to measure the spread of a variable by calculating the range, interquartile range (IQR), variance, and standard deviation. We will explore how to use Minitab to calculate such measures of central tendency and variability by working through numerous examples.

Most of this book is devoted to the practice of statistical inference. Recall that statistical inference relies on using sample statistics to estimate unknown population parameters. But before we begin a discussion on statistical inference, we first need to understand random variables along with their corresponding distributions. The notion of what sampling distributions are will be presented in addition to standard and non-standard normal distributions along with a discussion of how to calculate probabilities for random variables that have a normal distribution. We will also be introducing some of the more common discrete probability distributions such as the binomial distribution and the Poisson distribution, as well as some of the more common continuous probability distributions such as the t-distribution, the chi-square distribution, and the F-distribution. We will also learn how to use Minitab to graph probability distributions.

3.2 Descriptive Statistics

A *descriptive statistic* is a summary measure that is calculated from a sample variable. The purpose of calculating descriptive statistics is to summarize useful information or properties of a sample variable. For example, if we are interested in the first-year grade point averages (GPAs) for a sample of 100 students at a specific college, viewing the entire list of the first-year GPAs in order to get a good sense of some of the properties of this variable may not be very helpful. Instead of looking at every single observation, we could simply calculate the average GPA. Ideally, by using descriptive statistics, we can succinctly convey useful information by using a relevant numerical or summarized representation.

There are two basic types of descriptive statistics that we will discuss, namely, *measures of central tendency* and *measures of variability*.

3.3 Measures of Central Tendency

Measures of central tendency are descriptive statistics that represent the middle or center point of a quantitative sample variable. One common measure of central tendency for quantitative sample data is the *sample mean* (or the sample average). The sample mean of a variable is found by adding up all the sample observations and dividing by the sample size. The sample mean is denoted by $\bar{x}$ and can be found by using the following formula:

$$\bar{x} = \frac{\sum_{i=1}^{n} x_i}{n}$$

where n denotes the number of observations in the sample and $x_1, x_2, x_3, \ldots, x_n$ are the individual sample observations with an assigned position number.

The summation notation $\sum_{i=1}^{n} x_i$ is simply shorthand notation that is used for summing the values labeled as x_1–x_n as follows:

$$\sum_{i=1}^{n} x_i = x_1 + x_2 + x_3 + \ldots + x_n.$$

Example 3.1

Consider the data presented in Table 3.1 that gives the first-year GPAs for a random sample of five students.

To find the sample mean, we begin by labeling the individual observations, and this is done by assigning an arbitrary position number to each observation as follows:

$$x_1 = 2.56$$

$$x_2 = 3.21$$

$$x_3 = 3.56$$

$$x_4 = 2.10$$

$$x_5 = 1.87$$

When you are calculating the sample mean, the order in which the position numbers are assigned to the observations does not matter because addition is commutative.

Summing all these values, we have:

$$\sum_{i=1}^{5} x_i = x_1 + x_2 + x_3 + x_4 + x_5 = 2.56 + 3.21 + 3.56 + 2.10 + 1.87 = 13.30$$

TABLE 3.1

First-Year GPAs for a Random Sample of Five Students

2.56	3.21	3.56	2.10	1.87

The sample mean is then found by dividing this sum by the number of observations that are in the sample, which, for this example, is $n = 5$.

$$\bar{x} = \frac{\sum_{i=1}^{n} x_i}{n} = \frac{13.30}{5} = 2.66$$

Example 3.2

Table 3.2 gives the speeds (in miles per hour) for a random sample of seven cars driving on a particular stretch of highway.

The mean for this sample is calculated as follows:

$$\bar{x} = \frac{\sum_{i=1}^{n} x_i}{n} = \frac{68 + 72 + 73 + 84 + 67 + 62 + 74}{7} = \frac{500}{7} \approx 71.43$$

In addition to using descriptive measures to describe a sample variable, we can also use descriptive measures to describe characteristics or attributes of a population variable. For instance, the *population mean* is a parameter that represents the average value of a quantitative population variable. The population mean is represented by the symbol μ and is calculated as follows:

$$\mu = \frac{\sum_{i=1}^{N} x_i}{N}$$

where N denotes the total number of observations that are in the population and $x_1, x_2, x_3, \ldots, x_N$ denote the N individual population values with an assigned position number.

Example 3.3

Table 3.3 represents the population-level data that consists of the ten even numbers from 2 to 20.

The population mean can be found by adding up these values and dividing them by the total number of observations in the population ($N = 10$) as follows:

TABLE 3.2

Sample of Speeds of Seven Cars Driving on a Particular Stretch of the Highway

68	72	73	84	67	62	74

TABLE 3.3

Population of Data That Consists of the Ten Even Numbers from 2 to 20

2	4	6	8	10	12	14	16	18	20

$$\mu = \frac{\sum_{i=1}^{N} x_i}{N} = \frac{2+4+6+8+10+12+14+16+18+20}{10} = \frac{110}{10} = 11$$

Another measure of central tendency for a quantitative sample variable is called the *sample median* and is defined as the numeric value that partitions the sample in half. The median is a number such that at least 50% of the sample observations lie at or below this number and at least 50% of the sample observations lie at or above this number. You may recall from Chapter 2 that the median is also referred to as the second quartile or Q_2.

To find the median, we first need to find the *median position*. The median position is found by first putting the data in ascending order and then finding the observation that partitions the ordered data set in half. For a sample of size n, if there is an odd number of observations, the median is the observation located in position $(n + 1)/2$, and if there is an even number of observations, the median is the mean (or average) of the observations that are located in positions $n/2$ and $(n + 2)/2$.

Example 3.4

Table 3.4 consists of the number of hours per week that a random sample of five adults spend exercising.

The median of the sample in Table 3.4 can be found by first arranging the data in ascending order and assigning a position number, as is illustrated in Table 3.5.

Since there is an odd number of observations ($n = 5$), the median of this sample will correspond to the observation that is in the $(n + 1)/2 = (5 + 1)/2 = 3$rd position. Therefore, the median for this sample is the number 4.

Example 3.5

Consider the data set given in Table 3.6 that consists of the number of hours spent by a sample of ten homeowners maintaining their lawn each week (already in ascending order).

Since there is an even number of observations ($n = 10$), the median of this sample will be the average of the observations located in the $n/2 = 10/2 = 5$th

TABLE 3.4

Number of Hours Spent Exercising per Week for a Random Sample of Five Adults

2	4	7	3	8

TABLE 3.5

Data from Table 3.4 in Ascending Order with the Corresponding Position Number

Position number	1	2	3	4	5
Data	2	3	4	7	8

TABLE 3.6

Number of Hours That a Sample of Ten Homeowners Spend Working on Their Lawn Each Week

Position number	1	2	3	4	5	6	7	8	9	10
Data	2	4	5	7	9	13	15	17	28	30

and $(n + 2)/2 = (10 + 2)/2 = $ 6th positions. The observation located in the fifth position is 9, and the observation located in the sixth position is 13. The median is the average of these two values, or $(9 + 13)/2 = 11$. Notice that although the number 11 is not an actual observation in the data set, it still represents a numerical measure that cuts the data set in half because 50% (or five) of the observations lie at or below the number 11, and 50% (or five) of the observations lie at or above the number 11.

In some cases, the median is the preferred measure of central tendency as compared to the mean. This is because the median is less sensitive to outliers, as will be illustrated in the next example.

Example 3.6

Table 3.7 is a generic sample of 11 observations (already in ascending order). The mean and median for this sample are 58.55 and 22, respectively.

For the data in Table 3.7, the median may be the preferred measure of center when compared to the mean because the median is less affected by outliers. In this case, the median provides a more accurate representation of the central tendency of the underlying data set because most of the numbers are around 20 and the observations of 0 and 451 are more extreme values.

Another measure of central tendency is called the *mode*. The mode can be found for a quantitative or qualitative variable, and it is defined as the observation or observations that occur most often.

Example 3.7

The data set in Table 3.8 describes the number of hours a random sample of 12 college students spend studying for their classes each week.

The mode of the data in Table 3.7 is 8 because this is the observation that occurs most often, namely, twice.

The mode can also be found for qualitative variables, as is illustrated in the following example.

TABLE 3.7

Sample of 11 Observations

0	12	17	18	20	22	24	25	27	28	451

TABLE 3.8

Number of Hours a Random Sample of 12 College Students Spend Studying Each Week

5	6	7	8	8	9	10	12	15	18	19	40

TABLE 3.9

Political Affiliations for a Random Sample of Five Shoppers

Republican	Democrat	Democrat	Independent	Republican

Example 3.8

The data given in Table 3.9 consist of the political party affiliations for a random sample of five shoppers surveyed at a local shopping mall.

For this example, there will be two modes, namely, Democrat and Republican. This is because both of these observations occur most often (i.e., twice).

Although the mode can be found for both quantitative and qualitative variables, one limitation can arise when using the mode as a measure center for quantitative variables because many such variables may not have any observations that occur on more than one occasion. For instance, think of asking a random sample of students what their GPA is, you will probably not be able to find a mode because it is unlikely that two GPAs will be exactly the same.

3.4 Measures of Variability

Descriptive statistics that measure the variability or spread of a variable are numeric summaries that can be used to describe how the sample observations differ from each other. To illustrate the concept of variability, suppose I purchase ten 1-pound bags of my favorite brand of coffee from ten different stores around town. I would expect each of the 1-pound bags to weigh approximately the same. In other words, I would not expect much spread or variability in the weights of the different bags of the same brand of coffee purchased at various locations. However, if I asked ten of my friends to tell me how much money they spent to put gasoline in their cars this week, I would expect a significant amount of variation in the amount spent because of differences in the types of cars and distances driven, as well as differences in the prices of gasoline.

One simple measure of the spread for a quantitative variable is called the *range*. The range is found by subtracting the smallest value from the largest value.

Example 3.9

Consider the data previously given in Table 3.1, which consists of the first-year GPAs for a random sample of five students.

The range for this set of data can be found by taking the smallest observation and subtracting it from the largest observation as follows:

$$range = (largest\ value - smallest\ value) = (3.56 - 1.87) = 1.69$$

While the range is a very simple statistic to calculate, one concern when using the range to assess variability is that the range is based only on the two most extreme observations of the variable. And because of this, the range does not reflect the variability because no other observations are used that are not at the two extremes.

Another measure of variability for a quantitative variable that does not rely on the two most extreme observations is the *IQR*. The IQR is found by taking the difference between the third quartile and the first quartile, where the quartiles of a variable are values that partition the data into quarters:

$$IQR = Q_3 - Q_1$$

Recall from Chapter 2 that to find the quartiles of a data set, the data must first be put in ascending order and the quartiles are found such that a given percentage of the observations lie below or above the respective quartile. For instance, the first quartile, or Q_1, represents the value such that at least 25% of the data fall at or below this value and at least 75% of the data fall at or above this value. Similarly, the third quartile, or Q_3, represents the value such that at least 75% of the data fall at or below this value and at least 25% of the data fall at or above this value. Once the data is in ascending order, the value of Q_1 is the observation in the $(n + 1)/4$th position. Similarly, the value of Q_3 is the observation in the $3(n + 1)/4$th position. If these positions are not integer values, then interpolation is used, as described in detail in Section 2.8.

Example 3.10

Suppose we want to find Q_1 and Q_3 for the data provided in Table 3.6 (the data set of the number of hours a random sample of ten homeowners spend maintaining their lawn each week). Since this sample has an even number of observations ($n = 10$), Q_1 will be the value in the $(10 + 1)/4 = 2.75$th position. Because this value is not an integer, we then interpolate to find the value of Q_1:

$$Q_1 = x_2 + 0.75(x_3 - x_2) = 4 + 0.75(5 - 4) = 4.75$$

Similarly, to find Q_3, it will be the value in the $3(10 + 1)/4 = 8.25$th position, and since this is not an integer, we would interpolate as follows:

$$Q_3 = x_8 + 0.25(x_9 - x_8) = 17 + 0.25(28 - 17) = 19.75$$

Therefore, the IQR would be:

$$IQR = Q_3 - Q_1 = 19.75 - 4.75 = 15.00$$

Similar to using the range to assess variability, the IQR also tends not to be the best measure of variability because it also uses only a limited amount of the data.

Two additional measures of spread for a quantitative sample variable are the *sample variance* and the *sample standard deviation*. These measures provide information that is generally more useful than the range or the IQR because both the sample variance and the sample standard deviation use *all* of the individual observations in a given sample. These measures are typically used to summarize how the sample observations vary about the mean.

The sample variance, denoted as s^2, can be found using the following formula:

$$s^2 = \frac{\sum_{i=1}^{n}(x_i - \bar{x})^2}{n-1}$$

where n is the sample size, $\bar{x}$ is the sample mean, and x_i represents the ith observation.

The sample variance measures the squared difference between each observation and the sample mean. Notice that this statistic is found by dividing the sum of the squared differences between each observation and the mean by $n-1$ rather than by n. This adjustment is made so that the sample variance (which is a sample statistic) can provide a more accurate estimate of the population variance (which is a population parameter). We will be describing how the sample variance can be used to estimate a population variance in more detail in Chapter 4.

The *sample standard deviation*, denoted as s, is the square root of the sample variance, and it is calculated as follows:

$$s = \sqrt{s^2} = \sqrt{\frac{\sum_{i=1}^{n}(x_i - \bar{x})^2}{n-1}}$$

Typically, the sample variance and the sample standard deviation are better measures of variability because they both use all of the observations in a data set, as compared to the range and IQR that use only a few select values. The sample standard deviation is used more than the sample variance as a measure of variability because the units of the sample standard deviation are the same as those of the variable itself, whereas the units of the sample variance are in units squared.

Example 3.11

For the sample of five first-year GPAs as presented in Table 3.1, we found that the sample mean is 2.66. We can find the sample variance and sample standard deviation by first taking the difference between each sample observation and the sample mean, and then squaring this difference as illustrated in Table 3.10.

The first column in Table 3.10 gives the actual observations with an assigned position number, the second column is the difference between each of the sample observations and the mean $(x_i - \bar{x})$, and the third column is the squared difference between each of the sample observations and the sample mean $(x_i - \bar{x})^2$.

TABLE 3.10

Difference and Squared Difference Between Each Observation and the Sample Mean for the Data Given in Table 3.1

x_i	$(x_i - \bar{x})$	$(x_i - \bar{x})^2$
$x_1 = 2.56$	$2.56 - 2.66 = -0.10$	$(-0.10)^2 = 0.0100$
$x_2 = 3.21$	$3.21 - 2.66 = 0.55$	$(0.55)^2 = 0.3025$
$x_3 = 3.56$	$3.56 - 2.66 = 0.90$	$(0.90)^2 = 0.8100$
$x_4 = 2.10$	$2.00 - 2.66 = -0.56$	$(-0.56)^2 = 0.3136$
$x_5 = 1.87$	$1.87 - 2.66 = -0.79$	$(-0.79)^2 = 0.6240$
		$\sum_{i=1}^{5}(x_i - \bar{x})^2 \approx 2.0602$

To find the sample variance, we first need to add the values in the third column of Table 3.10 as follows:

$$\sum_{i=1}^{5}(x_i - \bar{x})^2 = 0.0100 + 0.3025 + 0.8100 + 0.3136 + 0.6240 \approx 2.0602$$

Then, we divide this sum by $n - 1$ as follows:

$$s^2 = \frac{\sum_{i=1}^{n}(x_i - \bar{x})^2}{n-1} = \frac{2.0602}{5-1} = 0.5151$$

To find the sample standard deviation, we simply take the square root of the variance:

$$s = \sqrt{s^2} = \sqrt{0.5151} \approx 0.7177$$

If we have data for an entire population, then we could calculate the *population variance* and the *population standard deviation* using the following formulas:

$$\sigma^2 = \frac{\sum_{i=1}^{N}(x_i - \mu)^2}{N}$$

$$\sigma = \sqrt{\frac{\sum_{i=1}^{N}(x_i - \mu)^2}{N}}$$

where μ is the symbol for the population mean, σ^2 is the symbol for the population variance, σ is the symbol for the population standard deviation, and N represents the size of the population.

Example 3.12

To find the variance and standard deviation for the population-level data presented in Table 3.3, we would take the sum of the squared difference between each observation and the population mean, and then divide by the population size, which is $N = 10$, to obtain the population variance of $\sigma^2 = 33.01$ and the population standard deviation of $\sigma = 5.745$.

3.5 Using Minitab to Calculate Descriptive Statistics

Minitab can be used to calculate many different types of descriptive statistics for quantitative sample data. Using the sample of five GPAs given in Table 3.1, we can have

Minitab calculate the descriptive statistics of interest by first entering the data into a Minitab worksheet column and then selecting **Stat** from the top menu bar and then **Basic Statistics** and **Display Descriptive Statistics**. This gives the **Display Descriptive Statistics** dialog box that is presented in Figure 3.1.

To select the variable, or variables for which you wish to calculate the descriptive statistics for, simply highlight the variables of interest and choose **Select**. This will place the variables or variables of interest in the **Variables** box.

Clicking on the **Statistics** tab in Figure 3.1 will bring up the descriptive statistics dialog box in Figure 3.2, which allows you to select the specific descriptive statistics you want Minitab to calculate for you.

Checking the boxes for the mean, standard deviation, variance, Q_1, median, Q_3, IQR, minimum, maximum, range, and the number of non-missing and missing data values provides the Minitab printout that is given in Figure 3.3.

3.6 More on Statistical Inference

Often it can be very difficult or even impossible to obtain data for an entire population. However, we can make inferences about various population parameters by using

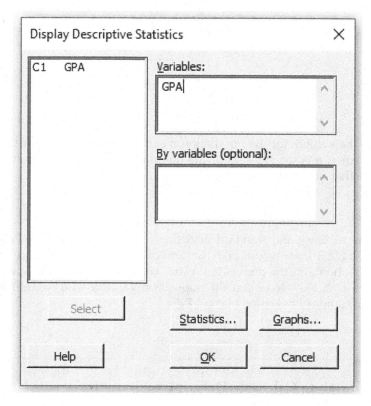

FIGURE 3.1
Minitab dialog box for calculating descriptive statistics for the GPA data presented in Table 3.1.

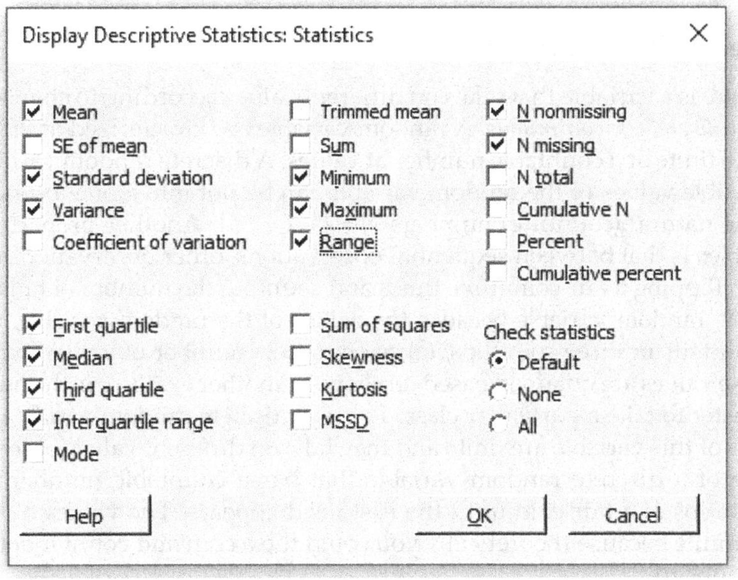

FIGURE 3.2
Minitab dialog box to select descriptive statistics.

Descriptive Statistics: GPA

Statistics

Variable	N	N*	Mean	StDev	Variance	Minimum	Q1	Median	Q3	Maximum	Range	IQR
GPA	5	0	2.660	0.718	0.515	1.870	1.985	2.560	3.385	3.560	1.690	1.400

FIGURE 3.3
Minitab output of selected descriptive statistics for the data presented in Table 3.1.

information from a representative sample that is obtained from the population of interest. For example, we can use the sample mean (a statistic) to estimate the population mean (a parameter), and similarly, we can use the sample variance (a statistic) and the sample standard deviation (a statistic) to estimate the population variance (a parameter) and the population standard deviation (a parameter). Thus, $\bar{x}$, the sample mean, can be used in combination with s, the sample standard deviation, to provide a good estimate of the population mean μ and the population standard deviation σ.

But before we can use sample statistics to make inferences about population parameters, we need to consider some properties of the sample statistic that we are using. In particular, we need to consider what are the possible values of the statistic in addition to how likely it is to observe these values. Similar to how a histogram can illustrate the distribution of a given variable along with its respective frequencies or probabilities, the distribution of a sample statistic can be used to display the values of the sample statistic along with their respective probabilities. In order to describe distributions for different sample statistics, we first need to discuss random variables and their associated probability distributions.

3.7 Discrete Random Variables

A *random variable* is a variable that takes on different values according to chance. A random variable can be *discrete* or *continuous*. A random variable is characterized as *discrete* if it can assume either a finite or a countable number of values. A discrete random variable is countable if the possible values of the random variable can be put into a one-to-one correspondence with the natural (counting) numbers $N = \{1, 2, 3, ...\}$. Another property of discrete random variables is that between sequential observations, other observations do not exist.

For example, flipping a fair coin three times and counting the number of tails is an example of a discrete random variable because the values of the random variable, which is the total number of tails in three coin flips, takes on a finite number of values, namely, 0, 1, 2, and 3, and these values are obtained based on chance. Another example is the number of students that register to take a particular class. This is a discrete random variable because the possible values of this variable are finite and they take on different values based on chance.

An example of a discrete random variable that has a countable number of values is the number of flips of a fair coin until the first head appears. The values of this random variable are infinite because theoretically you could flip a coin and count indefinitely until the first head appears.

The probability pattern that is associated with the possible values of a discrete random variable constitutes what is called the *probability distribution* of the random variable. For instance, if the random variable is the number of tails counted after flipping a fair coin three times, then the probability pattern that corresponds to each of the possible values of the random variable is presented in Table 3.11.

The probability distribution in Table 3.11 was constructed by first listing the set of all possible outcomes of the random variable and then finding the corresponding probabilities. The set of all of the outcomes obtained from flipping a fair coin three times is as follows:

$$\{HHH, HHT, HTH, HTT, THH, THT, TTH, TTT\}$$

Since there are a total of eight possible outcomes and each outcome has the same probability of occurring, the probability of observing 0 tails in three coin flips is 1/8 (this corresponds to observing HHH), the probability of observing one tail in three coin flips is 3/8 (this corresponds to observing HHT, HTH, or THH), the probability of observing two tails in three coin flips is 3/8 (this corresponds to observing HTT, THT, or TTH), and the probability of observing three tails in three coin flips is 1/8 (this corresponds to observing TTT).

Typically, random variables are represented as uppercase letters and the specific outcomes of a random variable as lowercase letters. In counting the number of tails from flipping a fair coin three times, the uppercase letter X would represent the random variable

TABLE 3.11

Probability Distribution for the Discrete Random Variable That Consists of Counting the Number of Tails after Flipping a Fair Coin Three Times

Number of Tails (X)	Probability $P(x_i)$
$x_1 = 0$	$P(x_1) = 1/8$
$x_2 = 1$	$P(x_2) = 3/8$
$x_3 = 2$	$P(x_3) = 3/8$
$x_4 = 3$	$P(x_4) = 1/8$

itself and lowercase letters would represent the specific values that the random variable X can take on, which in this case would be either $x_1 = 0$, $x_2 = 1$, $x_3 = 2$, or $x_4 = 3$.

There are two properties that must hold true for the probability distribution of a discrete random variable. First, the probabilities associated with each value of the random variable must be probabilities. In other words, they must be numbers that are between 0 and 1 inclusive. Symbolically, this means:

$$0 \le P(x_i) \le 1$$

where $P(x_i)$ is the probability of observing the specific value x_i of the random variable X.

Second, all of the probabilities from the probability distribution for a discrete random variable must add up to 1:

$$\sum P(x_i) = 1$$

For the probability distribution in Table 3.11, notice that all the probabilities are between 0 and 1 inclusive and they add up to 1.

We can also summarize discrete random variables by calculating measures of center and measures of spread.

One measure of center for a discrete random variable X is the *mean* or *expected value*. This can be found by summing the product of the values of the random variable with their corresponding probabilities as follows:

$$\mu_X = E(X) = \sum x_i \cdot P(x_i)$$

where x_i is the ith observation of the random variable X, and $P(x_i)$ is the probability that observation x_i will occur.

Measures of spread for a discrete random variable X are the variance $\left(\sigma_X^2\right)$ and standard deviation (σ_X). These can be found by using the following formulas:

$$\sigma_X^2 = \sum (x_i - \mu_X)^2 \cdot P(x_i)$$

$$\sigma_X = \sqrt{\sigma_X^2}$$

where x_i is the ith observation of the random variable, $P(x_i)$ is the probability that observation x_i will occur, and μ_X is the mean of the random variable X.

Example 3.13

To find the mean, or expected value, of the discrete random variable given in Table 3.11, we first find the products of each of the possible outcomes of the random variable with their corresponding probabilities and add them together as follows:

$$\mu_X = E(X) = \sum_{i=1}^{n} x_i \cdot P(x_i) = 0\left(\frac{1}{8}\right) + 1\left(\frac{3}{8}\right) + 2\left(\frac{3}{8}\right) + 3\left(\frac{1}{8}\right) = 1.50$$

The mean, or expected value, of the random variable X is $\mu_X = E(X) = 1.50$.

The variance of the discrete random variable X given in Table 3.11 is found by taking the squared difference between each value of the random variable and the mean of the random variable multiplied by the respective probability, and then adding them together as follows:

$$\sigma_X^2 = \sum_{i=1}^{n}(x_i - \mu_X)^2 \cdot P(x_i) = (0-1.5)^2\left(\frac{1}{8}\right) + (1-1.5)^2\left(\frac{3}{8}\right)$$

$$+ (2-1.5)^2\left(\frac{3}{8}\right) + (3-1.5)^2\left(\frac{1}{8}\right) = 0.75$$

The standard deviation of the discrete random variable X is found by taking the square root of the variance:

$$\sigma_X = \sqrt{\sigma_X^2} = \sqrt{0.75} \approx 0.866$$

The graph of the probability distribution of a random variable can be a useful way to visualize the probability pattern of the random variable. This graph can be created by plotting the possible values of the random variable X on the x-axis versus the respective probabilities for each of the values of the random variable on the y-axis. The graph of the probability distribution from Table 3.11 is illustrated in Figure 3.4.

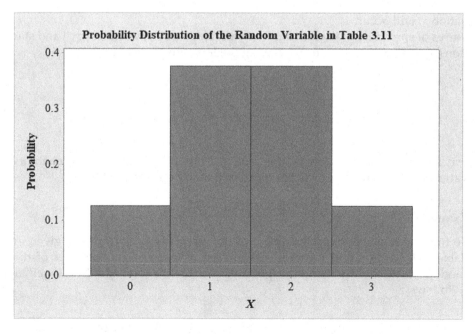

FIGURE 3.4
Graph of the probability distribution for the discrete random variable X given in Table 3.11.

3.8 Sampling Distributions

In order to use sample statistics, such as the sample mean and the sample standard deviation, to estimate the mean and standard deviation of a larger population, we need to consider the probability pattern of these sample statistics. In other words, we need to describe the possible values of the sample statistic along with their respective probabilities. The probability distribution of a sample statistic is referred to as the *sampling distribution* of the sample statistic.

The following example illustrates how the sampling distribution of the sample mean can be created.

Example 3.14

Consider the probability distribution for the discrete random variable X that is given in Table 3.12, and suppose that this random variable X represents population-level data.

We can graph this probability distribution by plotting the values of the random variable against their respective probabilities, as shown in Figure 3.5.

We can calculate the mean and standard deviation for this probability distribution as follows:

$$\mu_X = \sum x_i \cdot P(x_i) = 1(0.25) + 2(0.25) + 3(0.25) + 4(0.25) = 2.50$$

$$\sigma_X = \sqrt{\sum (x_i - \mu_X)^2 \cdot P(x_i)}$$

$$= \sqrt{(1-2.5)^2 (0.25) + (2-2.5)^2 (0.25) + (3-2.5)^2 (0.25) + (4-2.5)^2 (0.25)}$$

$$= \sqrt{1.25} \approx 1.12$$

So our population has a mean of 2.50 and a standard deviation of 1.12.

Now suppose that we draw random samples of size $n = 2$ from the random variable X that is given in Table 3.12 and we list every one of these samples. If we sample with replacement (meaning that we allow for the second outcome to be the same as the first), then every possible sample of size $n = 2$ would be those that are listed in Figure 3.6.

Assuming that each random sample of size $n = 2$ has the same probability of being selected as any other sample, and because there are 16 possible samples of size $n = 2$, then the probability of drawing any one sample of size $n = 2$ from those listed in Figure 3.6 would be $1/16 = 0.0625$.

TABLE 3.12

Probability Distribution for the Random Variable X

X	$P(x_i)$
$x_1 = 1$	0.25
$x_2 = 2$	0.25
$x_3 = 3$	0.25
$x_4 = 4$	0.25

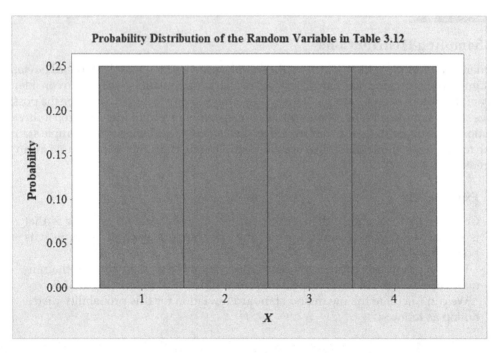

FIGURE 3.5
Graph of the probability distribution for the discrete random variable X given in Table 3.12.

$$\{1, 1\} \quad \{2, 1\} \quad \{3, 1\} \quad \{4, 1\}$$

$$\{1, 2\} \quad \{2, 2\} \quad \{3, 2\} \quad \{4, 2\}$$

$$\{1, 3\} \quad \{2, 3\} \quad \{3, 3\} \quad \{4, 3\}$$

$$\{1, 4\} \quad \{2, 4\} \quad \{3, 4\} \quad \{4, 4\}$$

FIGURE 3.6
All possible samples of size $n = 2$ for the population-level data given in Table 3.12.

We are now going to create a *new random variable* that represents the mean of all of these samples of size $n = 2$. The sample means for every sample given in Figure 3.6 is presented in Figure 3.7.

To create the probability distribution for this new random variable that represents the sample mean for all of the samples of size $n = 2$ (this probability distribution is referred to as the *sampling distribution of the sample mean*), we take each possible value of the random variable and assign the corresponding probability. For instance, since there are four ways to obtain a sample mean of 2.5, as can be seen in Figure 3.7, and the probability of selecting each sample is equally likely with a probability of 0.0625, then the probability of drawing a sample of size $n = 2$ and getting a sample mean of 2.5 would be:

$$4\left(\frac{1}{16}\right) = 4(0.0625) = 0.2500$$

$$\frac{1+1}{2} = 1 \qquad \frac{2+1}{2} = 1.5 \qquad \frac{3+1}{2} = 2 \qquad \frac{4+1}{2} = 2.5$$

$$\frac{1+2}{2} = 1.5 \qquad \frac{2+2}{2} = 2 \qquad \frac{3+2}{2} = 2.5 \qquad \frac{4+2}{2} = 3$$

$$\frac{1+3}{2} = 2 \qquad \frac{2+3}{2} = 2.5 \qquad \frac{3+3}{2} = 3 \qquad \frac{4+3}{2} = 3.5$$

$$\frac{1+4}{2} = 2.5 \qquad \frac{2+4}{2} = 3 \qquad \frac{3+4}{2} = 3.5 \qquad \frac{4+4}{2} = 4$$

FIGURE 3.7
Means for all the samples of size $n = 2$ listed in Figure 3.6.

Similarly, for all the other possible values of the sample mean for the samples of size $n = 2$, the means and their corresponding probabilities are presented in Table 3.13.

The sampling distribution of the sample mean as presented in Table 3.13 illustrates the probability pattern of all possible sample means for samples of size $n = 2$. The sampling distribution of the sample mean has the two properties of a probability distribution, namely, (1) all of the probabilities are between 0 and 1 inclusive, and (2) the sum of the probabilities is equal to 1.

We can graph the sampling distribution of the sample mean by plotting the possible values of the random variable (which in this case would be the values of the sample mean) on the x-axis, and the corresponding probabilities on the y-axis. Figure 3.8 gives the graph of the sampling distribution of the sample mean from the probability distribution that is given in Table 3.13.

In general, to use the sampling distribution of a sample statistic to estimate a population parameter, we would want the sampling distribution to center about the unknown population parameter of interest. Notice in Figure 3.8 that the sampling distribution of the sample mean is centered about the *true population mean of the original random variable* as given in Table 3.12 $(\mu_X = 2.50)$. We also want the standard deviation of the sampling distribution (this is also referred to as the standard error) to be as small as possible.

TABLE 3.13

Sampling Distribution of the Sample Mean for Drawing a Sample of Size $n = 2$ from the Probability Distribution of the Random Variable in Table 3.12

$\bar{x}$	$P(\bar{x})$	Percentage
1	$1/16 = 0.0625$	6.25
1.5	$2/16 = 0.1250$	12.50
2	$3/16 = 0.1875$	18.75
2.5	$4/16 = 0.2500$	25.00
3	$3/16 = 0.1875$	18.75
3.5	$2/16 = 0.1250$	12.50
4	$1/16 = 0.0625$	6.25

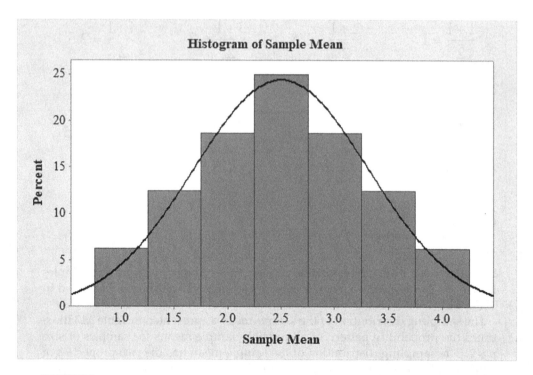

FIGURE 3.8
Graph of the sampling distribution of the sample mean from the probability distribution given in Table 3.13.

The *central limit theorem* states that as the sample size increases, the sampling distribution of the sample mean will approximately have the shape of a bell curve for *any* shape of the distribution of the underlying population that is being sampled from. The central limit theorem also states that the mean of the sampling distribution will center about the mean of the original population μ_X. It is because of this theorem that we will be able to make inferences about a population mean using the sample mean. But before we begin to use sampling distributions to make inferences about population parameters in Chapter 4, we first need to describe continuous random variables.

3.9 Continuous Random Variables

A *continuous random variable* is defined as a random variable such that between *any* two values of the random variable, other values can theoretically exist. Another way of describing a continuous random variable is a random variable that consists of an uncountable number of values. For example, consider the in-flight time it takes to fly from Boston, Massachusetts to New York City, New York. If it takes at least 75 minutes to fly from Boston to New York, we can call this the smallest possible flight time t_1, where $t_1 = 75$ minutes. It is impossible to precisely describe the next possible flight time, or flight time t_2, because it could be 83, 82, 82.5, 82.25 minutes, and so on. Thus, the flight time from Boston to New York City is a continuous random variable because between any two flight times, other flight times can possibly exist. Consequently, this means that there are an uncountable number of values for the random variable. As another example, consider the variable of blood alcohol content (BAC). It is a continuous random variable because between any two values of BAC, one could (at least theoretically) observe another BAC.

3.10 Standard Normal Distribution

Similar to discrete random variables, the relationship between a continuous random variable and its respective probability pattern can be displayed graphically. Perhaps the most common distribution of a continuous random variable is the *normal distribution*. A normal distribution has the shape of a bell curve, as can be seen in Figure 3.9.

A normal distribution is characterized by its shape, its center, and its spread about the center. The shape of a normal distribution is a bell curve. The center, or mean, of a random variable X that has the shape of a normal distribution is characterized by μ_X, and σ_X, the standard deviation of the random variable X, describes the spread or variability of the random variable about the mean, as illustrated in Figure 3.9. The horizontal axis of a normal distribution represents the possible values of the random variable X, and this axis goes indefinitely in each direction approaching but never crossing the horizontal axis. Figure 3.10 shows the comparison of two different normal distributions; one distribution that has a mean of 5 and a standard deviation of 10; and the other distribution that has a mean of –3 and a standard deviation of 4.

When a normal curve represents the probability distribution of a continuous random variable X, the area under the curve over an interval of possible values of this random variable corresponds to the probability that the random variable will take on a value within the given interval. We denote $P(a < X < b)$ as the probability that a random variable X takes on a value between a and b. This probability corresponds to the shaded area under the curve, as illustrated in Figure 3.11.

The total area under a normal curve must equal 1, and since the distribution is symmetric about the mean, the area to the left (or right) of the mean is equal to 0.50, as can be seen in Figure 3.12.

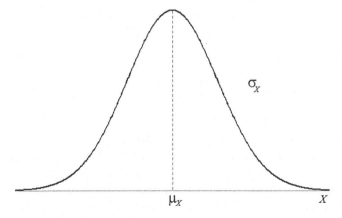

FIGURE 3.9
Normal distribution for the random variable X with a mean of μ_X and standard deviation of σ_X.

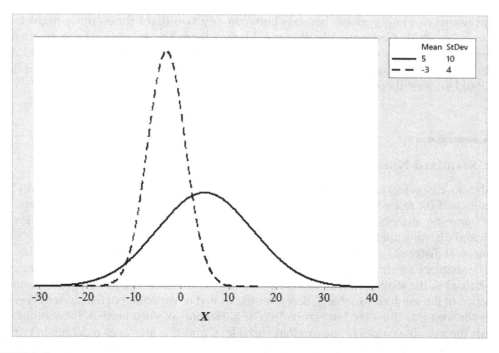

FIGURE 3.10
Comparison of two normal distributions, one with a mean of 5 and a standard deviation of 10 (solid curve), and the other with a mean of –3 and a standard deviation of 4 (dashed curve).

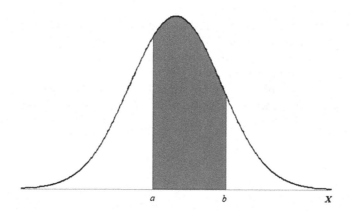

FIGURE 3.11
Shaded region that represents the probability that a normally distributed random variable X takes on values between a and b.

Because the area under a normal curve represents the probability that a continuous random variable X will take on a value within a given range, the probability that the random variable X takes on any *exact* value is 0. This is because the area under the curve for any exact value of a normally distributed random variable is equal to 0.

There is one special normal distribution, called the *standard normal distribution*. The standard normal distribution is a normal distribution that has a mean of 0 and a standard deviation of 1, as illustrated in Figure 3.13.

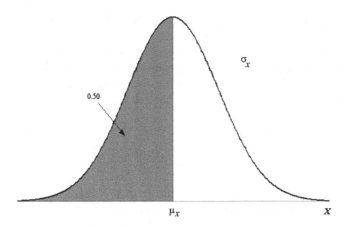

FIGURE 3.12
Shaded region of a normal distribution that represents the area to the left of the mean that is equal to 0.50.

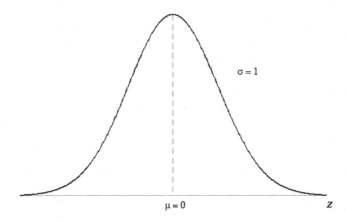

FIGURE 3.13
The standard normal distribution has a mean of 0 and a standard deviation of 1.

The standard normal distribution (sometimes referred to as a z-distribution) is a very important distribution because it allows us to find probabilities of random variables by using a *standard normal table*. The standard normal table is given in Table A.1. The standard normal table presents the area under the standard normal curve to the right of a given value of z. In other words, the shaded region in the table represents the probability that a random variable Z that follows a standard normal distribution will take on a value greater than z.

Example 3.15

Suppose we want to find the area under the standard normal curve that represents the probability that the random variable Z takes on a value less than $z = 1.25$. Because we want to find the probability that a random variable having the standard normal distribution takes on a value less than $z = 1.25$, this corresponds to the area under the standard normal curve, represented by the shaded region as illustrated in Figure 3.14.

In order to find this probability, we can use the standard normal table (Table A.1) because the mean is equal to 0 and the standard deviation is equal to 1. Notice that the standard normal table is referenced by a shaded area that is to the right of the value z, where this shaded area represents the probability that a random variable will take on a value greater than z. Since the area greater than $z = 1.25$ is 0.1056, the shaded area in Figure 3.14 that is less than $z = 1.25$ can be found by subtracting this area from 1 as follows:

$$1 - 0.1056 = 0.8944.$$

Therefore, the probability that the random variable Z takes on a value less than 1.25 is 0.8944 and the notation for this would be:

$$P(Z < 1.25) = 1 - 0.1056 = 0.8944$$

Example 3.16

To find the probability of a random variable Z that follows a standard normal distribution takes on a value between $z = -1.00$ and $z = 1.43$, we could use the standard normal tables to find the probability that is represented by the shaded region give in Figure 3.15.

To determine this probability, we use the standard normal tables to find the probability that the random variable falls below $z = -1.00$, which is 0.1587. Even though this area is in the left tail, it is the same as if we were considering the area in the right tail because the standard normal distribution is symmetric about 0. The probability that the random variable falls above $z = 1.43$ is 0.0764. Then adding these two probabilities and subtracting from 1 gives:

$$P(-1.00 < Z < 1.43) = 1 - (0.1587 + 0.0764) = 0.7650$$

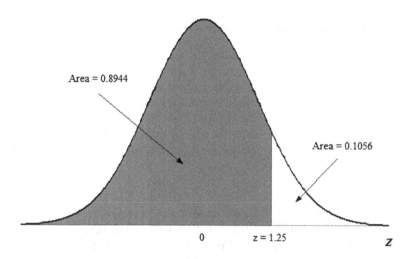

FIGURE 3.14

Standard normal distribution where the shaded area corresponds to the probability that the random variable Z takes on a value less than $z = 1.25$.

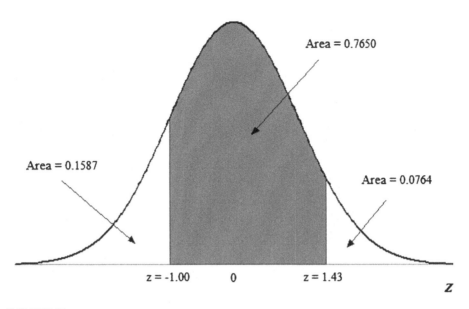

FIGURE 3.15
Area under the standard normal curve representing the probability that the random variable Z that follows the standard normal distribution falls between $z = -1.00$ and $z = 1.43$.

3.11 Non-Standard Normal Distributions

More often than not, normally distributed random variables tend not to have a mean of 0 and a standard deviation of 1. Because there is an infinite number of possible means and standard deviations that describe normally distributed random variables, we can convert (or standardize) any normally distributed random variable to the standard normal distribution. By standardizing a normal distribution, it means that a normal distribution with a mean different from 0 or a standard deviation different from 1 is converted to a standard normal distribution with a mean of 0 and a standard deviation of 1. Essentially, standardizing a non-standard normal distribution allows us to use the standard normal table to find probabilities (or the areas under the curve).

For a non-standard normal distribution, the shaded area under the curve, as illustrated in Figure 3.16, represents the probability that a random variable X takes on a value less than x.

To standardize, we can transform any x-value of a non-standard normally distributed random variable into a z-value that follows a standard normal distribution, and this requires the use of the following transformation formula:

$$z = \frac{x - \mu}{\sigma}$$

where x is the specific value of the non-standard normally distributed random variable, μ is the mean of the non-standard normally distributed random variable, and σ is the standard deviation of the non-standard normally distributed random variable.

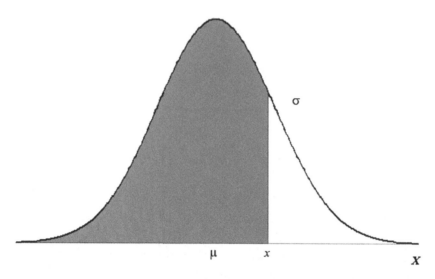

FIGURE 3.16
Non-standard normal distribution with a mean of μ and a standard deviation of σ where the shaded area represents the probability that a random variable X takes on a value less than x.

Standardizing converts a non-standard normal distribution to a standard normal distribution by converting the x-value to a z-value and converting the mean to 0 and the standard deviation to 1, as illustrated in Figure 3.17.

Example 3.17

Suppose we want to find the probability that the random variable X takes on a value less than 175, if X follows a normal distribution with a mean of 140 and a standard deviation of 40.

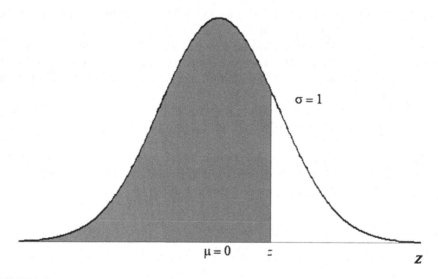

FIGURE 3.17
Standardized normal distribution from Figure 3.16 where the shaded area represents the probability that the random variable Z takes on a value less than z.

The shaded region in Figure 3.18 illustrates the area corresponding to the probability that the random variable takes on a value less than 175 if X is a normally distributed random variable with a mean of 140 and a standard deviation of 40.

By transforming the given x-value of the random variable X to a z-value, we get:

$$z = \frac{x-\mu}{\sigma} = \frac{175-140}{40} \approx 0.88$$

This now gives us the standard normal distribution such that $P(X < 175) = P(Z < 0.88)$, which is illustrated in Figure 3.19.

We can use the standard normal tables (Table A.1) to find the area of the shaded region that corresponds to the probability that the random variable X takes on a value less than 175. In referring to the standard normal table, the shaded area in the table corresponds to the probability that the random variable Z is greater than z. For this example, the shaded area that corresponds to the probability of interest can be found by subtracting the area found in the standard normal table from 1, as is illustrated in Figure 3.20.

Therefore, the probability that the random variable X takes on a value less than 175 is approximately:

$$P(X < 175) = P(Z < 0.88) = 1 - 0.1894 = 0.8106$$

Example 3.18

Suppose we want to find the probability that the random variable X takes on a value between 46 and 54, if X follows a normal distribution with a mean of 52 and a standard deviation of 2. Figure 3.21 shows the shaded region that corresponds to the probability that the random variable will take on a value between 46 and 54.

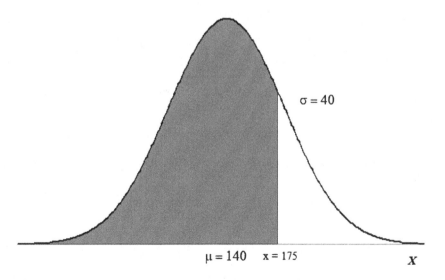

FIGURE 3.18
Shaded area that represents the probability that the random variable X that follows a non-standard normal distribution with mean 140 and standard deviation of 40 takes on a value less than 175.

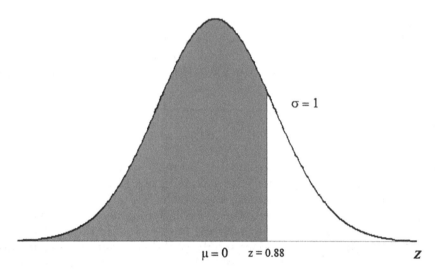

FIGURE 3.19
Standardized normal distribution from Figure 3.18.

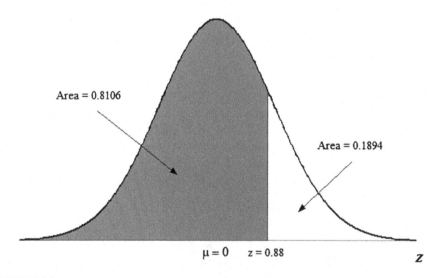

FIGURE 3.20
Probability under the standard normal curve corresponding to $P(Z < 0.88) = 0.8106$.

The standardized (or z) values are found by transforming each of the x-values as follows:

$$z = \frac{46 - 52}{2} = -3.00$$

$$z = \frac{54 - 52}{2} = 1.00$$

This converts the non-standard x-values of 46 and 54 to the standard z-values of −3.00 and 1.00, as is illustrated in Figure 3.22.

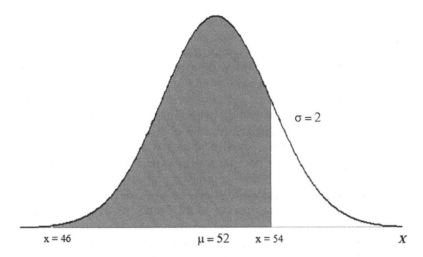

FIGURE 3.21
Shaded area that corresponds to the probability that the random variable X will take on a value between 46 and 54, if X is a normal distribution random variable with a mean of 52 and a standard deviation of 2.

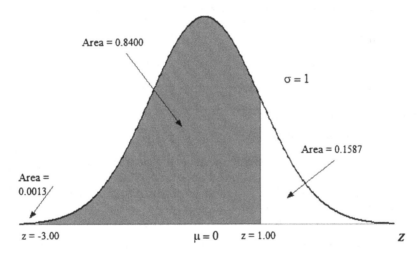

FIGURE 3.22
Standardized normal distribution for the random variable X in Figure 3.21.

To find the probability, we need to add the areas in both the tails and then subtract them from 1, as is shown in Figure 3.22.

$$P(46 < X < 54) = P(-3.00 < Z < 1.00) = 1 - (0.0013 + 0.1587) = 0.8400$$

3.12 Other Discrete and Continuous Probability Distributions

The backbone of making inferences about a population parameter relies on using information that is obtained from a sample statistic. For instance, we can make inferences about an unknown population mean using the sample mean. In addition to making inferences about population means, we can also make inferences about population proportions, population variances, as well as various other population parameters.

As we have seen earlier in this chapter, the probability pattern of the sample mean can give us some information about an unknown population mean. In Section 3.8, we discovered that the probability pattern (also known as the probability distribution) of the sample means has the shape of a normal distribution and is centered about the unknown population mean. The probability pattern that we obtain from a sample statistic is called the *sampling distribution* of the sample statistic. Thus, having sample information allows us to see the probability pattern of the sample statistic and then this sampling distribution can be used to make inferences about an unknown population parameter. However, not all sampling distributions will follow the shape of a normal distribution. The shape of sampling distributions will vary and depend on the properties of the statistic that is being considered.

In this section, we will describe some of the more common probability distributions that will be used throughout this book, as well as the relationship between different probability distributions. When probability distributions are generated from sample statistics, they provide us with the basic framework for making statistical inferences about unknown population parameters. Therefore, by understanding the probability distributions for different sample statistics, we can use these distributions to make generalizations about population parameters even when it is impossible or impractical to measure the population parameter itself.

3.13 The Binomial Distribution

A discrete distribution that can be used to model count data is called the *binomial distribution*. Count data is often described as the number of times that a given situation occurs. For instance, we can count the number of heads that appear in three coin flips.

With a binomial distribution, there are four criteria:

1. There are a fixed number of trials,
2. Each of the trials are independent of each other,
3. The outcome for each trial is either a success or failure,
4. The probability of a success for each trial is the same.

If these criteria hold true, then we can calculate the probability of observing x successes in n trials as follows:

$$P(X = x) = \frac{n!}{x!(n-x)!} \cdot p^x (1-p)^{n-x}$$

where p is the probability of a single success, and $x!$ (called "n-factorial") is the product of the integer x and all the numbers below it $\left(x! = x(x-1)(x-2)...(3)(2)(1)\right)$.

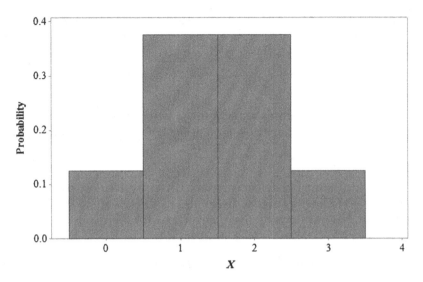

FIGURE 3.23
Graph of the binomial distribution for three trials and the probability of a single success is 0.50 or 50%.

Example 3.19

Suppose we flip a fair coin and count the number of heads. The probability of getting two heads in three coin flips could be found using the binomial distribution. Here is why:

1. For this example, there are a fixed number of trials, namely, three.
2. Each time we flip the coin, it has nothing to do with the other coin flips, so each of the trials are independent.
3. The outcome for each trial is either obtaining a head (success) or obtaining tails (failure).
4. The probability of a success for each trial is the same because the probability of getting a head for any of the three coin flips is 0.50 or 50%.

Thus, the probability of getting two heads ($x = 2$) in three coin flips ($n = 3$) can be calculated as follows:

$$P(X = 2) = \frac{3!}{2!(3-2)!} \cdot (0.50)^2 (1 - 0.50)^{3-2} = 3(0.25)(0.50) = 0.375$$

The graph of the binomial distribution with three trials and the probability of a single success being 0.50 or 50% is illustrated in Figure 3.23.

3.14 The Poisson Distribution

Another discrete distribution that can be used to model count data is called the *Poisson distribution*. The Poisson distribution is typically used when the binomial distribution has a large number of trials and the probability of a success is small. The possible values of a Poisson distribution are whole numbers 0, 1, 2, 3, 4, ..., and the probability of a count x can be found as follows:

$$P(X = x) = \frac{e^{-\mu} \cdot \mu^x}{x!}$$

where μ is the mean number of times that a given situation occurs, and e is the natural logarithm $(e \approx 2.7183)$. The standard deviation of the Poisson distribution is $\sqrt{\mu}$.

Example 3.20

Suppose that the number of defects produced at a given factory over the course of a 7-day period follows a Poisson distribution and we want to know the probability that (a) there are exactly five defects, and (b) there are less than three defects. If we know that the average number of defects is $\mu \approx 3.43$, then:

a. The probability that there are exactly five defects per day $(X = 5)$ can be calculated as follows:

$$P(X = 5) = \frac{e^{-3.43} \cdot (3.43)^5}{5!} \approx 0.1281$$

b. The probability that there are less than three defects per day $(X < 3)$ can be found as follows:

$$P(X < 3) = P(X = 0) + P(X = 1) + P(X = 2)$$

$$= \frac{e^{-3.43} \cdot (3.43)^0}{0!} + \frac{e^{-3.43} \cdot (3.43)^1}{1!} + \frac{e^{-3.43} \cdot (3.43)^2}{2!}$$

$$\approx 0.0324 + 0.1111 + 0.1905 \approx 0.3340$$

By definition, $0! = 1$.

The graph of the Poisson distribution with a mean of $\mu = 3.43$ is given in Figure 3.24.

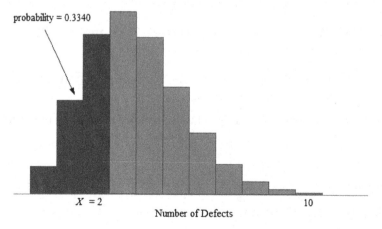

FIGURE 3.24
Graph of the Poisson distribution with a mean of $\mu = 3.43$.

3.15 The *t*-Distribution

The *t-distribution* is a continuous distribution that is an approximation of the standard normal distribution. One instance where the *t*-distribution is used is when inferences are made about a population mean using the sample mean. Like the standard normal distribution, the *t*-distribution is centered at 0 and goes infinitely in each direction, approaching but never touching the horizontal axis and the total area under the *t*-distribution is equal to 1. A comparison of the *t*-distribution with the standard normal distribution is illustrated in Figure 3.25.

What distinguishes the *t*-distribution from the standard normal distribution for making inferences about a population mean is that the standard normal distribution assumes that the population standard deviation is known. In contrast, the *t*-distribution does not rely on knowing the population standard deviation and it accounts for the uncertainty that arises when using the sample standard deviation to estimate the population standard deviation by using the *degrees of freedom*.

The degrees of freedom, often abbreviated as *df*, is a measure of the amount of information available in a sample that can be used to estimate a population parameter or parameters of interest. Typically, the degrees of freedom can be found by taking the sample size and subtracting from it the number of population parameters that are being estimated. As the degrees of freedom for the *t*-distribution increase, the shape of the *t*-distribution resembles that of the standard normal distribution. In addition to using the *t*-distribution to make inferences about a population mean, as will be described in Chapters 4 and 5, we will also use the *t*-distribution to make inferences about correlations and regression parameters in Chapters 6 and 7.

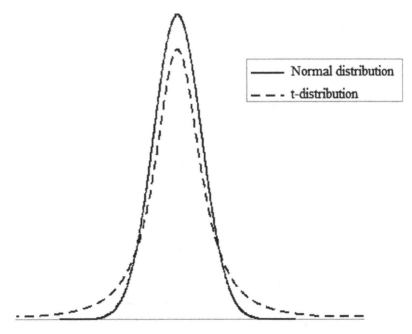

FIGURE 3.25
Comparison of the standard normal distribution and the *t*-distribution.

3.16 The Chi-Square Distribution

The sum of squared normal distributions is the *chi-square* distribution. Chi-square distributions are continuous distributions that are skewed to the right, as can be seen in Figure 3.26, and like the other probability distributions we have seen thus far, the total area under the distribution is equal to 1.

The tail of the chi-square distribution goes infinitely to the right, approaching but never touching the horizontal axis and like the *t*-distribution, the shape of the chi-square distribution depends on the degrees of freedom.

Notice that for the different chi-square distributions that are illustrated in Figure 3.27, as the degrees of freedom increase, the amount of skewness begins to decrease. Also notice

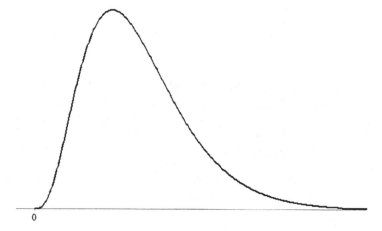

FIGURE 3.26
The chi-square distribution.

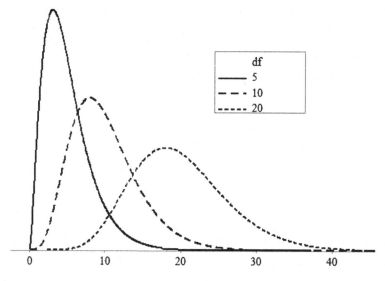

FIGURE 3.27
Comparison of chi-square distributions with 5, 10, and 20 degrees of freedom.

that as the degrees of freedom increase, the chi-square distribution starts to approach the shape of a normal distribution. You may also have noticed that the mean or center of a chi-square distribution is its degrees of freedom. We will be using the chi-square distribution to make inferences about variances in Chapter 4 and when testing for equal variances in Chapter 10.

3.17 The *F*-Distribution

The *F-distribution* is a continuous distribution that is the ratio of two chi-square distributions. Because this distribution is a ratio, there are two degrees of freedom to consider, the *numerator degrees of freedom* and the *denominator degrees of freedom*. In fact, the shape of the *F*-distribution begins to resemble the shape of the chi-square distribution as the degrees of freedom for the denominator approaches infinity. The *F*-distribution is also a right-skewed distribution that approaches but never crosses the horizontal axis and the total area under the *F*-distribution is equal to 1. Figure 3.28 gives a comparison of two different *F*-distributions. We will be using the *F*-distribution for a regression analysis in Chapter 8 and for an analysis of variance in Chapter 10.

3.18 Using Minitab to Graph Probability Distributions

Minitab can be used to graph many different probability distributions. To graph a probability distribution, first go to **Graph > Probability Distribution Plot**. This will bring up the dialog box that is shown in Figure 3.29.

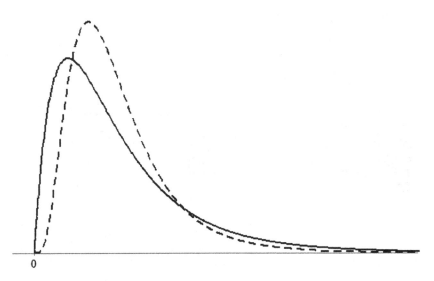

FIGURE 3.28
Comparison of two different *F*-distributions.

As you can see from Figure 3.29, you can graph a single distribution, vary the parameters of a single distribution, compare two different distributions, or view the probability of a single distribution.

For example, suppose we were interested in comparing a standard normal distribution with a *t*-distribution with two degrees of freedom. To do this, we would select **Two Distributions** and choose the distributions we are interested in comparing along with their characteristics, as is presented in Figure 3.30. The graph that would be generated is presented in Figure 3.31.

We can also use Minitab to calculate probabilities so we do not have to rely on using tables.

Example 3.21

Suppose we want Minitab to find the probability that the random variable X takes on a value between 46 and 54, if X follows a normal distribution with a mean of 52 and a standard deviation of 2 (Example 3.18).

We would need to select **Graph > Probability Distribution Plot > View Probability**, as is illustrated in Figure 3.32.

This brings up the dialog box in Figure 3.33 where we need to select the normal distribution and specify a mean of 52 and a standard deviation of 2.

We then need to click on the **Shaded Area** tab on the top of Figure 3.33. This will bring up the dialog box in Figure 3.34 where we need to specify three things: the area of the distribution we are interested in (which in this case will be the middle), whether a probability or *x*-values are being specified, and the given *x*-values or the probability. For our example, we would specify the *x*-values as 46 and 54.

This will generate the distribution in Figure 3.35 with the corresponding probability of 0.8400.

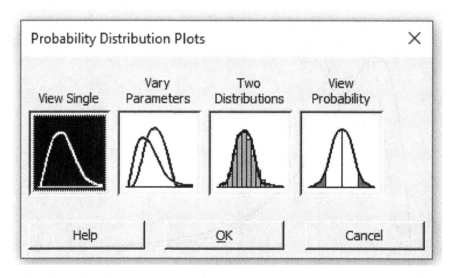

FIGURE 3.29
Dialog box to graph probability distributions.

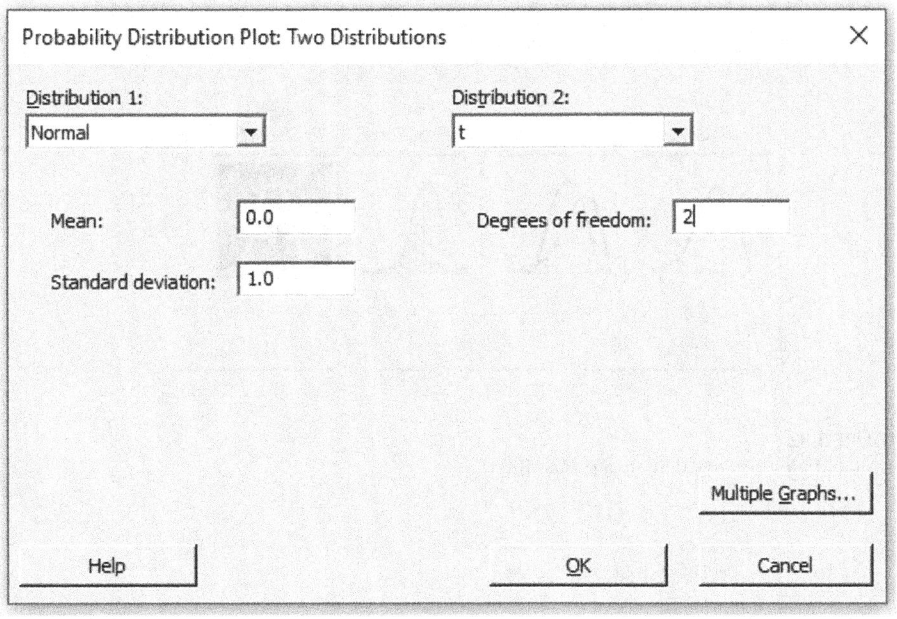

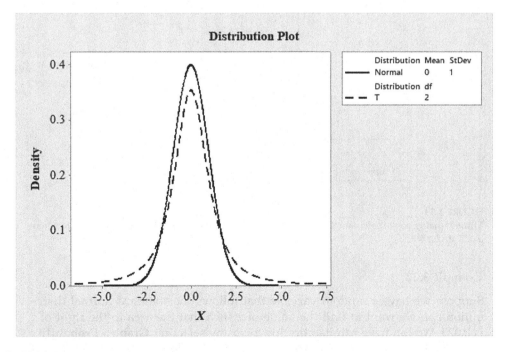

FIGURE 3.30
Dialog box for comparing a standard normal distribution (mean of 0.0 and a standard deviation of 1.0) to a *t*-distribution with two degrees of freedom.

FIGURE 3.31
Graph showing the standard normal distribution (solid line) and the *t*-distribution with two degrees of freedom (dashed line).

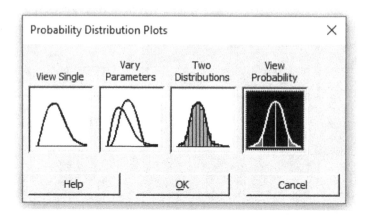

FIGURE 3.32
Minitab dialog box to calculate a probability.

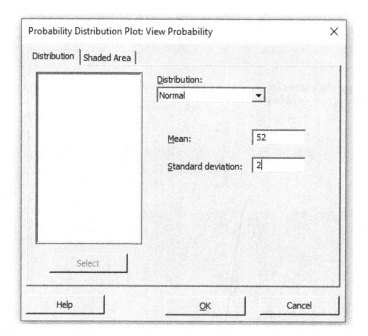

FIGURE 3.33
Minitab dialog box to calculate the probability for a normally distributed random variable where $\mu = 52$ and $\sigma = 2$.

Example 3.22

Suppose we have a random variable that follows the standard normal distribution and we want to find the value of z such that the area to the right of z is 0.025. We can have Minitab do this for us by selecting **Graph > Probability Distribution Plot > View Probability**. We can then specify the normal distribution with a mean of 0 and a standard deviation of 1 and click on **Shaded Area**. This would bring up the dialog box in Figure 3.36.

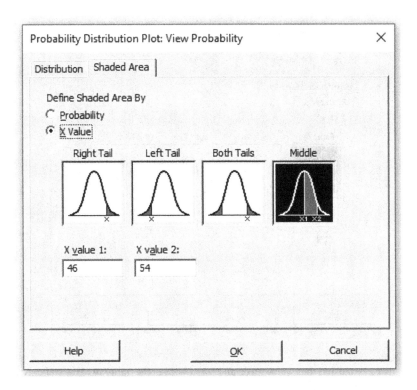

FIGURE 3.34
Minitab dialog box to define the shaded area of a distribution.

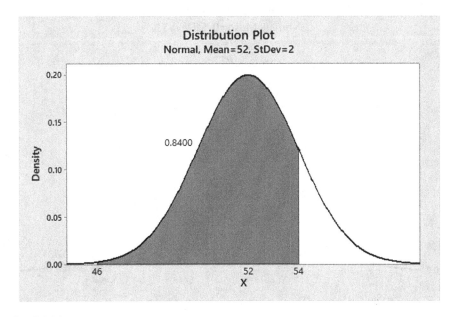

FIGURE 3.35
Minitab-generated shaded area that corresponds to the probability that the random variable X will take on a value between 46 and 54, if X is a normal distribution random variable with a mean of 52 and a standard deviation of 2.

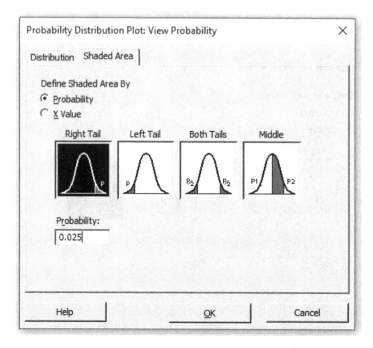

FIGURE 3.36
Minitab dialog box to find the *z*-value given an area of 0.025 in the right-tail of the standard normal distribution.

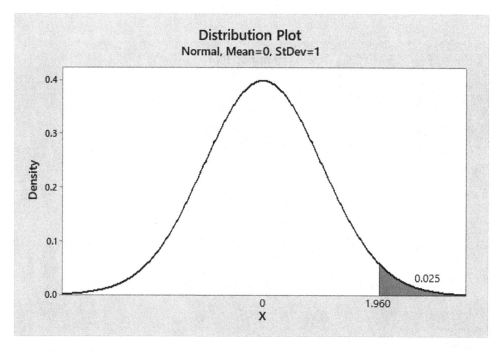

FIGURE 3.37
Minitab-generated graph of a random variable that follows the standard normal distribution where *z* = 1.96 corresponds to the area to the right of *z* is 0.025.

We then need to specify the probability (which, for this example, is 0.025), and select the shaded region to be in the right tail. This will bring up the graph of the distribution in Figure 3.37 where we can see that the value of $z = 1.96$ gives us an area of 0.025 in the right tail of a standard normal distribution.

Exercises

1. The data provided in Table 3.14 represents a random sample of the daily rainfall (in inches) over a 1-week period during the summer for a small town in Connecticut.
   Calculate the mean, median, mode, range, variance, and standard deviation for this set of data. Which measure of center do you think best represents this set of data?

2. The data in Table 3.15 is a random sample of the first-year salaries (in thousands of dollars) for a random sample of ten graduates from a university.
   a. Calculate the mean first-year salary and the median first-year salary. Which of these two measures do you think best represents the center of the underlying data set? Why?
   b. Calculate the standard deviation and the range for this set of data. Which of these two measures best represents the underlying data set? Why?

3. The data given in Table 3.16 are the speeds (in miles per hour) for a random sample of 12 cars driving down a given section of a highway that has a posted speed limit of 65 miles per hour.
   a. Calculate Q_1, Q_2, and Q_3 for these highway speeds.
   b. Find the range and the IQR.
   c. Draw and clearly label a boxplot of these highway speeds.

4. Find the value of z using the standard normal table or Minitab such that:
   a. The area to the right of z is 0.5557.
   b. The area to the left of z is 0.0116.

TABLE 3.14

Daily Rainfall for a Small Town in Connecticut

1.25	0.20	0.00	0.00	0.00	0.10	2.45

TABLE 3.15

Random Sample of the First-Year Salaries for a Random Sample of Ten University Graduates

25	34	26	95	45	52	34	51	29	38

TABLE 3.16

Speeds (in miles/hour) for a Random Sample of 12 Cars

72	70	63	68	75	80	65	67	68	70	68	65

c. The area between –z and z is 0.95.

d. The area to the right of z is 0.01.

5. Lifetimes of automobile parts often follow the pattern of a normal distribution. For example, the mean lifetime of car batteries tends to be normally distributed. Suppose that the mean number of years that a car battery lasts is 5 years and the standard deviation is 1.5 years. Find each of the following:

a. What is the probability that a randomly selected car battery will last more than 5 years?

b. What is the probability that a randomly selected car battery will last less than 2 years?

c. What is the probability that a randomly selected car battery will last between 4 and 6 years?

d. What is the probability that a randomly selected car battery will last more than 7 years?

e. What is the probability that a randomly selected car battery will last exactly 5 years?

6. Suppose you have a random variable Z that follows a standard normal distribution, find the following probabilities that correspond to the area under the standard normal curve:

a. $P(Z < 1.25)$

b. $P(Z < -0.49)$

c. $P(-1.39 < Z < 1.16)$

d. $P(Z > 1.25)$

e. $P(Z > -1.03)$

f. $P(Z > -0.21)$

g. $P(0.73 < Z < 1.54)$

h. $P(Z < -1.45)$

i. $P(Z > 1.96)$

7. Suppose you are given a random variable that has the standard normal distribution.

a. Find the value of z such that 5% of all values of this random variable will be greater.

b. Find the value of z such that 1% of all values of this random variable will be greater.

c. Find the value of z such that 95% of the standard normal distribution is between –z and +z.

8. Grades received on computerized examinations can often be approximated by a normal distribution. Assume that the average grade on a statistics examination is 73.25 points and the standard deviation is 12.58 points.

a. What is the probability that a grade selected at random will be greater than 80.50 points?

b. What is the probability that a grade selected at random will be between 60.25 and 70.75 points?

    c. What is the probability that a grade selected at random will be exactly 90.23?

10. Suppose a random variable X has a normal distribution with $\mu = 70$ and $\sigma = 12$. Using Minitab or the standard normal table, find each of the following probabilities:

    a. $P(X < 50)$

    b. $P(X > 70)$

    c. $P(40 < X < 60)$

    d. $P(120 < X < 160)$

    e. $P(60 < X < 95)$

    f. $P(X < 72.3)$

11. The sticker price for a new luxury SUV is $93,755, and most people who are interested in purchasing the vehicle will try to negotiate the price. To help in the negotiation process, there are many websites that provide information about what other consumers have paid for the vehicle. What is very interesting is that the price paid tends to follow the pattern of a normal distribution.

    If the average price paid for this vehicle is $92,450 and if the standard deviation is $1,275, find each of the following:

    a. What is the probability that a buyer will pay more than the average price?

    b. What is the probability that a buyer will pay less than $92,000?

    c. What is the probability that a buyer will pay between $91,500 and $93,000?

    d. At what price will 95% of the buyers pay more?

    e. What is the range of prices paid for the middle 95% of the buyers?

12. Justify whether or not each of the following tables represents the probability distribution of a random variable X:

X	P(X)
−1	0.26
−2	0.45
−3	0.22
−4	0.07

X	P(X)
0.21	0.20
0.22	0.20
0.23	0.20
0.24	0.20
0.25	0.20

X	P(X)
0.245	0.16
1.25	0.18
2.89	−0.12
3.47	0.39
0.18	0.39

13. You may have taken classes where the exam grades are scaled. One way to scale grades is to use a normal distribution. Suppose an instructor found the mean exam score of 64 and a standard deviation of 14.2 points and the distribution of the exam scores appeared to be normally distributed. Using this information, answer each of the following:

   a. If a student scores at least 1.5 standard deviations above the mean, they will receive the grade of A. What percentage of students will receive the grade of A?

   b. If a student scores from 0.5 to 1.5 standard deviations above the mean, they will receive the grade of B. What percentage of students will receive the grade of B?

   c. If a student scores from 0.5 standard deviations below the mean to 0.5 standard deviations above the mean, they will receive the grade of C. What percentage of students will receive the grade of C?

   d. If a student scores more than 1.5 standard deviations below the mean, they will receive the grade of F. What percentage of students will receive the grade of F?

14. Find the mean, variance, and standard deviation for the following probability distribution for the random variable X:

X	$P(X)$
1	0.20
2	0.27
3	0.19
4	0.34

15. For the probability distribution in Exercise 14, find the sampling distribution of the sample mean for all samples of size $n = 2$. (Hint: The sample means will be the same as given in Figure 3.7, but the probabilities will be different because each of the outcomes of the random variable X are not equally likely as they are in Table 3.12. What this means is that the probability of drawing a {1, 1} to give a sample mean of 1 would be $(0.20)(0.20) = 0.04$. And the probability of drawing a {1, 2} or a {2, 1} to give a sample mean of 1.5 would be $(0.20)(0.27) + (0.27)(0.20) = 0.108$, etc.).

16. Minitab can also be used to create a graphical summary of a given variable. The graphical summary can be found by selecting the Stat menu; under **Basic Statistics < Graphical Summary**. There are four graphs that appear in the summary: a histogram with a normal curve superimposed, a boxplot, and 95% confidence intervals for the mean and median, respectively.

   a. Using Minitab, create a graphical summary for the annual fuel cost for the data given in Table 3.17.

   b. Create a graphical summary for the annual fuel cost by Make.

   c. Create a graphical summary for the amount of greenhouse gas by Make.

17. Often we may be interested in comparing the amount of variability between two different data sets. This can be done using the sample *coefficient of variation* (*COV*), which is a percentage that can be used to describe the relative variability between two different samples. The COV is calculated by dividing the standard deviation by the mean and then multiplying by 100, as can be seen in the following formula:

TABLE 3.17

Make, City MPG, Highway MPG, Annual Fuel Cost, and Amount of Greenhouse Gas (in Tons) for a Random Sample of 25 Automobiles

Make	City MPG	Highway MPG	Annual Fuel Cost	Greenhouse Gas (Tons)
Chevrolet	27	37	1,084	6.1
Chevrolet	26	34	1,123	6.3
Chevrolet	27	37	1,084	6.1
Chevrolet	25	34	1,125	6.1
Chevrolet	25	34	1,280	6.5
Chevrolet	24	32	1,326	6.6
Chevrolet	24	36	1,204	6.9
Chevrolet	22	30	1,434	7.3
Chevrolet	23	30	1,253	7.1
Chevrolet	22	31	1,302	7.4
Chevrolet	23	30	1,253	7.1
Chevrolet	22	31	1,302	7.4
Honda	28	39	681	4.7
Honda	30	40	986	5.5
Honda	30	38	986	5.5
Honda	23	32	1,380	7
Honda	49	51	651	3.7
Honda	33	38	931	5.3
Honda	31	38	957	5.4
Honda	31	37	986	5.5
Mazda	28	35	1,051	6
Mazda	26	34	1,123	6.4
Mazda	26	33	1,162	6.5
Mazda	25	31	1,024	6.7
Mazda	20	28	1,559	8

$$COV = \frac{s}{\bar{x}} \cdot 100$$

The sample that has the larger value of the COV represents the sample that has more variability relative to the mean.

The data set provided in Table 3.18 gives the time (in minutes) that Phil and Barbara spend talking on their cell phones each day over the course of a 5-day period.

Using the COV, which of these two individuals shows more variability in the time spent (in minutes) talking on his or her cell phone?

18. The *weighted mean*, denoted as $\bar{x}_w$ is another measure of central tendency that can be used to assign weights (or measures of influence) to each of the individual observations in a sample. The formula for the weighted mean is as follows:

$$\bar{x}_w = \frac{\sum_{i=1}^{n} w_i \cdot x_i}{\sum_{i=1}^{n} w_i}$$

TABLE 3.18

Time (in Minutes) Spent by Phil and Barbara Talking on Their Cell Phones

Day	Phil	Barbara
Monday	52	32
Tuesday	127	45
Wednesday	285	29
Thursday	6	38
Friday	29	36

where w_i is the weight associated with each observation x_i.

Suppose you want to find the mean amount that you spent on gasoline (per gallon) over the course of a 12-week period. The data provided in Table 3.19 gives the number of gallons and the price paid per gallon over the 12-week period.

a. Find $\bar{x}$, the mean amount paid per gallon for gas over the 12-week period.

b. Find $\bar{x}_w$, the (weighted) mean amount paid per gallon for gas over the 12-week period.

c. Describe why you think there is a difference between the mean and the weighted mean.

19. The *trimmed mean* can also be a useful measure of center. It is found by trimming off a certain percentage of the smallest and largest observations and then calculating the mean. For instance, we can trim 5% of the smallest observations and 5% of the largest observations (rounded to the nearest integer), and then calculate the mean of the remaining values. The trimmed mean can be a good measure of center, a variable that may have pronounced outliers. For the data given in Table 3.6, which represents the number of hours that a sample of ten homeowners spend working on their lawn each week, find the trimmed mean by trimming off 10% of the largest and smallest observations.

TABLE 3.19

Number of Gallons of Gasoline, Price Per Gallon, and Total Cost Over a 12-Week Period

Number of Gallons	Price Per Gallon (in Dollars)	Total (in Dollars)
13.2	2.35	31.02
9.8	2.41	23.62
12.4	2.28	28.27
8.1	2.30	18.63
15.2	2.41	36.63
6.8	2.50	17.00
7.5	2.40	18.00
9.3	2.29	21.30
8.4	2.38	19.99
6.2	2.93	18.17
13.3	2.20	29.26
12.8	2.25	28.80

20. The *kurtosis* of a variable represents a measure of how the distribution of a variable is peaked (in other words, how "sharp" the data set appears when you consider its distribution). The kurtosis can be used to gauge whether the distribution of a variable differs from the normal distribution. For instance, the closer the kurtosis value is to 0, the more normally distributed the variable is; the closer the kurtosis value is to 1, the sharper the peak of the distribution; and a negative kurtosis value represents more of a flat distribution. The formula for kurtosis is as follows:

$$\text{Kurtosis} = \left[\frac{n(n+1)}{(n-1)(n-2)(n-3)} \right] \cdot \sum_{i=1}^{n} \left(\frac{x_i - \bar{x}}{s} \right)^4 - \frac{3(n-1)^2}{(n-2)(n-3)}$$

where n is the sample size, x_i are the individual observations, $\bar{x}$ is the sample mean, and s is the sample standard deviation for the given data.

   a. For the round-trip commuting distance data set in Table 2.1, draw a histogram and calculate the value of the kurtosis statistic.

   b. Does the measure of kurtosis make sense for the shape of the distribution for the commuting distance data set in Table 2.1?

21. The *skewness* of a variable can be used as a measure of how symmetrical a distribution is. If a variable is normally distributed, the skewness will be approximately equal to 0. If the skewness is positive, this indicates that more observations are below the mean than are above the mean (i.e., the data is right-skewed), and if the skewness is negative, this indicates that there are more observations above the mean than are below the mean (i.e., the data is left skewed). The formula to calculate the skewness is as follows:

$$\text{Skewness} = \frac{n}{(n-1)(n-2)} \cdot \sum_{i=1}^{n} \left(\frac{x_i - \bar{x}}{s} \right)^3$$

where n is the sample size, x_i are the individual observations, $\bar{x}$ is the sample mean, and s is the sample standard deviation for the given data.

   a. For the commuting distance data given in Table 2.1, calculate the value of the skewness.

   b. Does the measure of skewness make sense for this data set?

4

Statistical Inference for One Sample

4.1 Introduction

Statistical inference involves making an inference or prediction about an *unknown* population parameter of interest from information that is obtained by using a sample statistic. In this chapter, we will begin by describing two basic types of statistical inference: *estimation* and *hypothesis testing*.

In making statistical inferences using *estimation*, we try to obtain an estimate of a likely range of values for an unknown population parameter of interest. For example, we could use estimation techniques to infer the mean (or average) amount spent on advertising for the population of car dealers.

In making statistical inferences with *hypothesis testing*, we can test to see if an unknown population parameter differs enough from some hypothesized value. For instance, we may be interested in estimating if the entire population of students who take a final examination receive a mean score greater than 70%. For both estimation and hypothesis testing, we will be using sample data to make inferences or predictions about an unknown population parameter.

This chapter presents some basic statistical methods that can be used for estimation and hypothesis testing for a single mean, a one-sample proportion, a one-sample variance, and a one-sample count variable. This chapter also elaborates on the importance of a statistical power analysis and describes how to use Minitab for estimation and hypothesis testing.

4.2 Confidence Intervals

Estimation is concerned with using sample data to estimate a value or a likely range of values for an unknown population parameter of interest. One such way of estimating an unknown population parameter is by using a sample statistic. When an estimate of a population parameter consists of only a single sample statistic, it is called a *point estimate*. For example, the sample mean $\bar{x}$ can be used as a point estimate for an unknown population mean μ. However, using only a single sample statistic to estimate a population parameter is not a very good estimate because different samples will generate different sample statistics.

A better estimate is to obtain a likely range of values for a population parameter of interest by calculating a *confidence interval*. A confidence interval for a population parameter can be used to estimate a range of values where we have some degree of confidence that this interval contains the unknown population parameter of interest. For example, a confidence interval for a population mean is determined by using the sample mean and then calculating upper and lower bounds such that between these two boundary points, we have some degree of confidence that the interval contains the unknown population mean.

The theory behind confidence intervals for a population mean relies on the fact that for large samples, the sampling distribution of the sample mean is approximately normally distributed and is centered about the unknown population mean and the standard deviation can be estimated from the sample. Recall that this is what was shown in Section 3.8 when we described the central limit theorem. For small samples, we need to assume that the population being sampled from is approximately normally distributed.

Standardizing a given value of the sample mean gives the following *test statistic*, which is a random variable that follows the *t*-distribution:

$$T = \frac{\bar{x} - \mu}{\frac{s}{\sqrt{n}}}$$

where $\bar{x}$ is the sample mean, s is the sample standard deviation, n is the sample size, and μ is the unknown population mean. This test statistic is simply the standardized value of a given sample mean.

The *t*-distribution resembles a standard normal distribution in that it is a bell-shaped curve that is centered at 0. However, the shape of the *t*-distribution depends on what are called the *degrees of freedom*. Recall from Chapter 3 that the degrees of freedom (often abbreviated as *df*) is a measure of the amount of information available in the sample that can be used to estimate a population parameter or parameters of interest. Typically, the degrees of freedom is found by taking the number of sample observations minus the number of parameters that are being estimated. For instance, if we are using the *t*-distribution to compute a confidence interval for a single population mean, then the degrees of freedom would be equal to $n - 1$ for a sample of size n because we are estimating a single parameter, namely, the population mean μ, and this requires that we subtract 1 from the sample size.

For a random variable T that follows the *t*-distribution, we claim that the statistic T will fall between $-t_{\alpha/2}$ and $t_{\alpha/2}$ with probability $1 - \alpha$, as illustrated in Figure 4.1.

Because $T = \dfrac{\bar{x} - \mu}{\frac{s}{\sqrt{n}}}$ is a random variable and if $-t_{\alpha/2} < T < t_{\alpha/2}$, then:

$$-t_{\alpha/2} < \frac{\bar{x} - \mu}{\frac{s}{\sqrt{n}}} < t_{\alpha/2}$$

Multiplying through by $\left(\dfrac{s}{\sqrt{n}} \right)$ gives:

$$-t_{\alpha/2} \left(\frac{s}{\sqrt{n}} \right) < \bar{x} - \mu < t_{\alpha/2} \left(\frac{s}{\sqrt{n}} \right)$$

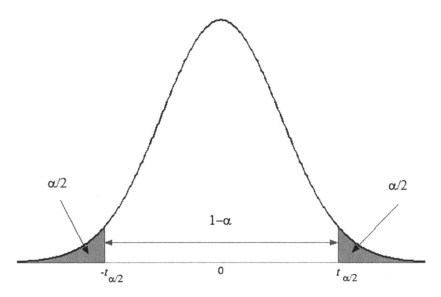

FIGURE 4.1
Probability distribution of the sample mean where the sample statistic T falls between $-t_{\alpha/2}$ and $t_{\alpha/2}$, with probability $1 - \alpha$.

Subtracting $\bar{x}$ from the inequality gives:

$$-\bar{x} - t_{\alpha/2}\left(\frac{s}{\sqrt{n}}\right) < -\mu < -\bar{x} + t_{\alpha/2}\left(\frac{s}{\sqrt{n}}\right)$$

Multiplying through by -1 requires changing the sign of the inequality as follows:

$$\bar{x} + t_{\alpha/2}\left(\frac{s}{\sqrt{n}}\right) > \mu > \bar{x} - t_{\alpha/2}\left(\frac{s}{\sqrt{n}}\right)$$

Correcting the direction of the inequality gives

$$\bar{x} - t_{\alpha/2}\left(\frac{s}{\sqrt{n}}\right) < \mu < \bar{x} + t_{\alpha/2}\left(\frac{s}{\sqrt{n}}\right)$$

Thus, the unknown population mean μ will fall between $\bar{x} - t_{\alpha/2}\left(\frac{s}{\sqrt{n}}\right)$ and $\bar{x} + t_{\alpha/2}\left(\frac{s}{\sqrt{n}}\right)$ with probability $1 - \alpha$.

We can also denote this confidence interval in the following manner:

$$\left(\bar{x} - t_{\alpha/2} \cdot \frac{s}{\sqrt{n}}, t_{\alpha/2} \cdot \frac{s}{\sqrt{n}}\right)$$

This interval is called a $100(1 - \alpha)\%$ confidence interval. If we take repeated samples of size n from the population of interest and calculate confidence intervals for all of these samples in the exact same manner, then the population mean μ will be contained in approximately $100(1 - \alpha)\%$ of all such intervals.

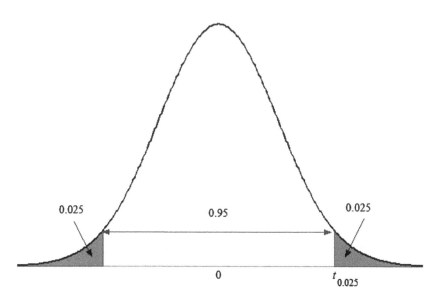

FIGURE 4.2
Probability distribution of the sample mean $\bar{x}$ for calculating a 95% confidence interval for the unknown population mean μ.

For instance, a $100(1 - 0.05)\% = 95\%$ confidence interval for a population mean μ obtained from a random sample of size n can be determined such that for all samples of size n, approximately 95% of all such confidence intervals calculated in the same manner are expected to contain the population parameter μ. In other words, for repeated samples of size n, 95% of all such intervals of the form:

$$\left(\bar{x} - t_{\alpha/2} \cdot \frac{s}{\sqrt{n}}, t_{\alpha/2} \cdot \frac{s}{\sqrt{n}} \right)$$

will contain the unknown population mean where $\alpha = 0.05$ and $t_{0.025}$ is the value that gives an area of 0.025 in the right tail of a t-distribution with $n - 1$ degrees of freedom. Thus, we can claim that we are 95% confident that any given interval will contain the unknown population mean. Such a confidence interval is illustrated in Figure 4.2.

In order to find the value of $t_{\alpha/2}$, we need to reference a t-table (see Table A.2). Notice that the t-table is different from the standard normal table in that the degrees of freedom are given in the first column, values of the area in the right tail are given in the top row, and the reference area is the area to the right of t. For instance, if we wanted to find $t_{0.025}$ for 25 degrees of freedom, we would locate the value of 0.025 in the top row of the t-table, and find 25 degrees of freedom in the leftmost column of the t-table. This would give a value of $t_{0.025} = 2.060$.

Example 4.1

A random sample of 18 car dealers found that this particular sample of dealers spent an average amount of $5,500 per month on the cost of advertising and the sample standard deviation was calculated to be $500. Suppose that we want to find a 95% confidence interval for the unknown mean (or average) amount of money that the *population* of car dealers spends monthly on advertising.

Assuming that the population we are sampling from is normally distributed, we can use the sample mean ($\bar{x} = \$5,500$) and the sample standard deviation

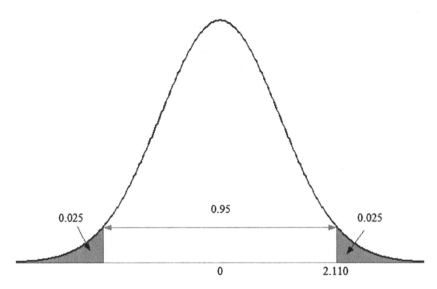

FIGURE 4.3
Probability distribution of the sample mean with the value of $t = 2.110$ that defines the upper boundary point for a 95% confidence interval with 17 degrees of freedom.

($s = 500$) to calculate a 95% confidence interval. We need to use the t-table (Table A.2) to find the value for $t_{\alpha/2} = t_{0.05/2} = t_{0.025}$. Since $n = 18$, we have $n - 1 = 17$ degrees of freedom; then the values for $t_{\alpha/2} = t_{0.05/2} = t_{0.025}$ are the values that describe the upper and lower boundary points of our confidence interval as follows:

$$t_{\alpha/2} = t_{0.05/2} = 2.110$$

This is illustrated in Figure 4.3.

Therefore, a 95% confidence interval for the unknown population mean amount spent on advertising would be:

$$\left(\bar{x} - t_{\alpha/2} \cdot \frac{s}{\sqrt{n}}, \bar{x} + t_{\alpha/2} \cdot \frac{s}{\sqrt{n}} \right)$$

$$= \left(5500 - 2.110 \cdot \frac{500}{\sqrt{18}}, 5500 + 2.110 \cdot \frac{500}{\sqrt{18}} \right)$$

$$= \left(5500 - 248.67, 5500 + 248.67 \right) \approx \left(5251.33, 5748.67 \right)$$

Thus, if repeated samples of size 18 are drawn from the amount that the population of car dealers spends per month on advertising, and the confidence intervals are calculated in a manner similar to that given, then 95% of all such intervals will contain the unknown population mean. In other words, we say we are 95% confident that the mean amount that the population of car dealers spend on advertising falls between $5251.33 and $5748.67 per month.

Another way to look at this is as follows:

$$5251.33 < \mu < 5748.67$$

where μ is the unknown *population mean*.

TABLE 4.1

Sample of Final Examination Grades for a Random Sample of 20 Students

71	93	91	86	75
73	86	82	76	57
84	89	67	62	72
77	68	65	75	84

Example 4.2

The final examination grades for a sample of 20 statistics students selected at random are presented in Table 4.1.

Assuming that the population we are sampling from is normally distributed, calculating the necessary descriptive statistics for this sample gives:

$$\bar{x} \approx 76.65$$

$$s \approx 10.04$$

$$n = 20$$

Suppose we want to find a 99% confidence interval for the mean final examination grade for the population of all students taking the statistics examination at our site. Since $1 - \alpha = 0.99$, this means that $\alpha = 0.01$. So for a sample size of 20 with 19 degrees of freedom, the upper boundary point $t_{\alpha/2} = t_{0.01/2} = t_{0.005}$ would be described by the value of $t = 2.861$, as illustrated in Figure 4.4.

On the basis of the t-distribution with $20 - 1 = 19$ degrees of freedom, we have $t_{\alpha/2} = t_{0.005} = 2.861$, so a 99% confidence interval would be:

$$\left(76.65 - 2.861 \cdot \frac{10.04}{\sqrt{20}}, 76.65 + 2.861 \cdot \frac{10.04}{\sqrt{20}}\right) = \left(76.65 - 6.42, 76.65 + 6.42\right) \approx \left(70.23, 83.07\right)$$

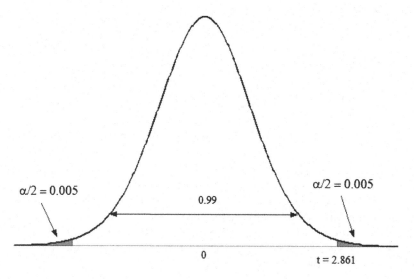

FIGURE 4.4

Probability distribution of the sample mean that defines the upper boundary point of a 99% confidence interval with 19 degrees of freedom.

This confidence interval suggests that if we repeat the process of taking samples of size $n = 20$ from the population of all statistics students taking the final examination at our site and calculate confidence intervals in the exact same manner, then 99% of all such intervals will contain the unknown population mean final examination grade. In other words, we can claim that we are 99% confident that the *population* of statistics students will score a mean (or average) final examination score between 70.23 and 83.07 points.

4.3 Using Minitab to Calculate Confidence Intervals for a Population Mean

Minitab can be used to calculate confidence intervals for a population mean either by entering the raw data in a worksheet column or by entering the relevant descriptive statistics.

To use Minitab to calculate a 99% confidence interval for the population mean score on the statistics final examination for the data given in Table 4.1 using either the raw data or a collection of descriptive statistics, select **Stat > Basic Statistics > 1-Sample *t*** from the pull-down menu.

The summarized data is entered into Minitab, as can be seen from the one-sample *t*-test dialog box in Figure 4.5.

By clicking on the **Options** box, we can input the appropriate confidence level, as shown in Figure 4.6. We also need to make sure that the alternative hypothesis is always set to "not equal" whenever we are calculating a confidence interval (the reason for doing so will become more apparent when we cover hypothesis tests in the next section).

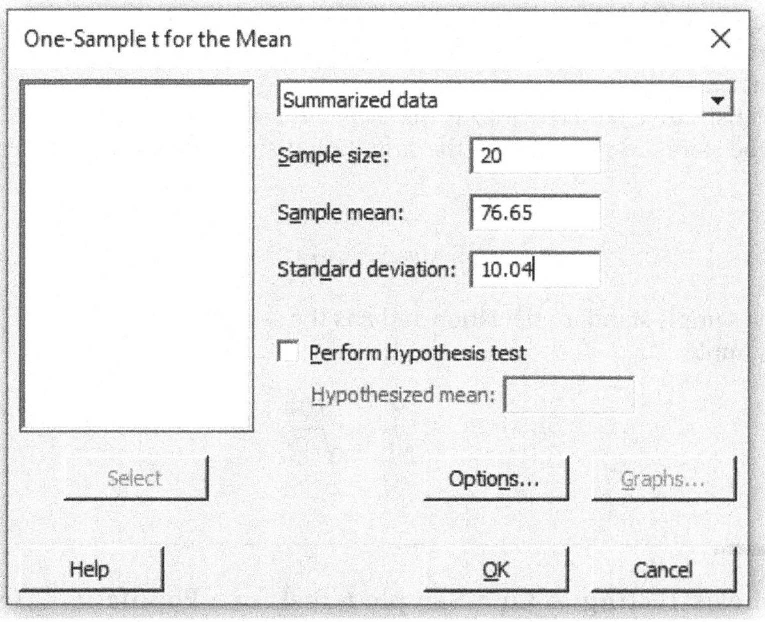

FIGURE 4.5
Minitab dialog box for a confidence interval for a population mean using summarized data.

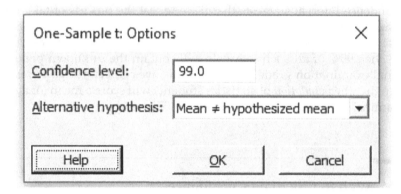

FIGURE 4.6
Minitab options dialog box for calculating for a 99% confidence interval for a population mean.

One-Sample T

Descriptive Statistics

N	Mean	StDev	SE Mean	99% CI for μ
20	76.65	10.04	2.25	(70.23, 83.07)

μ: mean of Sample

FIGURE 4.7
Minitab printout for a 99% confidence interval for the unknown mean score on the statistics final examination for the entire population of students.

Then selecting **OK** gives the Minitab printout illustrated in Figure 4.7. Notice that Figure 4.7 gives a 99% confidence interval for the true population mean score on the statistics final examination.

Another statistic given in Figure 4.7 is the *standard error of the mean* (SE Mean). This is an estimate of the standard deviation of the sampling distribution and can be calculated as follows:

$$\text{SE Mean} = \frac{s}{\sqrt{n}}$$

where s is the sample standard deviation and n is the sample size.

For this example:

$$\text{SE Mean} = \frac{s}{\sqrt{n}} = \frac{10.04}{\sqrt{20}} \approx 2.25$$

4.4 Hypothesis Testing: A One-Sample *t*-Test for a Population Mean

A confidence interval for a population mean describes a range of values based on sample data where we have a certain degree of confidence that the unknown population parameter can be found within this interval. A *hypothesis test*, which is also based on sample data, can tell us whether an unknown population parameter is different enough from some hypothesized value. For instance, we can conduct a hypothesis test that would allow us to determine whether the unknown population mean is significantly different from some specific or hypothesized value of interest. The symbol for the hypothesized population mean is denoted as μ_0.

In order to perform a hypothesis test about a population mean, we first have to set up a *null hypothesis* and an *alternative hypothesis*. The alternative hypothesis is established based on what is being investigated and is found by considering what it is that we are looking to accept that is *different* from a given hypothesized population parameter. The null hypothesis is the hypothesized population parameter that is set up for the purpose of being rejected. For instance, in using the data from Table 4.1, which provides the final examination grades for a sample of 20 statistics students, suppose we want to investigate whether the unknown population mean grade on the statistics final examination is greater than 75. Since we want to know whether the population mean is greater than the hypothesized value of 75, we would state our alternative hypothesis as H_1: $\mu > 75$ (this is the hypothesis that we are interested in testing) and the null hypothesis would then be, H_0: $\mu = 75$ (this is the hypothesis that we are interested in rejecting).

What we are trying to infer with hypothesis testing is whether it is likely that a given sample statistic comes from a population whose mean is significantly greater than 75. Because we are using the sample mean to make an inference about the unknown population mean, we would expect there to be some amount of variability in the sample data that is due to sampling error. From our last example, recall that we calculated the sample mean to be 76.65. In performing a hypothesis test, what we are really interested in showing is whether a sample mean of 76.65 is a reasonable sample mean that would likely be obtained from a population whose true mean is equal to 75. Or is it the case that the population we are sampling from really does have a mean that is actually greater than 75? We want to make sure that the difference between the sample mean and the hypothesized mean is not simply due to sampling variability.

Recall from Chapter 3 that for large samples (typically when $n \geq 30$), the distribution of all possible sample means is approximately normally distributed and is centered about the unknown population mean. If our sample size is less than 30 and the population we are sampling from is approximately normally distributed, then we can assume that the distribution of the sample mean will also be normally distributed and centered about the unknown population mean.

One way of deciding whether the sample we selected comes from a population with a mean of μ_0 is by expecting that any given sample mean will fall within two standard deviations of the population mean with a high degree of certainty. Therefore, if the sample mean falls more than two standard deviations away from the population mean, then we may believe that the population we are sampling from does not have a mean that is equal to μ_0. Thus, if the mean of a random sample is not likely to be drawn from a population that has a hypothesized mean of μ_0, we may infer that the mean of the true population we are sampling from is different from the hypothesized mean of μ_0.

The graph in Figure 4.8 illustrates the sampling distribution of the sample mean centered about the true population mean μ_0 that is specified in the null hypothesis. The small shaded area in the right tail of Figure 4.8 represents an area of the sampling distribution of the sample mean where there is only a small chance that a sample mean from this small area would be drawn from a population with a hypothesized mean of μ_0 and this is because these sample means fall more than two standard deviations away from the population

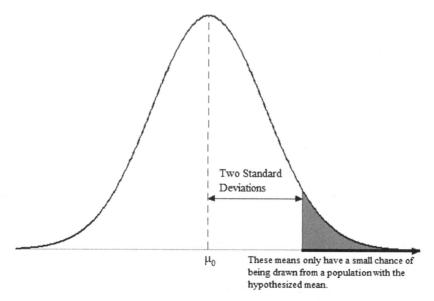

Two Standard
Deviations

μ_0

These means only have a small chance of
being drawn from a population with the
hypothesized mean.

FIGURE 4.8
Distribution of the sample mean centered about the true population mean, μ_0, where the shaded area in the right
tail represents the area under the standard normal curve, which corresponds to the values of the sample mean
that only have a small chance of coming from a population with a true mean of μ_0.

mean. If it were the case that a given sample mean had only a very small chance of coming
from a population with a mean of μ_0, then we may tend to believe that the population we
are sampling from may not have a true mean of μ_0. This shaded region of the sampling
distribution is often referred to as the *rejection region* and it represents the probability that
a given sample mean will be drawn from a population with a mean of μ_0.

In order to accept the alternative hypothesis that the mean of the population we are
sampling from is significantly *greater than* the hypothesized mean of μ_0 and reject the null
hypothesis that the hypothesized mean is equal to μ_0, we would expect a sample value to
fall far away from the hypothesized mean somewhere in this shaded region.

Hypothesis testing is basically a decision process that is used to decide between two
competing claims, the null hypothesis and the alternative hypothesis. If the sample data
does not significantly differ from what is stated in the null hypothesis beyond a reasonable doubt, then this suggests that the null hypothesis cannot be ruled out. Only when
the sample data strongly contradicts the null hypothesis do we reject the null hypothesis and accept the alternative hypothesis. If it were the case that a sample mean is not
likely to come from a population whose mean is μ_0, then we would accept the alternative
hypothesis and reject the null hypothesis.

In order to define the rejection region, we need to determine the area under the sampling
distribution of the sample mean that corresponds to the collection of sample means that
only have a small chance of coming from a population whose true mean is μ_0. The *level
of significance*, denoted as α, represents the area of the sampling distribution of the sample mean that describes the probability of observing a sample mean that comes from a
population with a mean of μ_0. The level of significance is established prior to beginning a
hypothesis test and it corresponds to how much we are willing to risk that the true population we are sampling from actually does have a mean of μ_0, even though the sample mean
obtained may be different enough from μ_0 to suggest otherwise.

It is for this reason that typical values of α tend to be fairly small, such as $\alpha = 0.05$ or $\alpha = 0.01$. This is because if a sample mean falls far enough away from the hypothesized mean into the rejection region, then for a significance level of either 5% or 1%, we have either a 5% or 1% chance that a given sample mean can actually be drawn from a population whose true mean is μ_0. The rejection region for testing a single population mean against some hypothesized value is determined by the level of significance, the sampling distribution of the test statistic, and the direction of the alternative hypothesis.

However, it could be the case that the population we are sampling from does indeed have a mean of μ_0, even though it is very unlikely. We say that a *Type I* error is committed when we accept the alternative hypothesis and reject the null hypothesis when in fact the null hypothesis is actually true. In other words, a Type I error is committed when we claim that the mean of the population we are sampling from has a mean that is different enough from μ_0 (this is when we accept the alternative hypothesis), even though it is the case that the population we are sampling from really does have a mean of μ_0, and this can happen α% of the time.

There are three types of alternative hypotheses that can be used for testing whether the true population we are sampling from has a mean different from some hypothesized value μ_0: a *two-tailed test*, a *right-tailed test*, and a *left-tailed test*. These three alternative hypotheses and their appropriate rejection regions for any given value of α are represented in Figures 4.9–4.11.

The shaded rejection regions in Figures 4.9–4.11 represent that for a given hypothesized mean of μ_0, the probability is α that the population mean is consistent with what was stated in the null hypothesis. The area of the rejection region is part of the sampling distribution for the test statistic under consideration, and for these small regions, the sample means are more likely to support what is stated in the alternative hypothesis.

For our last example, if we could accept the alternative hypothesis of $H_1: \mu > 75$ and reject the null hypothesis of $H_0: \mu = 75$, then we would infer that the population we are sampling from has a mean that is significantly greater than the hypothesized mean of $\mu = 75$. If the given sample mean does not fall in the shaded region as illustrated in Figure 4.10, then we

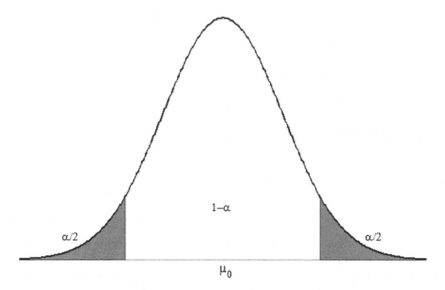

FIGURE 4.9
Sampling distribution for a two-tailed test where the rejection region is located in both the left and the right tails that corresponds to the following null and alternative hypotheses: $H_0: \mu = \mu_0$, $H_1: \mu \neq \mu_0$.

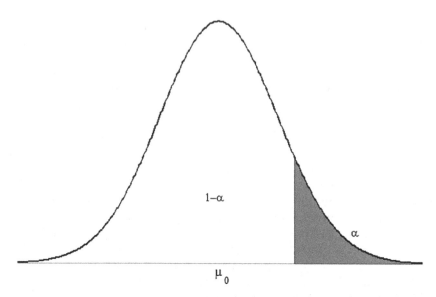

FIGURE 4.10
Sampling distribution for a right-tailed test where the rejection region is located in the right tail that corresponds to the following null and alternative hypotheses: $H_0: \mu = \mu_0$, $H_1: \mu > \mu_0$.

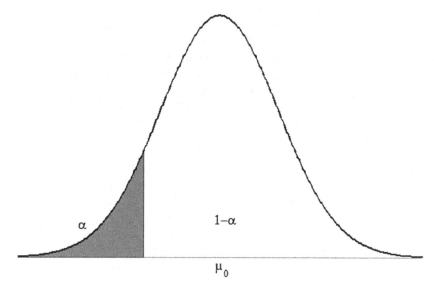

FIGURE 4.11
Sampling distribution for a left-tailed test where the rejection region is located in the left tail that corresponds to the following null and alternative hypotheses: $H_0: \mu = \mu_0$, $H_1: \mu < \mu_0$.

have reason to believe that the distribution we are sampling from does not have a mean that is significantly greater than $\mu = 75$.

What we are trying to do with this hypothesis test is to decide whether a sample mean of $\bar{x} = 76.65$ is likely to be drawn from a population whose true mean is $\mu = 75$. Recall that the distribution of all possible sample means is a bell-shaped curve centered about the true population mean, as is illustrated in Figure 4.12.

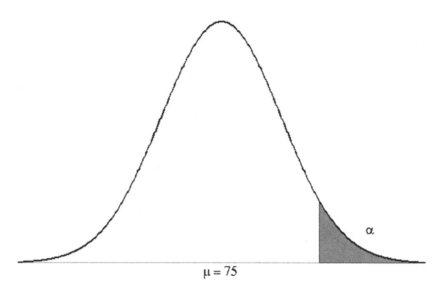

$\mu = 75$

FIGURE 4.12
Sampling distribution of the sample mean drawn from a population whose true mean is $\mu = 75$.

After initially establishing the appropriate rejection region based on the level of significance and the direction of the alternative hypothesis, we can then determine whether the test statistic $T = \dfrac{\bar{x} - \mu_0}{\dfrac{s}{\sqrt{n}}}$, which is the sample mean standardized, is consistent with what is specified in the null hypothesis.

Example 4.3

For the data given in Table 4.1, suppose we want to know if the population mean score on the statistics examination is greater than 75. If the significance level is $\alpha = 0.05$, then a right-tailed hypothesis test would be used $H_0: \mu = 75$, $H_1:$ $\mu > 75$. Since the size of our sample is $n = 20$ (with 19 degrees of freedom), then the rejection region can be defined as presented in Figure 4.13.

Notice in Figure 4.13 that the rejection region is located in the right tail. This is because the alternative hypothesis is specified as a strict inequality, which means that a given sample statistic would need to fall far in the right tail to be inconsistent with what is specified in the null hypothesis.

Since $n < 30$ and assuming that the population we are sampling from is normally distributed, by using the sample mean of $\bar{x} = 76.65$, a sample standard deviation of $s = 10.04$, and a sample size of $n = 20$, the test statistic would be calculated as follows:

$$T = \frac{\bar{x} - \mu_0}{\dfrac{s}{\sqrt{n}}} = \frac{76.65 - 75}{\dfrac{10.04}{\sqrt{20}}} \approx 0.735$$

If this test statistic falls into the rejection region in the right tail, this would correspond to a value greater than $t = 1.729$ and then we could reject the null

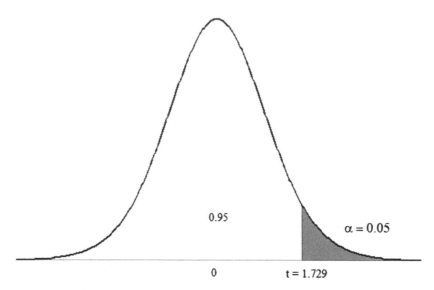

FIGURE 4.13
A right-tailed test of a population mean using a sample of size $n = 20$ with 19 degrees of freedom.

hypothesis (that the population mean is equal to 75) and accept the alternative (that the population mean is greater 75). However, notice in Figure 4.13, the value of the test statistic T *does not* fall into the shaded region because it is not greater than $t = 1.729$. Thus, we cannot claim that *the true population mean* is significantly greater than 75. To interpret this finding within the context of our example, we could claim that we do not have reason to believe that the mean of the population we are sampling is significantly greater than 75.

One thing to keep in mind when we do not have enough evidence to accept the alternative hypothesis is that this *does not* mean that we can accept the null hypothesis. We set up the null hypothesis only for the purpose of rejecting it. Because we are using a sample statistic to make an estimate of a population parameter, we can never obtain an exact estimate of an unknown population parameter. The best we can do is to show whether a sample statistic is different enough to come from a population whose parameter is specified in the null hypothesis. So for our last example, we do not have enough evidence to accept the alternative hypothesis $H_1: \mu > 75$, but this does not necessarily imply that the null hypothesis of $H_0: \mu = 75$ is true. We can never make an exact estimate of a population parameter.

4.5 Using Minitab for a One-Sample t-Test

Using Minitab to perform a one-sample t-test for the data given in Table 4.1 requires some of the exact same steps that are used when calculating confidence intervals for a population mean. We still use **Stat > Basic Statistics > 1-Sample t**. One difference is that the mean being tested under the null hypothesis needs to be specified, as illustrated in Figure 4.14.

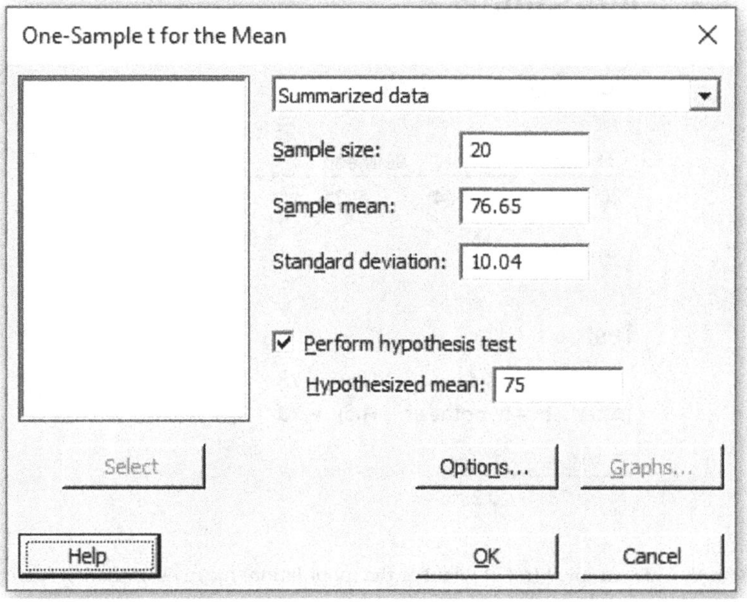

FIGURE 4.14
Minitab dialog box for a one-sample *t*-test.

In addition to providing the hypothesized mean, the confidence level and the direction of the alternative hypothesis also have to be specified. This is done by selecting the **Options** tab, as can be seen in Figure 4.15.

Figure 4.16 provides the Minitab printout for testing whether the true population mean score received on the final examination is significantly greater than 75.

Note that the portion of the Minitab printout in Figure 4.16 provides the descriptive statistics along with the specification of null and alternative hypotheses and the value of the test statistic *T*.

However, the information in the Minitab printout does not explicitly state whether to accept the alternative hypothesis and reject the null hypothesis. Instead, the printout provides what is called a *p-value*. We can use the *p*-value to assess whether we can accept

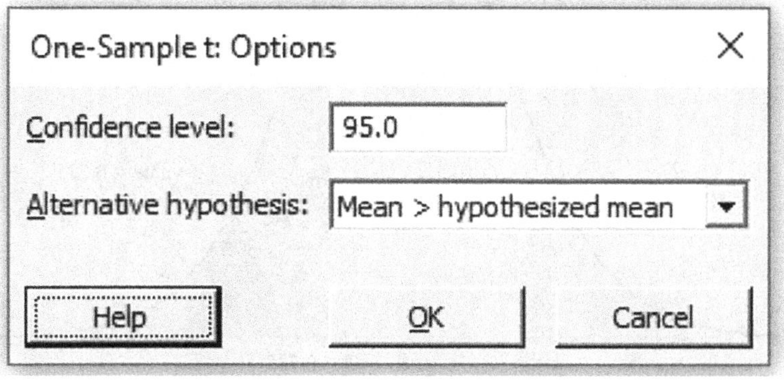

FIGURE 4.15
Options box to specify the confidence level and alternative hypothesis for a one-sample *t*-test.

One-Sample T

Descriptive Statistics

N	Mean	StDev	SE Mean	95% Lower Bound for μ
20	76.65	10.04	2.25	72.77

μ: mean of Sample

Test

Null hypothesis	H_0: $\mu = 75$
Alternative hypothesis	H_1: $\mu > 75$

T-Value	P-Value
0.73	0.236

FIGURE 4.16
Minitab printout for the one-sample *t*-test of whether the (population) mean examination score is greater than 75, using the random sample of data provided in Table 4.1.

the alternative hypothesis or not. The *p*-value is defined as the area under the sampling distribution that is based on the value of the test statistic. The *p*-value can also be described as the observed level of significance because it indicates the likelihood of observing an extreme value as it represents the smallest value of α that can be expected from a given test statistic that will lead to rejecting the null hypothesis.

From the Minitab printout in Figure 4.16, notice that the *p*-value is 0.236. This *p*-value tells us that the observed level of significance is 0.236. In other words, a *p*-value of 0.236 describes the smallest value of α that we can expect from our given test statistic, as illustrated in Figure 4.17.

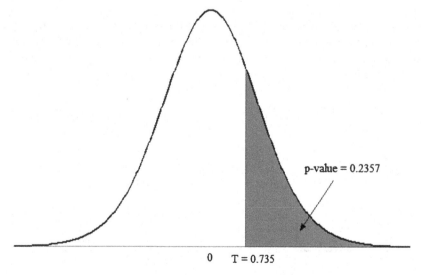

p-value = 0.2357

0   T = 0.735

FIGURE 4.17
The rejection region as defined by the test statistic $T = 0.73$ that corresponds to a *p*-value of 0.2360.

The *p*-value of 0.2360 represents the probability of observing a standardized sample mean greater than $T = 0.735$ from a population with a mean of $\mu = 75$ and there is a 23.60% chance of this actually happening. In other words, there is a 23.60% chance of observing a sample mean greater than 76.65 from a population whose true mean is 75. Clearly, this does not make for a very strong argument and that is why we are almost always concerned with significance levels that are less than 10%.

When the *p*-value is less than our predetermined level of significance of α, we can reject the null hypothesis and accept the alternative hypothesis because when this happens, the test statistic falls in the rejection region. If the *p*-value is greater than our predetermined level of significance of α, then this implies that the test statistic does not fall in the rejection region. If this happens, we can only claim that we do not have enough evidence to reject the null hypothesis and thus cannot accept the alternative hypothesis. For our example, a *p*-value of 0.2360 is greater than the predetermined level of significance $\alpha = 0.05$, which indicates that the test statistic T does not fall in the rejection region and so the true mean of the population we are sampling from is not significantly different from the hypothesized mean of $\mu = 75$. The top panel in Figure 4.18 shows the area of the *T*-distribution for the desired level of significance ($\alpha = 0.05$), and the bottom panel in Figure 4.18 illustrates the area to the right of the test statistic $T = 0.735$ (the *p*-value). Notice that the area represented as the *p*-value in the bottom panel of Figure 4.18 is greater than the area that illustrates the level of significance $\alpha = 0.05$ in the top panel of Figure 4.18.

We will now review confidence intervals and hypothesis tests about an unknown population mean by working through an additional example.

Example 4.4

Suppose we are interested in finding a confidence interval for the population mean time, in hours, it takes a manufacturing plant to produce a certain type of part

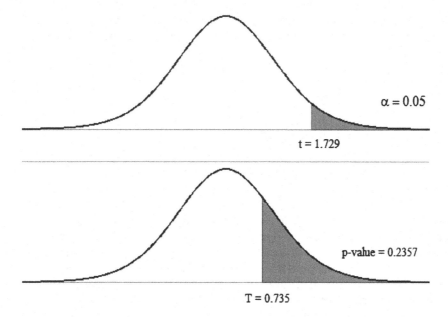

FIGURE 4.18
Top panel: *T*-distribution with an area of $\alpha = 0.05$ in the right tail for 19 degrees of freedom. Bottom panel: *T*-distribution showing the area to the right of the test statistic $T = 0.735$ as 0.02357 (the *p*-value).

TABLE 4.2

Sample of the Time in Hours Taken to Manufacture a Certain Type of Part from Start to Finish at a Given Plant ($n = 25$)

55	53	46	37	52
42	45	55	50	70
45	65	50	51	30
37	45	42	55	46
60	61	30	32	46

from start to finish ($\alpha = 0.05$). The data in Table 4.2 gives the time, in hours, taken from the plant to produce this specific type of part for a sample of size $n = 25$.

Recall that to find a 95% confidence interval for the population mean time taken to manufacture this part, we would need to calculate the sample mean and the sample standard deviation.

The sample mean is calculated as follows:

$$\bar{x} = \frac{\sum_{i=1}^{n} x_i}{n} = \frac{1200}{25} = 48.00$$

The sample standard deviation can be calculated using the following formula:

$$s = \sqrt{\frac{\sum_{i=1}^{n}(x_i - \bar{x})^2}{n-1}} \approx 10.30$$

Assuming that the population we are sampling from is approximately normally distributed, then a $100(1 - \alpha)\% = 100(1 - 0.05)\% = 95\%$ confidence interval for the unknown population mean time taken to manufacture the part would be calculated as follows:

$$\left(\bar{x} - t_{\alpha/2} \cdot \frac{s}{\sqrt{n}}, \ \bar{x} + t_{\alpha/2} \cdot \frac{s}{\sqrt{n}} \right)$$

$$= \left(48 - 2.064 \cdot \frac{10.30}{\sqrt{25}}, \ 48 + 2.064 \cdot \frac{10.30}{\sqrt{25}} \right) \approx \left(43.75, 52.25 \right)$$

where $t_{\alpha/2} = t_{0.25} = 2.064$ is the value from the t-distribution with $n - 1 = 25 - 1 = 24$ degrees of freedom.

This confidence interval suggests that if we repeatedly collect random samples of size 25 from the time taken to manufacture the given part for the entire population and calculate confidence intervals in the exact same manner, then 95% of all such intervals will contain the mean time taken to manufacture this part for the *entire population*. Thus, we would claim to be 95% confident that the population mean time taken to manufacture the part from start to finish falls between 43.75 and 52.25 hours.

To summarize, a confidence interval for a population mean provides a way of arriving at an estimate of a possible range of values for a population mean by using data that was collected from a representative sample.

We can also conduct a hypothesis test if we are interested in testing some hypothesis about a population mean by using sample data.

Example 4.5

Using the data from Table 4.2 for the time taken to manufacture a specific type of part at a certain factory, suppose that we are interested in whether the mean (or average) time it takes to manufacture this part is significantly *less than* 55 hours ($\alpha = 0.05$). In order to do this test, the appropriate null and alternative hypotheses are set up as follows:

$$H_0 : \mu = 55$$

$$H_1 : \mu < 55$$

On the basis of the direction of the alternative hypothesis, a level of significance of $\alpha = 0.05$, and 24 degrees of freedom, the rejection region would be defined as presented in Figure 4.19.

In Figure 4.19, the rejection region is defined in the left tail because we are testing whether the mean of the true population we are sampling from is *less than* the hypothesized mean of 55 hours.

The test statistic would then be calculated as follows:

$$T = \frac{\bar{x} - \mu}{\frac{s}{\sqrt{n}}} = \frac{48.00 - 55.00}{\frac{10.30}{\sqrt{25}}} \approx -3.40$$

Since the value of the test statistic falls into the rejection region ($T = -3.40 < t = -1.711$), as illustrated in Figure 4.20, we can accept the alternative hypothesis

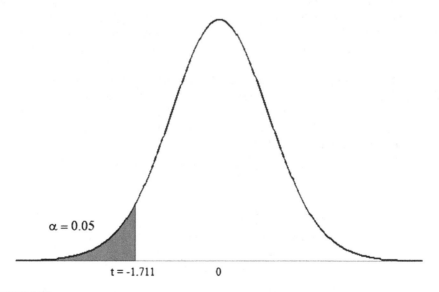

$\alpha = 0.05$

$t = -1.711$ 0

FIGURE 4.19
Rejection region for a left-tailed one-sample *t*-test with 24 degrees of freedom and a significance level of $\alpha = 0.05$.

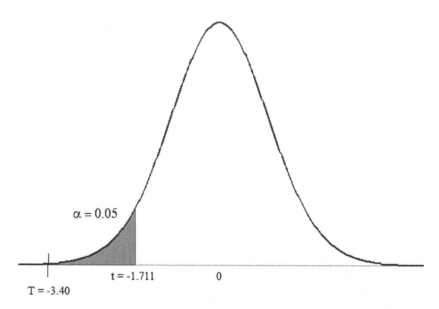

FIGURE 4.20
Rejection region and the corresponding value of test statistic to determine if the population mean time to manufacture the part is significantly less than 55 hours.

and reject the null hypothesis. This suggests that the true population mean time taken to manufacture this specific part is significantly less than 55 hours.

Figure 4.21 provides the Minitab printout for finding a 95% confidence interval for the population mean time taken to manufacture the part, and Figure 4.22 provides the Minitab printout of the hypothesis test to determine whether or not the true population average time it takes to manufacture the part is less than 55 hours.

From the Minitab printout in Figure 4.22, the value of the test statistic $T = -3.40$ corresponds to a p-value of 0.001. The p-value is the area to the left of the test statistic T that describes the minimum level of significance that we can expect from the given test statistic with the given alternative hypothesis. In other words, there is only a 0.1% chance of observing a sample mean less than 48 from a population whose true mean is H_0: $\mu = 55$. Figure 4.23 illustrates how the p-value describes the area that is to the left of T, which is clearly less than our predetermined level of significance of $\alpha = 0.05$. Because it is very unlikely that this sample mean comes from a population whose true mean is H_0: $\mu = 55$, this leads to accepting the alternative hypothesis and rejecting the null hypothesis.

One-Sample T

Descriptive Statistics

N	Mean	StDev	SE Mean	95% CI for μ
25	48.00	10.30	2.06	(43.75, 52.25)

μ: mean of Sample

FIGURE 4.21
Minitab printout for the 95% confidence interval for the mean time taken to manufacture the part.

One-Sample T

Descriptive Statistics

N	Mean	StDev	SE Mean	95% Upper Bound for μ
25	48.00	10.30	2.06	51.52

μ: mean of Sample

Test

Null hypothesis	$H_0: \mu = 55$
Alternative hypothesis	$H_1: \mu < 55$

T-Value	P-Value
-3.40	0.001

FIGURE 4.22
Minitab printout for the hypothesis test of whether, on average, it takes less than 55 hours to manufacture the part.

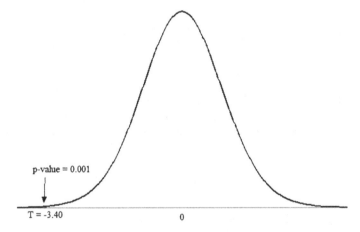

FIGURE 4.23
Illustration of the *p*-value as being the area to the left of the test statistic $T = -3.40$.

Example 4.6

Historically, students tend to work on an average of 22 hours/week. An educational researcher wants to know if the mean number of hours that the students work is different from the historical average ($\alpha = 0.10$). A sample of 18 students reported an estimate of the number of hours they work each week and these results are given in Table 4.3.

Since we want to know if the mean number of hours worked is *different from* the historical average, we use a two-tailed test and hence the null and alternative hypotheses would be:

$$H_0 : \mu = 22$$

$$H_1 : \mu \neq 22$$

TABLE 4.3

Number of Hours Worked Each Week for a Sample of 18 Students

25	28	15	20	25	24	10	8	30
15	27	30	25	22	26	30	19	27

For a sample of size 18, the degrees of freedom would be $n - 1 = 18 - 1 = 17$ and the level of significance of $\alpha = 0.10$ would be split in the two tails of the *t*-distribution, as is illustrated in Figure 4.24.

We can then calculate the necessary descriptive statistics as follows:

$$\bar{x} \approx 22.56$$

$$s \approx 6.73$$

$$n = 18$$

The test statistic is calculated as follows:

$$T = \frac{\bar{x} - \mu}{\frac{s}{\sqrt{n}}} = \frac{22.56 - 22}{\frac{6.73}{\sqrt{18}}} \approx 0.35$$

Since this value does not fall in the shaded regions as illustrated in Figure 4.24, we claim that the population mean number of hours that college students work is not significantly different from the historical mean of 22 hours/week.

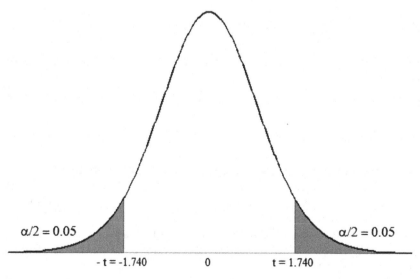

FIGURE 4.24
T-distribution for 17 degrees of freedom for a two-tailed test ($\alpha = 0.10$).

4.6 Power Analysis for a One-Sample *t*-Test

There are two types of conclusions that can be made when doing a one-sample hypothesis test about a population mean:

1. If the test statistic *does fall* in the rejection region (this occurs when the *p*-value is less than the predetermined level of significance), we accept the alternative hypothesis and reject the null hypothesis and claim that there is enough evidence to suggest that we are sampling from a population whose mean is significantly different from, less than, or greater than what is specified in the null hypothesis.

2. If the test statistic *does not fall* in the rejection region (this occurs when the *p*-value is greater than the predetermined level of significance) and we do not reject the null hypothesis, we cannot claim that we are sampling from a population that has a mean that is significantly different from, or less than, or greater than, what is specified in the null hypothesis.

For the second conclusion to be true, we need to determine whether the population we are sampling from, in fact, does not have a true population mean that is different from what is specified in the null hypothesis. We need to decide whether the true population mean is not different enough from a specific hypothesized value, and that it is not simply the case that there is not a large enough sample to be able to detect whether or not the true population mean is different enough from the hypothesized value.

This is where the notion of a *power analysis* comes into play. Essentially, a power analysis provides you with the necessary sample size to detect a minimum difference for a given population mean. The *power of a test* is defined as the probability of finding a difference of a specific size provided that such a difference actually exists. The power of a test tends to range from 70% to 90%, with 80% being the typical power value.

A power analysis is often a very important first step in basic statistical inference because when there is not enough evidence to reject the null hypothesis and thus the claim that the population mean is different enough from some hypothesized value, you may want to maximize the probability of making the correct inference by deciding whether the collected sample is large enough to detect such a difference provided that such a difference were to actually exist.

So when do you need to conduct a power analysis? There are two basic scenarios that you may want to consider—a *prospective power analysis* and a *retrospective power analysis*.

A *prospective power analysis* is done *before data collection begins*. By conducting a power analysis before collecting your data, you can gain some insight as to the size of the sample that is needed to find a reasonable difference. For example, if you want to estimate whether the average final exam score is greater than 75 (see Example 4.3), you may want to first conduct a power analysis before you collect your sample. So for instance, if you want a score of at least 85 (i.e., at least a 10-point difference) to test out as significant, a prospective power analysis could give you some idea of how large of a sample you would need to find a difference of this size.

A *retrospective power analysis* is done *after you have conducted your analysis* and did not find any significant difference. This happens when the test statistic does not fall in the rejection region (when the *p*-value is greater than the predetermined level of significance) and we do not reject the null hypothesis. A retrospective power analysis will tell you whether or not the size of the sample you collected was large enough to see a specified difference. Not having a sample large enough to find a given difference would suggest that your analysis was *underpowered*. In other words, your sample size was not large enough to find an effect of a certain size, if such an effect were to actually exist.

4.7 Using Minitab for a Power Analysis for a One-Sample *t*-Test

Consider the data given in Table 4.1, where we were interested in determining whether the true population mean grade on a statistics final examination is significantly greater than 75. Recall that since we were interested in testing whether the population mean is greater than 75, we stated our alternative hypothesis to be H_1: $\mu > 75$ and the null hypothesis as H_0: $\mu = 75$.

Recall that we did not have enough evidence to suggest that the population mean grade on the examination was significantly greater than 75 (the *p*-value was greater than our predetermined level of significance, α). But even though we did not find a significant difference, we may want to investigate whether the sample we collected was large enough to be able to detect a difference if such a difference were to actually exist. This is an example of a retrospective power analysis since we already collected our data and did a one-sample *t*-test.

Suppose we wanted to detect a minimum difference of at least 10 points on the examination score with 80% power. In other words, we want a mean difference of at least 10 points between the hypothesized mean and the standardized sample mean to be significant 80% of the time. We can use Minitab to perform a power analysis to tell us how large a sample would be needed to detect this minimum difference for a one-sample *t*-test. This can be done by clicking on **Stat**, then **Power and Sample Size**, and then **1-Sample *t***, as illustrated in Figure 4.25.

This brings up the dialog box that is presented in Figure 4.26, where we need to provide the difference we are interested in finding, the desired power level (which is typically 80%), and an estimate of the standard deviation. We also need to click on the **Options** tab to select the level of significance and the direction of the alternative hypothesis (which would be **Greater than** for this example).

Notice in Figure 4.26 that we need to specify two of the three values of sample size, differences, and power values, along with an estimate of the standard deviation. By specifying a difference of 10 points, a power value of 80%, and the sample standard deviation of 10.04, this gives the Minitab printout that appears in Figure 4.27.

According to the Minitab printout in Figure 4.27, a sample size of 8 would be needed to find a difference of 10 points between the true and hypothesized population means 80% of the time. So our sample size of $n = 20$ was, in fact, large enough to find a 10-point difference if such a difference were to exist, but it appears that such a difference did not exist.

Suppose that we wanted to detect a smaller difference of 3 points (also at 80% power). Then we could run a similar power analysis in Minitab by specifying the difference, power level, and standard deviation as described in Figure 4.28.

The Minitab printout for this power analysis is given in Figure 4.29.

Notice from Figure 4.29 that we would need a sample size of at least 71 to detect if the population mean is greater than the hypothesized mean by at least 3 points. For our sample of size $n = 20$, this was not a large enough sample to detect a 3-point difference, so the test

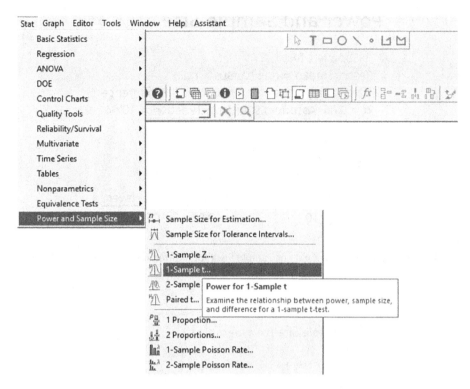

FIGURE 4.25
Minitab commands for running a power analysis for a one-sample *t*-test.

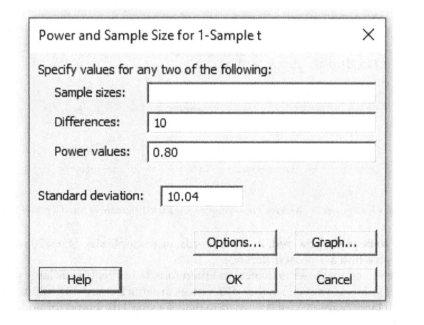

FIGURE 4.26
Minitab dialog box for testing a difference of more than 10 points with a power level of 80% for a one-sample *t*-test.

Power and Sample Size

1-Sample t Test
Testing mean = null (versus > null)
Calculating power for mean = null + difference
α = 0.05 Assumed standard deviation = 10.04

Results

| | Sample | Target | |
Difference	Size	Power	Actual Power
10	8	0.8	0.812334

FIGURE 4.27
Minitab printout for a power analysis for a one-sample *t*-test for a difference of at least 10 and a power level of 80%.

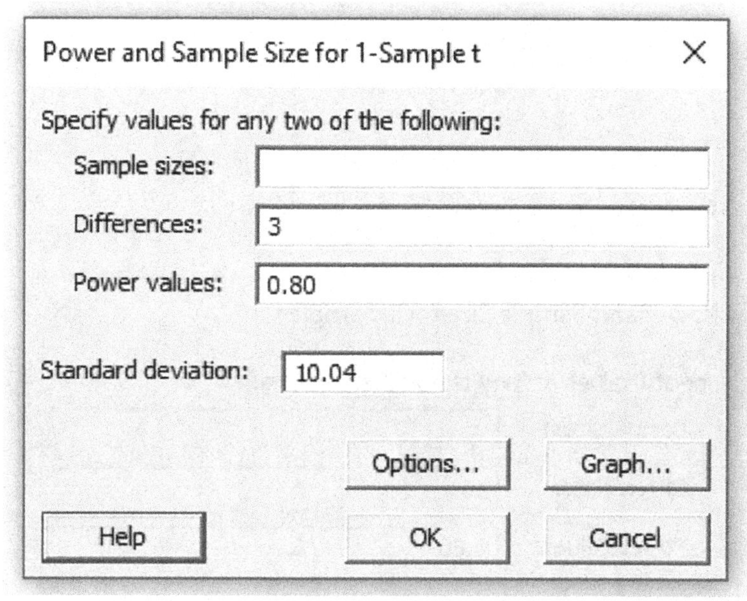

Power and Sample Size for 1-Sample t   ✕

Specify values for any two of the following:

Sample sizes: [   ]

Differences: [3   ]

Power values: [0.80   ]

Standard deviation: [10.04   ]

[Options...]   [Graph...]

[Help]   [OK]   [Cancel]

FIGURE 4.28
Minitab dialog box for a power analysis for a one-sample *t*-test for a difference of 3 and a power level of 80%.

is considered to be underpowered. In other words, our sample size of $n = 20$ was not large enough to be able find a 3-point difference.

Larger differences are easier to detect, and therefore do not require as large a sample for a given power value. However, smaller differences are much more difficult to find, which is why a much larger sample size is needed to detect a small difference provided that such a difference were to actually exist.

We can also determine what the power would be for a given sample size and difference. For the sample given in Table 4.1, we have a sample of size 20.

Power and Sample Size

1-Sample t Test
Testing mean = null (versus > null)
Calculating power for mean = null + difference
α = 0.05 Assumed standard deviation = 10.04

Results

Difference	Sample Size	Target Power	Actual Power
3	71	0.8	0.801910

FIGURE 4.29
Minitab printout for a power analysis for a one-sample *t*-test for a difference of size 3 and a power level of 80%.

Suppose we want to see what the power of our hypothesis test would be if we have a difference of at least 5. In other words, how likely is it that we will find a difference of at least 5 with a sample of size $n = 20$? Then we need to specify the sample size and the difference as presented in Figure 4.30.

The Minitab printout for this power analysis is presented in Figure 4.31.

The printout in Figure 4.31 suggests that we only have (approximately) a 69% chance of detecting a difference of 5 points with a sample of size $n = 20$ (provided that such a difference actually exists). Since this does not make for a very strong argument, we may want to consider increasing the sample size in order to detect such a difference with a higher probability. Similarly, we could also specify a given sample size and power value and Minitab will determine what the minimum difference would be.

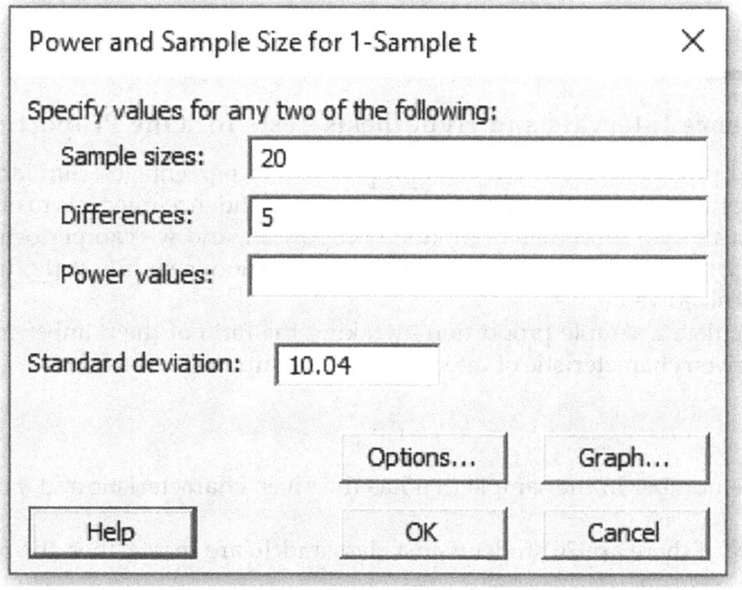

Power and Sample Size for 1-Sample t	✕

Specify values for any two of the following:

Sample sizes: 20

Differences: 5

Power values:

Standard deviation: 10.04

Options... Graph...

Help OK Cancel

FIGURE 4.30
Minitab dialog box for a power analysis for a one-sample *t*-test for a difference of size 5 with a sample size $n = 20$.

Power and Sample Size

1-Sample t Test
Testing mean = null (versus > null)
Calculating power for mean = null + difference
α = 0.05 Assumed standard deviation = 10.04

Results

Difference	Sample Size	Power
5	20	0.692139

FIGURE 4.31
Minitab results for a power analysis for a one-sample *t*-test for a difference of 5 with a sample size $n = 20$.

To summarize, in order to perform a power analysis, two of the three values of sample size, minimum difference, and power level must be provided in order for Minitab to calculate the other value that is left blank.

In general, there are no hard-and-fast rules for conducting a power analysis and the importance of using a power analysis varies for different fields of study, and most such analyses tend to be guided by generally accepted rules of thumb. Also, simply collecting a given sample of the appropriate size needed to detect a specified difference does not guarantee that such a difference will actually be detected. All that a power analysis can do is to provide you with some estimate as to the sample size needed to find such a difference if one were to actually exist. Furthermore, collecting the appropriate sample size for a given difference does not guarantee that any measures used to collect your data are valid. For an intuitive description of power analysis and validity, see Light et al. (1990).

4.8 Confidence Intervals and Hypothesis Tests for One Proportion

Many applied problems often deal with proportions or percentages. Similar to the work we have done thus far with means, we can calculate confidence intervals to obtain a likely range of values for an unknown population proportion, and we can perform hypothesis tests to infer whether an unknown population proportion is significantly different from some hypothesized value.

We can calculate a sample proportion by taking the ratio of the number in the sample that has the given characteristic of interest and the sample size as follows:

$$\hat{p} = \frac{x}{n}$$

where x is the number in the sample that has the given characteristic and n is the sample size.

For instance, if there are 20 students in a class and 16 are males, then the proportion of males is:

$$\hat{p} = \frac{x}{n} = \frac{16}{20} = 0.80$$

We can also easily obtain a percentage by simply multiplying a proportion by 100 to get:

$$\hat{p} = \frac{x}{n} = \frac{16}{20} = 0.80(100) = 80\%$$

A confidence interval for a population proportion takes the following form:

$$\left(\hat{p} - z_{\alpha/2} \cdot \sqrt{\frac{\hat{p}(1-\hat{p})}{n}}, \hat{p} + z_{\alpha/2} \cdot \sqrt{\frac{\hat{p}(1-\hat{p})}{n}} \right)$$

where $\hat{p}$ is the sample proportion, n is the sample size, and $z_{\alpha/2}$ is the value that defines the upper $\alpha/2$ portion of the standard normal distribution.

For large samples, the sampling distribution of the sample proportion $\hat{p}$ follows a normal distribution and is centered about the unknown population proportion p.

Example 4.7

Suppose that in a taste test given to a random sample of 350 shoppers at a local grocery store, 210 of these shoppers preferred the taste of a new brand of soft drink as compared to a competitor's brand.

To find a 95% confidence interval for the unknown population proportion of shoppers who would prefer this new brand of soft drink, we would need to find the sample proportion. The sample proportion is found by taking the number who prefer the taste of the new soft drink and dividing it by the total number of shoppers who participated in the test. This can be calculated as follows:

$$\hat{p} = \frac{210}{350} = 0.60$$

We then need to find the value of z that defines the upper 2.5% of the standard normal distribution, as is illustrated in Figure 4.32.

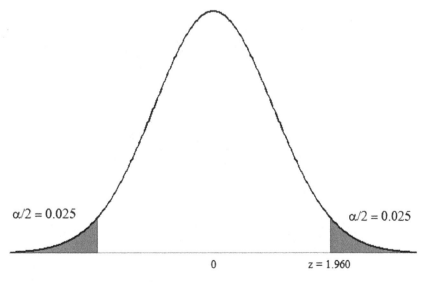

$\alpha/2 = 0.025$ $\alpha/2 = 0.025$

0         $z = 1.960$

FIGURE 4.32
Distribution of the sample proportion for a 95% confidence interval.

Thus, a 95% confidence interval would be

$$\left(\hat{p} - z_{\alpha/2} \cdot \sqrt{\frac{\hat{p}(1-\hat{p})}{n}}, \hat{p} + z_{\alpha/2} \cdot \sqrt{\frac{\hat{p}(1-\hat{p})}{n}} \right)$$

$$= \left(0.60 - 1.96 \cdot \sqrt{\frac{0.60(1-0.60)}{350}}, 0.60 + 1.96 \cdot \sqrt{\frac{0.60(1-0.60)}{350}} \right)$$

$$\approx (0.60 - 0.051, 0.60 + 0.051) \approx (0.549, 0.651) \approx (54.9\%, 65.1\%)$$

Interpreting this confidence interval suggests that we are 95% confident that between 54.9% and 65.1% of the *population* of shoppers would prefer this particular brand of soft drink over that of the competitor's.

We can also perform hypothesis tests to compare the true population proportion against some hypothesized value. For large samples, we can use the following test statistic:

$$Z = \frac{x - n \cdot p_0}{\sqrt{n \cdot p_0(1 - p_0)}}$$

where x is the number in the sample who has the given characteristic of interest, n is the sample size, and p_0 is the population proportion being tested under the null hypothesis. This test statistic for the sample proportion follows the standard normal distribution and it is centered about the unknown population proportion p_0, as illustrated in Figure 4.33.

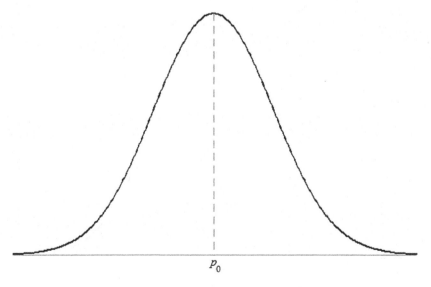

FIGURE 4.33
Distribution of the sample proportion if H_0: $p = p_0$ were true.

Example 4.8

Suppose a soft drink company wants to test whether at least 55% (0.55) of all shoppers would prefer their brand of soft drink as compared to that of the competitor's. A random sample of 350 shoppers found that 210 preferred the particular brand of soft drink. To test whether or not the true population proportion of shoppers that would prefer their particular brand of soft drink is significantly greater than 55% ($\alpha = 0.05$), we would set up the null and alternative hypotheses using proportions as follows:

$$H_0 : p = 0.55$$

$$H_1 : p > 0.55$$

Figure 4.34 shows the rejection region that is defined by the value of $z_{0.05} = 1.645$. The test statistic would then be calculated as follows:

$$Z = \frac{x - n \cdot p_0}{\sqrt{n \cdot p_0 (1 - p_0)}} = \frac{210 - (350)(0.55)}{\sqrt{(350)(0.55)(1 - 0.55)}} \approx 1.88$$

Since this test statistic falls in the rejection region, as illustrated in Figure 4.34, we can accept the alternative hypothesis and reject the null hypothesis. Thus, the unknown population proportion of shoppers who would prefer this particular brand of soft drink is significantly greater than 0.55

We can go one step further and actually calculate the p-value by hand. Recall that the p-value is the minimum level of significance that is determined based on where the test statistic falls on the given sampling distribution. We can find the p-value by using either the standard normal tables since the test statistic for the population proportion follows a standard normal distribution, or we can use Minitab to generate the sampling distribution. To find the p-value, we need to find the area to the right of the test statistic Z, as is illustrated in Figure 4.35.

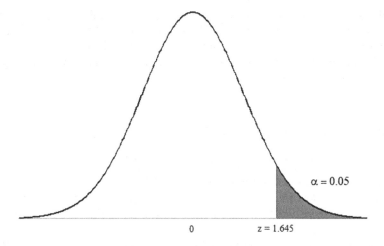

FIGURE 4.34
Rejection region for testing if the true population proportion of shoppers who prefer the new brand of soft drink is significantly greater than 55% (0.55).

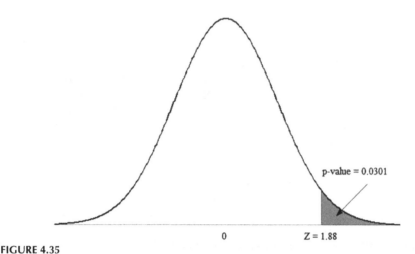

FIGURE 4.35

Area to the right of the test statistic $Z = 1.88$ corresponding to a p-value of 0.0301.

Figure 4.35 shows that the area to the right of $Z = 1.88$ is equal to 0.0301. This area corresponds to a p-value of 0.0301, which supports the conclusion that the true population proportion of shoppers who prefer the given brand of soft drink over that of the competitor's is significantly greater than 0.55 (i.e. 55%). This is because the p-value is less than the predetermined level of significance of 0.05.

It is fairly easy to find an exact p-value for a test statistic that follows the standard normal distribution using the standard normal table. For a right-tailed test, the p-value represents the area to the right of the test statistic on the standard normal curve. For a left-tailed test, the p-value represents the area to the left of the test statistic on a standard normal curve. For a two-tailed test, the p-value represents twice the area to the right of a positive test statistic on a standard normal curve or twice the area to the left of a negative test statistic on a standard normal curve.

It is much more difficult to find an exact p-value using tables for test statistics that are not normally distributed, such as for test statistics that follow the t-distribution. This is because the t-table only provides a few values of the area, whereas the standard normal table can be used to find the exact p-value. But we can always use Minitab to draw a distribution and calculate the p-value as was presented in Section 3.18.

4.9 Using Minitab for a One-Sample Proportion

Minitab can be used to calculate confidence intervals and perform hypothesis tests for a one-sample proportion. To calculate a confidence interval, under the **Stat < Basic Statistics** option on the menu bar you can select **1 Proportion** to test one proportion, as illustrated in Figure 4.36.

The dialog box for a one-sample proportion confidence interval and hypothesis test is presented in Figure 4.37, where we can enter either actual sample data or summarized data.

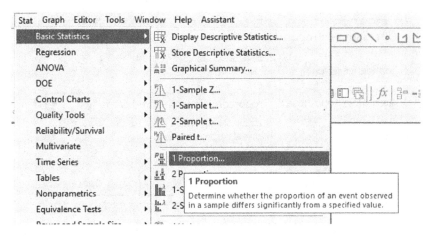

FIGURE 4.36
Minitab commands to calculate a confidence interval and conduct hypothesis tests for a one-sample proportion.

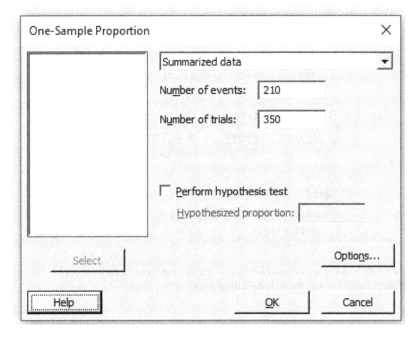

FIGURE 4.37
Minitab dialog box for a one-sample proportion using summarized data.

Under the **Options** tab, you can then select the confidence level and specify that the alternative has to be set to "not equal" to find a confidence interval, as can be seen in Figure 4.38. Also, notice that the method used is set to the **Normal approximation**. This will allow you to conduct hypothesis tests and calculate confidence intervals based on the normal distribution. This is done so that Minitab will use the same formulas that were introduced in the calculations that we did manually.

This gives the Minitab printout in Figure 4.39.

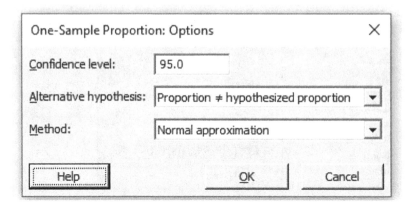

FIGURE 4.38
Minitab options box for a one-sample proportion.

Test and CI for One Proportion

Method

p: event proportion
Normal approximation method is used for this analysis.

Descriptive Statistics

N	Event	Sample p	95% CI for p
350	210	0.600000	(0.548676, 0.651324)

Test

Null hypothesis	H_0: p = 0.5
Alternative hypothesis	H_1: p ≠ 0.5

FIGURE 4.39
Minitab printout for the confidence interval for a one-sample proportion of those shoppers that prefer the particular brand of soft drink as compared to that of the competitor's.

Similarly, we could use Minitab to run a hypothesis test by specifying the proportion being tested under the null hypothesis, the direction of the alternative hypothesis, and the level of significance, as can be seen in Figures 4.40 and 4.41.

In Figure 4.41, we specified the method to be the **Normal approximation**. This is done so that the formulas used by Minitab will be the same as those we used when we did the calculations manually.

The Minitab printout is given in Figure 4.42. Notice in Figure 4.42 that the p-value of 0.030 represents the area to the right of the test statistic $Z = 1.88$. Thus, it is very unlikely that the given sample would be drawn from a population with a true proportion of $p = 0.55$. Since this p-value is less than the predetermined level of significance, we can claim that the unknown population proportion is significantly greater than 0.55 (which is 55%).

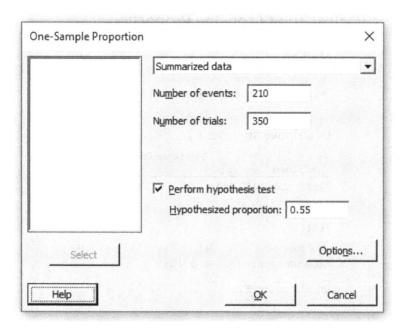

FIGURE 4.40
Minitab dialog box to perform a hypothesis test for a hypothesized proportion of 0.55.

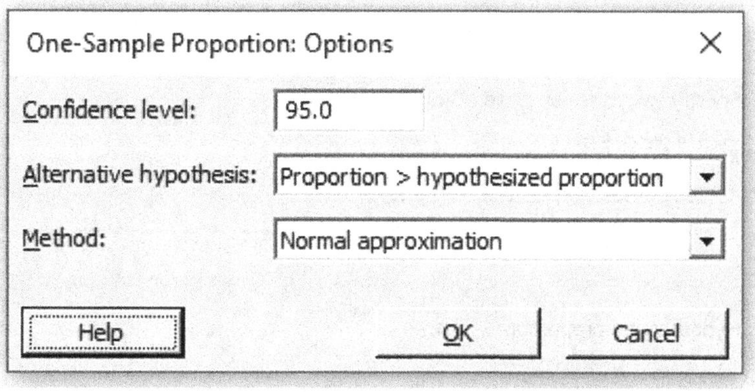

FIGURE 4.41
Minitab options box to test if the true population proportion is greater than 55% ($\alpha = 0.05$).

4.10 Power Analysis for a One-Sample Proportion

Similar to the power analyses we have done thus far, we can use Minitab to perform a power analysis for a one-sample proportion. The Minitab dialog box for a one-sample proportion is given in Figure 4.43.

Notice that Figure 4.43 requires that you specify any two of the three values for sample size, comparison proportion, or the power level, along with the proportion being tested under the null hypothesis. For our last example, suppose we want to know the sample

Test and CI for One Proportion

Method

p: event proportion
Normal approximation method is used for this analysis.

Descriptive Statistics

N	Event	Sample p	95% Lower Bound for p
350	210	0.600000	0.556928

Test

Null hypothesis	H_0: p = 0.55
Alternative hypothesis	H_1: p > 0.55

Z-Value	P-Value
1.88	0.030

FIGURE 4.42
Minitab printout for the one-sample test of a population proportion of whether at least 55% of shoppers prefer the particular brand of soft drink as compared to that of the competitor's.

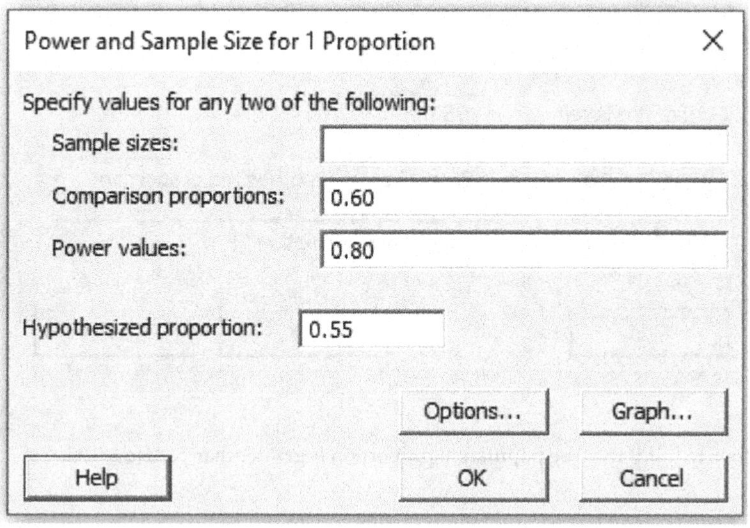

FIGURE 4.43
Minitab dialog box for a power analysis for a one-sample proportion.

size that would be needed to detect a population proportion of 0.60 with a power level of 80%. Note that 0.60 is 0.05 more than the hypothesized proportion of 0.55. We could specify 0.60 as the comparison proportion and specify the value being tested under the null hypothesis as the hypothesized value of *p*. Because the direction of the alternative hypothesis was "greater than," the significance level and the appropriate direction of

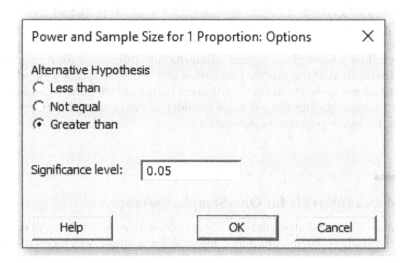

FIGURE 4.44
Minitab options box for a one-sample proportion.

Power and Sample Size

Test for One Proportion
Testing p = 0.55 (versus > 0.55)
α = 0.05

Results

Comparison p	Sample Size	Target Power	Actual Power
0.6	606	0.8	0.800137

FIGURE 4.45
Minitab printout for a power analysis to determine what sample size is needed to detect a difference of 0.05 over the hypothesized proportion of 0.55 with 80% power.

the alternative hypothesis needs to be specified, as can be seen in the **Options** box in Figure 4.44. Essentially, this power analysis allows us to obtain the sample size needed to find a difference of at least 0.05 more than the hypothesized proportion 0.55 to test out as significant at the given power level.

This gives the Minitab printout in Figure 4.45.

As can be seen in Figure 4.45, a sample size of at least 606 is needed to detect at least a 0.05 difference at 80% power, if such a difference were to actually exist.

4.11 Confidence Intervals and Hypothesis Tests for One-Sample Variance

Similar to how we calculated confidence intervals and performed hypothesis tests about the population mean and population proportion, we can also calculate confidence intervals and perform hypothesis tests about other population parameters such as the *population variance.*

What underlies all of the techniques we have used thus far is that we can describe the distribution (in other words, the probability pattern) of the sample statistic. For instance, when making inferences about the population mean, the distribution of the sample mean was found to follow a *t*-distribution, and when making inferences for a population proportion, the distribution of the sample proportion was found to follow a standard normal distribution. Once we understand the distribution pattern of the sample statistics, we can then use these sample statistics to calculate confidence intervals and perform hypothesis tests about an unknown population parameter.

4.12 Confidence Intervals for One-Sample Variance

In order to calculate confidence intervals for a population variance, we need to rely on a sampling distribution. For a confidence interval for a single variance (assuming that we are sampling from a normally distributed population), the sample statistic $\frac{(n-1)s^2}{\sigma^2}$ follows a chi-square distribution with $n-1$ degrees of freedom. This random variable falls between $\chi^2_{(1-\alpha/2)}$ and $\chi^2_{\alpha/2}$ with probability $1-\alpha$. With a little algebraic manipulation, the expression for a $100(1-\alpha)\%$ confidence interval for a population variance σ^2 will take the following form:

$$\frac{(n-1)s^2}{\chi^2_{\alpha/2}} < \sigma^2 < \frac{(n-1)s^2}{\chi^2_{(1-\alpha/2)}} \qquad df = n-1$$

where s^2 is the sample variance, n is the sample size, and $\chi^2_{\alpha/2}$ is the value of the chi-square distribution for an area of $\alpha/2$ in the right tail, and $\chi^2_{(1-\alpha/2)}$ is the value of the chi-square distribution for an area of $(1-\alpha/2)$ in the right tail. As with all the other confidence intervals that we have calculated so far, if we take repeated samples of size n from the population of interest and calculate confidence intervals in the exact same way, then the unknown population variance would be contained in approximately $100(1-\alpha)\%$ of these intervals.

When making inferences about a population variance, we first need to calculate the sample variance (denoted as s^2). Recall from Chapter 3 that the sample variance can found by taking the sum of the squared difference between each data value (x_i) and the sample mean ($\bar{x}$) and then dividing by $n-1$ as follows:

$$s^2 = \frac{\sum\limits_{i-1}^{n}(x_i - \bar{x})^2}{n-1}$$

Example 4.9

Table 4.2 presents a sample of the time taken (in hours) to manufacture a certain part from start to finish ($n = 25$). Suppose we are interested in finding a 99% confidence interval for the population variance using this set of data. Once we calculate the sample variance of $s^2 \approx 106.17$, we then need to find the values of

χ^2 that corresponds to areas of 0.005 and 0.995 on the chi-square distribution with 24 degrees of freedom. This is illustrated in Figure 4.46.

Then, a 99% confidence interval for the population variance would be:

$$\frac{(n-1)s^2}{\chi^2_{\alpha/2}} < \sigma^2 < \frac{(n-1)s^2}{\chi^2_{(1-\alpha/2)}}$$

$$\frac{(25-1)(106.17)}{45.558} < \sigma^2 < \frac{(25-1)(106.17)}{9.886}$$

$$55.93 < \sigma^2 < 257.75$$

If we take repeated samples of size 25 and calculate confidence intervals in the exact same way, then approximately 99% of these intervals will contain the unknown population variance. Thus, we are 99% confident that the unknown population variance will fall between 55.93 and 257.75 hours squared.

We can easily obtain a confidence interval for an unknown *population standard deviation* by simply taking the square root of the confidence interval for an unknown population variance. Recall that the standard deviation is found by taking the square root of the variance.

$$55.93 < \sigma^2 < 257.75$$

$$7.48 < \sigma < 16.05$$

We can then state that we are 99% confident that the unknown population standard deviation will fall between 7.48 and 16.05 hours.

One reason that confidence intervals for standard deviations may be more useful than that for variances is that the units for the standard deviation are the same as the units of the original data. The units for variance are squared units of the original data.

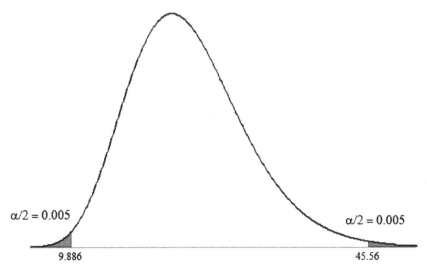

FIGURE 4.46
Chi-Square distribution for a 99% confidence interval with 24 degrees of freedom.

4.13 Hypothesis Tests for One-Sample Variance

In order to conduct a hypothesis test about an unknown population variance, we need to transform the sample variance in the form of a test statistic as follows:

$$\chi^2 = \frac{(n-1)s^2}{\sigma_0^2} \quad df = n-1$$

Where n is the sample size, s^2 is the sample variance, and σ_0^2 is the hypothesized population variance that we are interested in testing. The sampling distribution of this test statistic also follows the chi-square distribution with $n-1$ degrees of freedom.

Similar to other hypothesis tests we have done in the past, there are three different forms for the null and alternative hypotheses for a one-sample variance—a two-tailed test, a right-tailed test, and a left-tailed test:

$$H_0 : \sigma^2 = \sigma_0^2 \quad H_0 : \sigma^2 = \sigma_0^2 \quad H_0 : \sigma^2 = \sigma_0^2$$
$$H_1 : \sigma^2 \neq \sigma_0^2 \quad H_1 : \sigma^2 > \sigma_0^2 \quad H_1 : \sigma^2 < \sigma_0^2$$

Example 4.10

Historically, in Massachusetts, the standard deviation for high temperatures in the month of December is 8°F. December 2015 was considered one of the warmest months on record, with temperatures ranging from freezing to almost 70°F. Table 4.4 provides a random sample of 22 temperatures taken across Massachusetts during December 2015.

Suppose that we are interested in determining whether the variability of the temperatures in Massachusetts during December 2015 was more than what is known historically ($\alpha = 0.05$). We first begin by calculating any relevant descriptive statistics:

$$\bar{x} \approx 52.77$$

$$n = 22$$

$$s \approx 8.474$$

$$s^2 \approx 71.80$$

TABLE 4.4

Random Sample of Temperatures (in Degrees Fahrenheit) Taken Across Massachusetts During December 2015 ($n = 22$)

59	56	44	46	58	58
53	58	61	48	52	52
41	53	60	56	69	
57	35	38	45	62	

Since we were given the historical standard deviation of 8°F, this would yield a historical variance of 64 (recall that the variance is just the square of the standard deviation). Using this information, the null and alternative hypothesis can be stated as follows:

$$H_0 : \sigma^2 = 8^2$$

$$H_1 : \sigma^2 > 8^2$$

Now calculating the test statistic:

$$\chi^2 = \frac{(n-1)s^2}{\sigma_0^2} \qquad df = n-1$$

$$\chi^2 = \frac{(22-1)(8.474)^2}{8^2} \approx 23.56 \qquad df = 22-1 = 21$$

Given that the alternative hypothesis is a right-tailed test, we can describe the rejection region of the chi-square distribution with 21 degrees of freedom, as is illustrated in Figure 4.47.

Since the test statistic is less than the value that defines the rejection region on the chi-square distribution (32.67), we do not have enough evidence to suggest that the variability in temperatures during December 2015 in Massachusetts was more than the historical variance of 64.

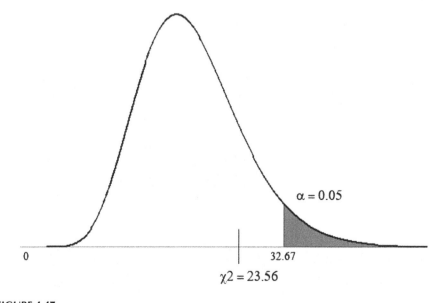

FIGURE 4.47
The rejection region for a right-tailed test statistic for the chi-square distribution with a significance level of 0.05 for 21 degrees of freedom.

4.14 Using Minitab for One-Sample Variance

Minitab can easily calculate confidence intervals and perform hypothesis tests for one-sample variances. Begin by selecting **Stat > Basic Statistics > 1-Variance**. This will bring up the dialog box as is seen in Figure 4.48.

Similar to what we have seen with other analyses using Minitab, we can enter either the raw data or the summarized data and we can select either the sample standard deviation or the sample variance. The summarized data to calculate the confidence interval for the population variance from Example 4.8 can be entered into Minitab, as is illustrated in Figure 4.49.

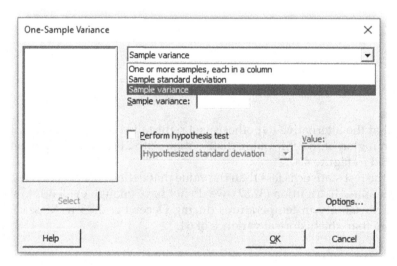

FIGURE 4.48
Minitab dialog box for confidence intervals and hypothesis tests for a single variance.

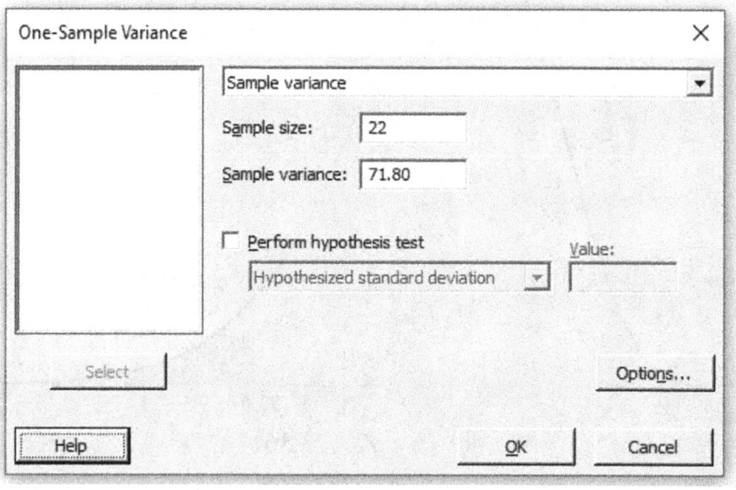

FIGURE 4.49
Minitab dialog box to find the confidence interval for the population variance that is presented in Example 4.8.

Since we are interested in finding a 99% confidence interval for the population variance, we need to select the confidence level under the **Options** tab, as is illustrated in Figure 4.50.

The results can be found in Figure 4.51.

Similarly, to perform the hypothesis test as described in Example 4.9, we could enter the raw or summarized data and the hypothesized standard deviation or variance, as is described in Figure 4.52.

Selecting the **Options** tab, we could specify the confidence level as well as the direction of the alternative hypothesis, as can be seen in Figure 4.53.

This gives the results presented in Figure 4.54. Notice that there may be a small amount of round-off error between the manual results as compared to those generated by Minitab with respect to the chi-square statistic.

The confidence intervals and hypothesis tests that we just performed rely on the population being normally distributed. When this is not the case, approximate confidence intervals can be obtained and hypothesis tests performed using the Bonett method (2006). These approximation methods only require that the population being sampled from is continuous but it does not necessarily need to be normally distributed. Furthermore, the raw data has to be used in order to perform the Bonett method. You may have noticed that the Bonett method was not provided when we wanted to find the confidence interval from Example 4.9. This is because we used the summarized data instead of the raw data. Whenever a single variance is tested, Minitab gives the results for both the chi-square

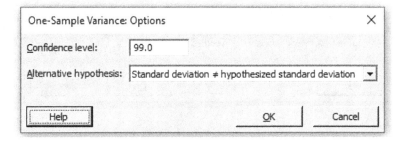

FIGURE 4.50
Options dialog box for finding a 99% confidence interval for a one-sample variance.

Test and CI for One Variance

Method

σ^2: variance of Sample
The Bonett method cannot be calculated for summarized data.
The chi-square method is valid only for the normal distribution.

Descriptive Statistics

N	StDev	Variance	99% CI for σ^2 using Chi-Square
25	10.3	106	(55.9, 257.7)

FIGURE 4.51
Minitab printout for finding a confidence interval for a population variance as given in Example 4.8.

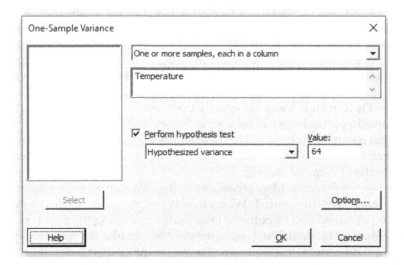

FIGURE 4.52
Minitab dialog box to perform a hypothesis test for a one-sample variance.

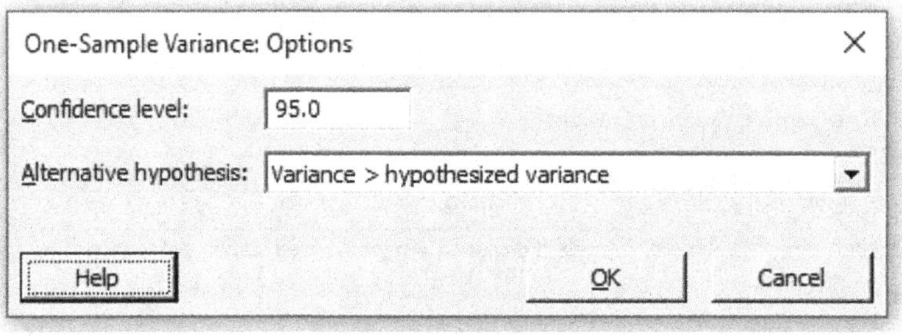

FIGURE 4.53
Minitab dialog box to select the confidence level and a right-tailed hypothesis test for a single variance, as is described in Example 4.9 ($\alpha = 0.05$).

method and the Bonett method only when the raw data is used. Just keep in mind that the method chosen depends on whether or not the population being sampled from is normally distributed.

4.15 Power Analysis for One-Sample Variance

Similar to all of the power analyses we have done thus far, we can use Minitab to perform a power analysis for a one-sample variance. The Minitab dialog box for a one-sample variance is given in Figure 4.55, and just like the other power analyses we have done thus far, we need to enter two of the three values for sample sizes, ratios, and power values. The value of the ratios represents the ratio between the comparison variance and the hypothesized variance. If your alternative hypothesis is specified as "less than," then you must

Test and CI for One Variance: Temperature

Method

σ: standard deviation of Temperature
The Bonett method is valid for any continuous distribution.
The chi-square method is valid only for the normal distribution.

Descriptive Statistics

N	StDev	Variance	95% Lower Bound for σ using Bonett	95% Lower Bound for σ using Chi-Square
22	8.47	71.8	6.76	6.79

Test

Null hypothesis	$H_0: \sigma^2 = 64$
Alternative hypothesis	$H_1: \sigma^2 > 64$

Method	Test Statistic	DF	P-Value
Bonett	—	—	0.328
Chi-Square	23.56	21	0.315

FIGURE 4.54
Minitab printout for testing a single variance as described in Example 4.9.

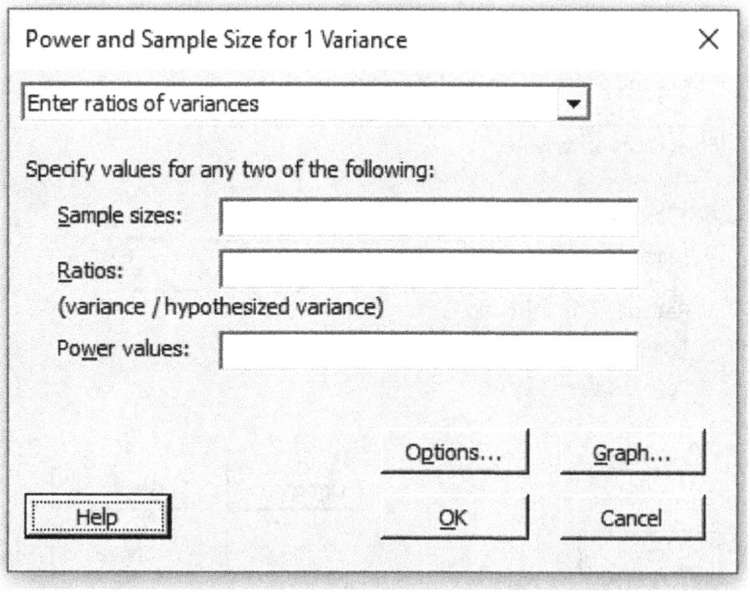

FIGURE 4.55
Minitab dialog box to do a power analysis for one variance.

enter a ratio that is less than 1. If your alternative hypothesis is specified as "greater than," then you must enter a ratio that is greater than 1. We can also enter ratios of standard deviations (recall that the standard deviation is the square root of the variance).

Example 4.11

For the data given in Example 4.10, suppose we want to find the sample size needed to detect a ratio of 1.25 between the variance and the hypothesized variance at 80% power. In other words, we want to know how big of a sample would be needed for a variance of 1.25 times our hypothesized variance to test out as significant (if such a difference were to actually exist).

Notice that since our hypothesis test is a right-tailed test (i.e., the direction of the alternative hypothesis is strictly greater than), we need to specify a ratio that is greater than 1. We would then enter 1.25 as the ratio and a power level of 0.80 (or 80%), as can be seen in Figure 4.56.

We also need to click on the **Options** tab in order to specify the direction of the alternative hypothesis (which in our case will be greater than) and the respective significance level (which is 0.05 or 5%), as can be seen in Figure 4.57.

The results of this power analysis are provided in Figure 4.58.

To interpret this finding in the context of our example, we would need a sample of size $n = 245$ in order to detect a difference of 1.25 times the variance that is specified in the null hypothesis. Since we specified a variance of 64 in the alternative hypothesis, we would need a sample size of at least 245 in order for a variance of 64(1.25) = 80 to test out as significant. In other words, for a difference of 80 − 64 = 16 to test out as significant, a sample size of at least 245 would be needed to find such a difference if such a difference were to actually exist.

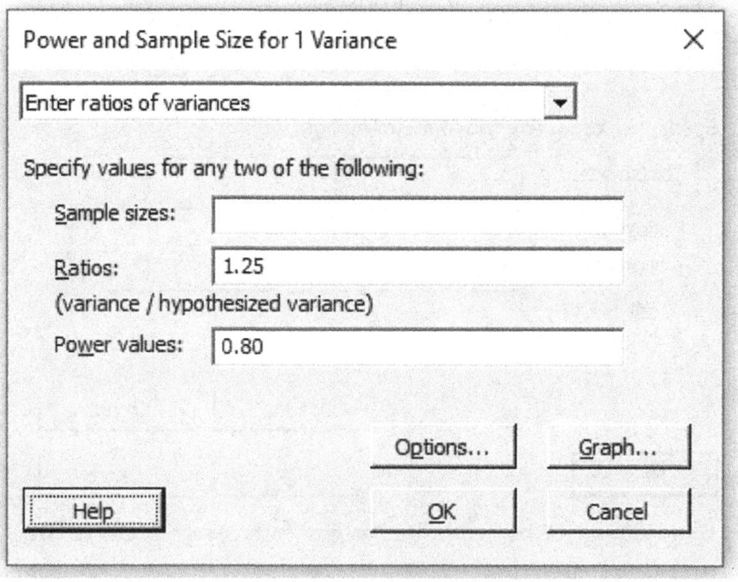

FIGURE 4.56
Minitab dialog box to do a power analysis for one variance for a ratio of 1.25 at 80% power.

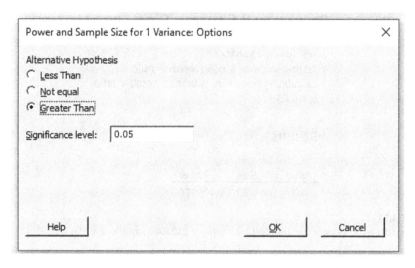

FIGURE 4.57
Minitab dialog box to specify the direction of the alternative hypothesis and the significance level for a power analysis for a one-sample variance test.

Power and Sample Size

Test for One Variance
Testing variance = null (versus > null)
Calculating power for (variance / null) = ratio
$\alpha = 0.05$

Results

Ratio	Sample Size	Target Power	Actual Power
1.25	245	0.8	0.801211

FIGURE 4.58
Power analysis for a single variance test for a ratio of 1.25 at 80% power.

We can also do a power analysis to see what power level we have achieved with our given sample size. Recall in Example 4.10 that for a sample of size $n = 22$, a sample variance of 71.8 was obtained, and we wanted to test if we were sampling from a population whose variance is greater than 64. The ratio of these variances will then be calculated as follows:

$$\frac{\text{variance}}{\text{hypothesized variance}} = \frac{71.8}{64} \approx 1.12$$

By entering a sample size of 22 and a ratio of 1.12, we get the Minitab printout that is provided in Figure 4.59.

Power and Sample Size

Test for One Variance
Testing variance = null (versus > null)
Calculating power for (variance / null) = ratio
α = 0.05

Results

Ratio	Sample Size	Power
1.12	22	0.109976

FIGURE 4.59
Power analysis for a single variance test for a ratio of 1.12 for a sample size of $n = 22$.

Recall that there was not enough evidence to suggest that the variability in temperatures for December 2015 in Massachusetts was more than the historical variance of 64 (i.e., that the hypothesized variance is significantly greater than 64). Thus, we can claim that our study is underpowered because a sample of size $n = 22$ would only achieve about 11% power to find a difference of at least $71.8 - 64 = 7.8$, if such a difference were to exist.

4.16 Confidence Intervals for One-Sample Count Data

We can also calculate confidence intervals for a *count variable*. Recall that a count variable is a variable that takes on only the nonnegative integer values $\{0, 1, 2, 3, 4,...\}$. We can calculate a $100(1 - \alpha)\%$ confidence interval for the *population mean count* as follows:

$$\left(\frac{\hat{\lambda} - z_{\alpha/2}\sqrt{\frac{\hat{\lambda}}{n}}}{t}, \frac{\hat{\lambda} + z_{\alpha/2}\sqrt{\frac{\hat{\lambda}}{n}}}{t} \right)$$

where $\hat{\lambda}$ is the mean of the sample count variable, n is the sample size, t is the observation length, and $z_{\alpha/2}$ is the value on the standard normal distribution for an area of $\alpha/2$ in the right tail. In other words,

$$\frac{\hat{\lambda} - z_{\alpha/2}\sqrt{\frac{\hat{\lambda}}{n}}}{t} < \lambda < \frac{\hat{\lambda} + z_{\alpha/2}\sqrt{\frac{\hat{\lambda}}{n}}}{t}$$

where λ is the unknown mean count for the population.

Example 4.12

Suppose a computer engineer wants to determine if the mean number of computer glitches is more than five per day. A random sample of computer glitches that occurred over a 13-day period is given in Table 4.5.

To find a 97% confidence interval for the population mean number of glitches, we first calculate the mean of the sample counts as follows:

$$\hat{\lambda} = \frac{4+7+2+0+5+7+1+0+0+3+2+5+8}{13} \approx 3.385$$

The observation length $t = 1$ since the mean number of computer glitches per day are being counted and the units of the sample data are the daily number of glitches.

Then a 97% confidence interval would be as follows:

$$\left(\frac{\hat{\lambda} - z_{\alpha/2}\sqrt{\frac{\hat{\lambda}}{n}}}{t}, \frac{\hat{\lambda} + z_{\alpha/2}\sqrt{\frac{\hat{\lambda}}{n}}}{t} \right) = \left(\frac{3.385 - 2.170\sqrt{\frac{3.385}{13}}}{1}, \frac{3.385 + 2.170\sqrt{\frac{3.385}{13}}}{1} \right)$$

$$= \left(\frac{3.385 - 1.107}{1}, \frac{3.385 + 1.107}{1} \right) \approx (2.278, 4.492)$$

We claim to be 97% confident that the mean number of computer glitches per day is between 2.278 and 4.492.

TABLE 4.5

Random Sample of the Number of Computer Glitches Over a 13-Day Period

Day	Number of Glitches
1	4
2	7
3	2
4	0
5	5
6	7
7	1
8	0
9	0
10	3
11	2
12	5
13	8

If we wanted to calculate a confidence interval for the number of glitches *per hour*, we let $t = 24$ and so the confidence interval would be:

$$\left(\frac{\hat{\lambda} - z_{\alpha/2}\sqrt{\frac{\hat{\lambda}}{n}}}{24}, \frac{\hat{\lambda} + z_{\alpha/2}\sqrt{\frac{\hat{\lambda}}{n}}}{24} \right) = \left(\frac{3.385 - 2.170\sqrt{\frac{3.385}{13}}}{24}, \frac{3.385 + 2.170\sqrt{\frac{3.385}{13}}}{24} \right)$$

$$= \left(\frac{3.385 - 1.107}{24}, 24 \right) \approx (0.095, 0.187)$$

4.17 Using Minitab to Calculate Confidence Intervals for a One-Sample Count Variable

We can have Minitab calculate a confidence interval for the population mean count rate by selecting **Stat < Basic Statistics < 1-Sample Poisson Rate,** as is illustrated in Figure 4.60.

This brings up the dialog box in Figure 4.61. We enter the number of glitches in the **Sample columns**.

By clicking on the **Options** tab we can specify the confidence level and choose the method as the **Normal approximation**, as is illustrated in Figure 4.62.

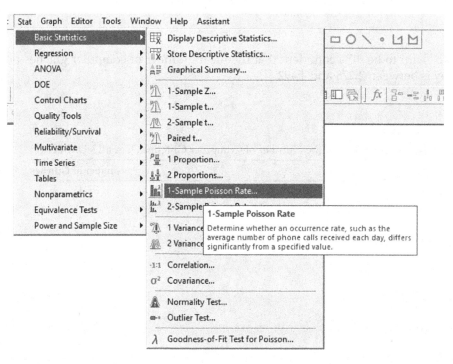

FIGURE 4.60
Minitab commands to find a 1-Sample Poisson Rate confidence interval and hypothesis test.

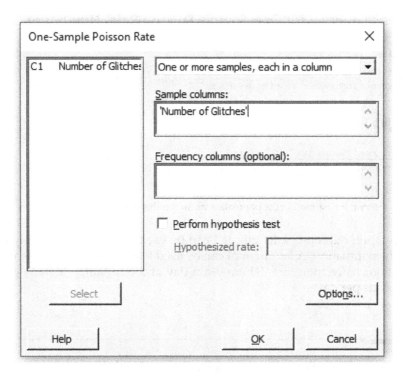

FIGURE 4.61
Minitab dialog box to calculate a confidence interval for the mean number of glitches.

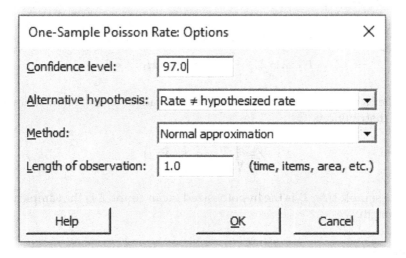

FIGURE 4.62
Minitab dialog box for a 97% confidence interval for a One-Sample Poisson based on the normal approximation.

The results are given in Figure 4.63.

The dialog box in Figure 4.62 also provides the opportunity to describe the length of the population mean rate. When just the mean counts are being considered, the length needs to be set equal to 1. For our example, since each sample observation counts the daily

Confidence Interval for One-Sample Poisson Rate: Number of Glitches

Method

λ: Poisson rate of Number of Glitches
Normal approximation method is used for this analysis.

Descriptive Statistics

	Total		
N	Occurrences	Sample Rate	97% CI for λ
13	44	3.38462	(2.27733, 4.49190)

FIGURE 4.63
97% Confidence Interval for estimating the population mean number of glitches per day.

number of computer glitches, a length of 1 can be used to estimate the daily number of glitches for the population. A length of 24 can be used to estimate the number of computer glitches per hour since there are 24 hours in a day and our sample data is given as the number of counts per day.

4.18 Hypothesis Test for a One-Sample Count Variable

We can also perform a hypothesis to estimate if the population mean count is different from some hypothesized value. To do this, one of the following null and alternative hypotheses can be set up:

$$H_0 : \lambda = \lambda_0 \quad H_0 : \lambda = \lambda_0 \quad H_0 : \lambda = \lambda_0$$

$$H_1 : \lambda \neq \lambda_0 \quad H_1 : \lambda > \lambda_0 \quad H_1 : \lambda < \lambda_0$$

where λ_0 is the hypothesized mean count. The test statistic given below follows the standard normal distribution:

$$Z = \sqrt{\frac{n}{\lambda_0 \cdot t}} \left(\hat{\lambda} \cdot t - \lambda_0 \cdot t \right)$$

where n is the sample size, λ_0 is the hypothesized mean count, $\hat{\lambda}$ is the sample mean count, and t is the length.

Example 4.13

Suppose we want to know if the mean number of computer glitches per day is more than 3 ($\alpha = 0.05$). The null and alternative hypothesis can be stated as follows:

$$H_0 : \lambda = 3$$

$$H_1 : \lambda > 3$$

The test statistic can then be calculated as:

$$Z = \sqrt{\frac{n}{\lambda_0 \cdot t}} \left(\hat{\lambda} \cdot t - \lambda_0 \cdot t \right) = \sqrt{\frac{13}{3 \cdot 1}} \left(3.385 \cdot 1 - 3 \cdot 1 \right) \approx 0.801$$

Comparing the test statistic to the value of $z = 1.645$ that defines the rejection region as illustrated in Figure 4.64, we conclude that there is not enough evidence to support the claim that the population mean number of computer glitches is more than 3 per day.

The p-value can be obtained by considering the area to the right of the test statistic $Z = 0.801$, as shown in Figure 4.65.

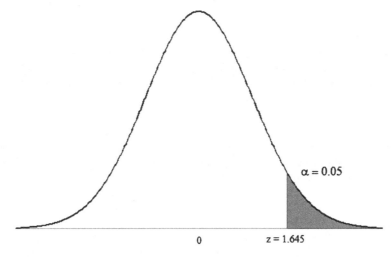

FIGURE 4.64
Normal approximation for the hypothesis test of whether the mean number of computer glitches per day is more than 3 ($\alpha = 0.05$).

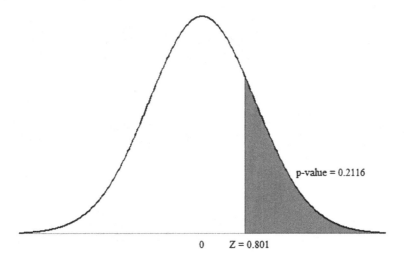

FIGURE 4.65
Standard normal distribution showing the p-value of 0.2116 for the test statistic $Z = 0.801$ for a one-sample count variable.

4.19 Using Minitab to Conduct a Hypothesis Test for a One-Sample Count Variable

Just as we have seen throughout this chapter, the same dialog box can be used for both confidence intervals and hypothesis tests. Figure 4.66 gives the dialog box for the hypothesis test in Example 4.12.

Clicking on the **Options** tab, the level of significance and the direction of the alternative hypothesis need to be specified, as is shown in Figure 4.67.

Using the **Normal approximation** gives the results that are presented in Figure 4.68.

Since $p = 0.212$, as illustrated in Figure 4.68 is greater than a level of significance of $\alpha = 0.05$, there is not enough evidence to claim that the population mean number of defects per day is greater than 3.

One aspect to keep in mind when making inferences with a one-sample Poisson is that the normal approximation of the Poisson distribution is only appropriate when there are larger samples or if the mean count is away from 0. In case of small samples or if the count data is around 0, then you may want to consider using an exact method. The reason is that the Poisson distribution is bounded below by 0, whereas the normal distribution goes infinitely in each direction. If the mean counts are larger and not around 0, a normal approximation can give you reasonable estimates. If the mean counts are smaller and are closer to 0, a normal approximation may not be appropriate. And while the calculations for exact methods are a bit more cumbersome to do out manually and are not presented in this book, the inferences will be the same.

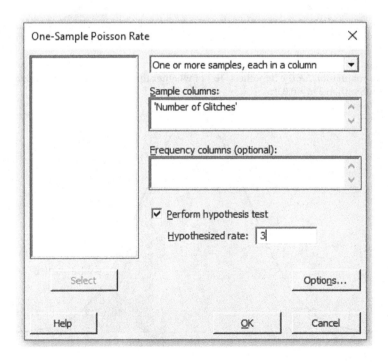

FIGURE 4.66
Minitab dialog box to perform a hypothesis test to estimate if the population mean number of glitches is more than 3 per day.

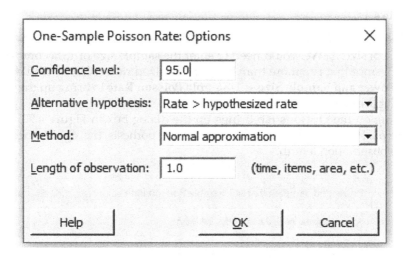

FIGURE 4.67
Minitab dialog box to specify the level of significance, the direction of the alternative hypothesis, and the normal approximation for a one-sample count variable.

Test and CI for One-Sample Poisson Rate: Number of Glitches

Method

λ: Poisson rate of Number of Glitches
Normal approximation method is used for this analysis.

Descriptive Statistics

N	Total Occurrences	Sample Rate	95% Lower Bound for λ
13	44	3.38462	2.54533

Test

Null hypothesis	$H_0: \lambda = 3$
Alternative hypothesis	$H_1: \lambda > 3$

Z-Value	P-Value
0.80	0.212

FIGURE 4.68
Minitab output of the hypothesis test of whether the population mean number of defects is greater than 3 per day.

4.20 Using Minitab for a Power Analysis for a One-Sample Poisson

Like many of the power analyses we have seen thus far, we can perform a power analysis to see whether a given sample size has enough power to detect an effect of a certain size, provided that such an effect actually exists. Similar to performing a power analysis for other types of hypothesis tests, we only need to enter two of the following: sample size, comparison rate, or power level.

Example 4.14

Suppose we want to find the power to detect a mean defective rate of 2 items for a sample of size 13. We would need to enter the sample size of 13, a comparison rate of 5 (since that is 2 more than the hypothesized rate of 3). We would select **Stat < Power and Sample Size < 1-Sample Poisson Rate** to bring up the dialog box in Figure 4.69.

Clicking on the **Options** tab brings up the dialog box in Figure 4.70, where we can specify the direction of the alternative hypothesis, the significance level, and the observation length.

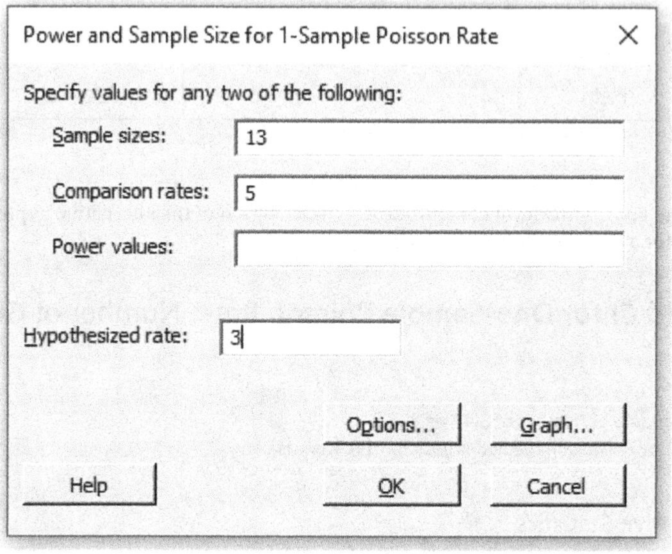

FIGURE 4.69
Minitab dialog box to perform a power analysis for a One-Sample Poisson Rate.

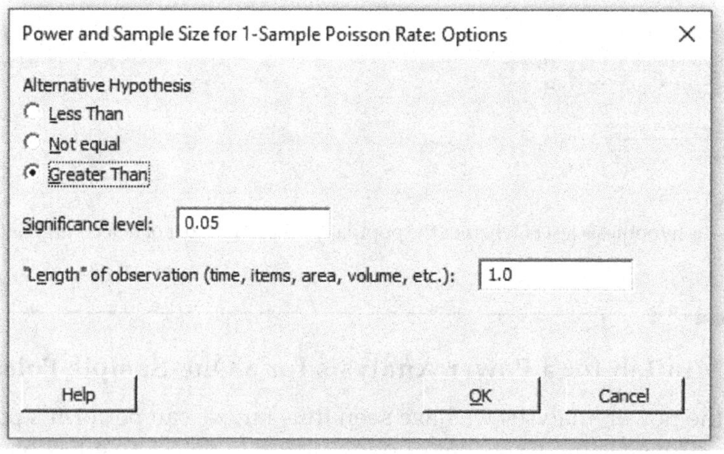

FIGURE 4.70
Minitab dialog box for a One-Sample Poisson Rate for a right-tailed test with a significance level of 0.05 and the length of observation as 1 day.

Power and Sample Size

Test for 1-Sample Poisson Rate
Testing rate = 3 (versus > 3)
α = 0.05
"Length" of observation = 1

Results

Comparison Rate	Sample Size	Power
5	13	0.974460

FIGURE 4.71
Minitab output for a power analysis for a 1-Sample Poisson Rate for a sample size of 13 and a comparison rate of 5.

The results are provided in Figure 4.71.

Figure 4.71 shows that a sample of size 13 would have approximately a 97% chance of detecting a difference of at least 2 mean items. The difference between the comparison rate of 5 items and hypothesized rate of 3 items is the desired effect size that we believe has practical meaning. Therefore, a sample of size 13 can detect such a difference, if such a difference were to actually exist. In other words, a sample of size 13 is large enough to detect if the population rate is greater than the hypothesized mean by at least 2 defective items per day.

4.21 A Note About One- and Two-Tailed Hypothesis Tests

You may have noticed that when a significant effect is found while conducting a two-tailed hypothesis test, this automatically allows you to make inferences for a one-tailed test. This is because the rejection region for a two-tailed test is much smaller than that for a one-tailed test.

Recall for a two-tailed test that the level of significance gets split between the two tails. For instance, consider a significance level of 0.05, where the rejection region in each tail would be 0.025. If you were able to accept the alternative hypothesis and reject the null hypothesis for a two-tailed test, then the test statistic of interest would fall in the rejection region that corresponds to *half of the original level of significance*. If the sign of the test statistic were positive, this would suggest acceptance of a right-tailed alternative hypothesis, and similarly, if the sign of the test statistic were negative, this would suggest acceptance of a left-tailed alternative hypothesis. Recall that for a one-tailed test, the area of the rejection region corresponds to the entire amount of the level of significance. Therefore, because a two-tailed test corresponds to a smaller rejection region in either tail, clearly a significant two-tailed test would correspond to a significant one-tailed test and the direction of that test would be based on the sign of the test statistic. This is illustrated in Figure 4.72 for a two-tailed test where the test statistic falls in the right tail.

It is for this reason that Minitab conducts many hypothesis tests using a two-tailed alternative hypothesis. But keep in mind that if there is not enough evidence to accept the

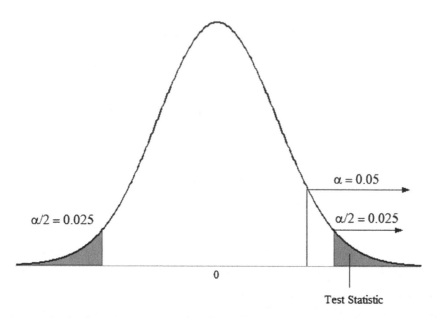

FIGURE 4.72
Sampling distribution of a test statistic for a two-tailed test ($\alpha/2 = 0.025$) with the area for a right-tailed test ($\alpha = 0.05$).

alternative hypothesis and reject the null hypothesis for a two-tailed test, you may still be able to infer significance for a one-tailed test. This is because the test statistic could fall in the rejection region that corresponds to the total level of significance for a one-tailed test (for instance, if the p-value is 0.045) but not fall into ½ of the level of significance (or 0.025). This is illustrated in Figure 4.73.

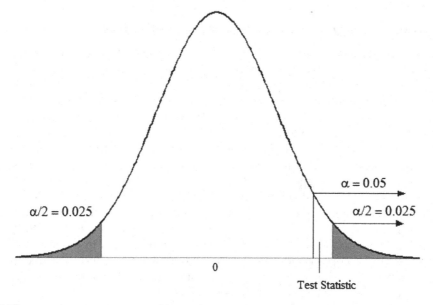

FIGURE 4.73
Sampling distribution of a test statistic for a one-tailed test ($\alpha = 0.05$) with the area for a two-tailed test ($\alpha/2 = 0.025$).

Exercises

Unless otherwise specified, use a significance level of $\alpha = 0.05$ and a power level of 80%.

1. Historically, the average age to get married (for the first time) has been 29 years old. A researcher was interested in whether millennials tend to get married later than did generations of the past. A random sample of 29 millennials found that the mean age at first marriage was 29.3 years, with a standard deviation of 2.5 years. Can you support the claim that millennials tend to marry later than generations of the past?

2. It has been established that the average time it takes for students to complete a bachelor's degree is approximately 4.8 years. In order to help students graduate earlier, a university wants to implement a first-year seminar. But before the program is put in place for all of their students, the university wants to test the effectiveness of such a program. To test the program, a pilot study was conducted by the university by first obtaining a random sample of incoming freshmen and assigning them to the program. From the original random sample, when a total of 25 students graduated from the university, they found that the mean time to graduation was 4.3 years, with a standard deviation of 1.3 years. Does participating in the first-year seminar help students to graduate in less than 4.8 years?

3. A news organization wants to estimate the wait times to vote in the current election. A reporter asked a random sample of 19 voters how long their wait time was and found their mean wait time to be 28.9 minutes and the standard deviation was calculated to be 23 minutes. If $\alpha = 0.01$, estimate the wait times for the population of voters who will vote in the current election.

4. Suppose that a bottled water plant wants to determine if the bacteria count in their water supply exceeds the safety threshold of 100 cfu/mL (colony-forming units per milliliter). Ten different samples were taken at random points in the water supply and the bacterial count was measured in cfu/mL. The data is presented in Table 4.6.

   a. Can you conclude whether the plant water supply exceeds the safety threshold for the bacteria count?

   b. How big of a sample would be needed to detect at least a 2-cfu/mL difference?

   c. The plant wants to be able to detect a bacteria count of at least 105 cfu/mL. How big of a sample would be needed to determine if the bacteria count exceeds 105 cfu/mL?

5. A machine that makes hose fittings is set for a tolerance of 0.25 cm (i.e., the standard deviation) and a mean diameter of 12 cm. A random sample of 12 fittings (in cm) is given in Table 4.7.
   Find and interpret 99% confidence intervals for both the population mean diameter and the population variance of the hose fittings.

TABLE 4.6

Bacteria Count (in cfu/mL)

104	102	100	97	93	105	102	97	96	102

TABLE 4.7

Mean Diameter (in Centimeters) for a Random Sample of 12 Hose Fittings

12.24	12.64	12.15	12.36	12.60	12.02	12.18	11.95	11.95	12.08	12.23	11.89

6. A recent political poll was given to 1,000 likely voters before the presidential election. Of these voters, only 475 said they would vote for the candidate who is currently in office. Determine if there is enough evidence to suggest that the incumbent candidate may lose the election by not getting at least 50% of the votes. Find and interpret the *p*-value.

7. The service department at a car dealership serviced 528 cars over the course of a 1-week period. These customers were asked to complete a survey regarding their experience with the service department. Of those 528 customers who serviced their cars during the course of this week, 484 said they were very satisfied with the service they received.

   a. Find and interpret a 95% confidence interval for the true population proportion of customers who would respond that they are very satisfied with the service department at this dealership.

   b. Estimate if more than 90% of all customers will respond that they are very satisfied with the service department at this dealership ($\alpha = 0.01$).

8. A recent study suggests that more than 80% of all high school mathematics teachers use calculators in their classroom. To test this theory, a national sample of 1,286 high school mathematics teachers were asked if they used calculators in their classroom and 1,274 responded that they did.

   a. On the basis of the information collected from this sample, do you agree with the results of this study that suggest that more than 80% of all high school mathematics teachers use calculators in their classroom?

   b. What size sample would be needed to detect at least a 5% difference at 70% power?

9. A weight loss program claims that its clients have lost, on average, more than 40 pounds over the course of a 4-month period. A random sample of 30 participants in the weight loss program shows an average loss of 44 pounds in a 4-month period, with a standard deviation of 11 pounds.

   a. For a significance level of $\alpha = 0.01$, test whether you would support the program's claims.

   b. What sample size is needed in order to find a mean difference of at least 3 pounds over a 4-month period?

   c. For a sample of size $n = 30$, determine the minimum mean difference in weight loss that can be detected provided that such a difference exists.

10. You may have heard of the Beck Depression Inventory (BDI). This is a 21-question survey where higher total scores are indicative of more severe depressive symptoms. In particular, scores greater than 30 may suggest severe depression. Suppose a psychologist wants to investigate whether retired adults are more likely to suffer from depression. A random sample of 18 retired adults was given the BDI and their total scores are presented in Table 4.8.

   a. Find and interpret a 95% confidence interval for the population mean BDI.

TABLE 4.8

BDI Scores for a Random Sample of 18 Retired Adults

28	29	30	30	31	32	39	21	29
29	32	33	28	28	28	25	39	38

    b. Do you think that retired adults have a mean BDI total score greater than 30?

    c. How many adults would have to be sampled to detect a difference of at least 33 on the BDI?

11. A machine makes pipe fittings to have a mean length of approximately 2 in. and the standard deviation of the lengths must be less than 0.04 in. To make sure the machine is making the fittings to spec, a random sample of 30 pipe fittings were selected and the sample mean was calculated to be 1.98 in. and the sample standard deviation was calculated to be 0.03 in. Is the machine manufacturing the pipe fittings such that the standard deviation of the lengths is less than 0.04 in.?

12. A random sample of 1,463 households across the United States found that 938 of these households watch television on more than one night per week. Find and interpret a 99% confidence interval for the true population proportion of households in the United States that watch television on more than one night per week.

13. A random sample of 1,463 households across the United States found that the mean amount of television watched per week was 15.36 h, with a standard deviation of 4.27 h. Find and interpret a 99% confidence interval for the average number of hours that households in the United States watch television per week.

14. It has been claimed that most car accidents happen less than 25 miles from home. The data set presented in Table 4.9 gives a random sample of the mileage away from home for 36 car accidents in a given state.

    a. Find the mean and standard deviation of this data set.

    b. Test the claim that on average, accidents happen less than 25 miles from home.

    c. How large of a sample size would be needed to find a difference of at least 5 miles?

15. In 2013, it was found that 10% of all fatal car crashes were caused by distracted driving. A recent study was done to estimate whether the percentage of fatal car crashes caused by distracted driving has recently changed. The causes for a random sample of 2,550 recent accidents were examined and it was found that 290 of these accidents were caused by distracted driving. On the basis of the information collected from this sample, explain whether or not the percentage of fatal car crashes caused by distracted driving has changed.

TABLE 4.9

Distances from Home for a Random Sample of 36 Car Accidents

2	18	27	47	29	24
18	46	38	36	12	8
26	38	15	57	26	21
37	18	5	9	12	28
28	17	4	7	26	29
46	20	15	37	29	19

16. A manufacturer of wood flooring wants to estimate the mean number of defective floor planks in each box of wood flooring. For a random sample of 12 boxes of planks, find the number of defective planks in each box, as given in Table 4.10.

   a. Find and interpret a 98% confidence interval for the mean number of defective planks (use the normal approximation and an exact method for a one-sample Poisson and compare the difference).

   b. Estimate if the mean number of defects per box is more than 1.

   c. How many boxes would you need to sample to be able to find an average of at least 1 defective plank per box?

17. A school district wants to estimate the mean number of snow days that they can expect each year. A sample of the number of snow days taken over the last 10 years is given in Table 4.11.

   a. Find and interpret a 98% confidence interval for the mean number of snow days.

   b. Estimate if the mean number of snow days per year is less than 3 ($\alpha = 0.01$).

   c. How big of a sample would be needed to detect the mean number of snow days to be no more than 2?

18. Acidity and alkalinity of soil are typically measured on the pH scale, which has a range from 0–14. A pH of 7 is considered neutral, a pH of 0–7 is acidic, and a pH of 7–14 is alkaline. Lawns should have a pH of between 6.3 and 6.7, where a pH of 6.5 is considered optimal. A landscaper needs to test a customer's lawn pH to see if the lawn needs to be treated to adjust the pH balance. Random soil samples were taken from nine different locations and the pH readings are presented in Table 4.12.

   a. Is the mean soil pH significantly different from the measure of 6.5 that is considered optimal?

   b. Is the standard deviation of the soil pH significantly different from 0.20?

   c. Find and interpret a 99% confidence interval for the population standard deviation.

19. A machine pours soda into 12-ounce bottles. If the standard deviation of the amount of soda poured exceeds 0.23 ounces, then this is an indication that the

TABLE 4.10

Number of Defects Found in a Random Sample of 12 Boxes of Planks

1	0	1	2	0	2	1	0	3	1	0	1

TABLE 4.11

Random Sample of the Number of Snow Days Taken Over the Past 10 Years

2	1	0	0	0	3	1	2	0	5

TABLE 4.12

pH Readings for a Random Sample of Nine Locations

6.2	6.8	6.3	6.9	6.2	6.5	6.0	6.9	6.3

TABLE 4.13

Number of Ounces in a Random Sample of 20 12-Ounce Bottles of Soda

12.35	12.41	12.30	12.04	12.33	12.06	12.04	12.05	12.33	12.09
12.22	12.05	12.00	12.06	12.85	12.31	12.22	12.04	12.25	12.36

machine is not working correctly. To see if the machine is working correctly, a random sample of 20 bottles of soda is taken and the number of ounces in each of these samples is given in Table 4.13.

a. Does the standard deviation exceed 0.23 ounces?

b. How big of a sample would be needed to find at least a 0.30-ounce difference in the standard deviation?

c. Is a sample size of $n = 20$ large enough to find at least a 0.07 difference in the standard deviation?

References

D.G. Bonett (2006). Approximate confidence interval for standard deviation of non-normal distributions, *Computational Statistics and Data Analysis*, 50(3): 775–782.

R. J. Light, J. D. Singer, and J. B. Willett (1990). *By Design: Planning Research on Higher Education.* Cambridge, MA: Harvard University Press.

5

Statistical Inference for Two-Sample Data

5.1 Introduction

In this chapter, we will describe how to calculate confidence intervals and perform hypothesis tests when sample data is collected from *two different populations*. We will present how to calculate confidence intervals and perform hypothesis tests for two means, two proportions, two variances, and two counts when the two populations we sample from are independent of each other. We will also illustrate how to perform before- and after-comparisons when the populations are dependent.

5.2 Confidence Interval for the Difference Between Two Means

Similar to how we calculated confidence intervals for a single mean, we can also calculate confidence intervals and perform hypothesis tests for comparing the difference between means from *two independent populations*. Two populations are said to be *independent* if any observation in the first population has nothing to do with any observation in the second population. For instance, we could test whether there is a difference in the mean lifetimes for two different brands of cell phone batteries. We can assume that these populations are independent of each other as the observations from one population would likely have nothing to do with the observations from the other population.

To find a $100(1 - \alpha)\%$ confidence interval for the difference between two independent population means, $\mu_1 - \mu_2$, we use the following formula:

$$\left((\bar{x}_1 - \bar{x}_2) - t_{\alpha/2} \cdot s_p \cdot \sqrt{\frac{1}{n_1} + \frac{1}{n_2}}, \ (\bar{x}_1 - \bar{x}_2) + t_{\alpha/2} \cdot s_p \cdot \sqrt{\frac{1}{n_1} + \frac{1}{n_2}} \right)$$

where $s_p = \sqrt{\dfrac{(n_1 - 1)s_1^2 + (n_2 - 1)s_2^2}{n_1 + n_2 - 2}}$ is the pooled sample standard deviation and $t_{\alpha/2}$ is the upper $\alpha/2$ portion of the t-distribution with $n_1 + n_2 - 2$ degrees of freedom. The sample statistics are: $\bar{x}_1$, the sample mean of the first population; $\bar{x}_2$, the sample mean of the second population; n_1, the sample size taken from the first population; n_2, the sample size taken from the second population; and s_1^2 and s_2^2 are the sample variances found from the first and second populations, respectively.

The pooled standard deviation is used if the samples are drawn from two populations that follow a normal distribution and if we assume that the *two population variances are equal*. However, if we have unequal variances, we will need to use the following statistic:

$$s = \sqrt{\frac{s_1^2}{n_1} + \frac{s_2^2}{n_2}}$$

Also, for unequal variances, the number of degrees of freedom would be found by using the following formula and *rounding it down* to an integer:

$$df = \frac{\left(\frac{s_1^2}{n_1} + \frac{s_2^2}{n_2}\right)^2}{\frac{\left(s_1^2/n_1\right)^2}{n_1 - 1} + \frac{\left(s_2^2/n_2\right)^2}{n_2 - 1}}$$

where s_1^2 is the variance of the first sample, s_2^2 is the variance of the second sample, n_1 and n_2 are the sample sizes for the first and second sample, respectively.

Then, a $100(1 - \alpha)\%$ confidence interval assuming unequal variances would be:

$$\left((\bar{x}_1 - \bar{x}_2) - t_{\alpha/2} \cdot s, \ (\bar{x}_1 - \bar{x}_2) + t_{\alpha/2} \cdot s \right)$$

where $\bar{x}_1$ is the sample mean of the first population, $\bar{x}_2$ is the sample mean of the second population, and $t_{\alpha/2}$ is the upper $\alpha/2$ portion of the t-distribution with $n_1 + n_2 - 2$ degrees of freedom.

Example 5.1

Suppose we are interested in estimating the mean difference between the lifetimes for two brands of cell phone batteries. If we collected a sample of 15 lifetimes in hours from each of the two different brands of cell phone batteries, we could record the mean lifetime, standard deviation, and sample variance for each of these samples as follows:

Brand 1: $n_1 = 15$ $\bar{x}_1 = 74.2\,\text{h}$ $s_1 = 6.85\,\text{h}$ $s_1^2 \approx 46.92$

Brand 2: $n_2 = 15$ $\bar{x}_2 = 73.9\,\text{h}$ $s_2 = 7.05\,\text{h}$ $s_2^2 \approx 49.70$

If we are assuming equal variances, the pooled standard deviation can be calculated as vfollows:

$$s_p = \sqrt{\frac{(n_1 - 1)s_1^2 + (n_2 - 1)}{n_1 + n_2 - 2}}$$

$$= \sqrt{\frac{14(6.85)^2 + 14(7.05)^2}{15 + 15 - 2}}$$

$$\approx \sqrt{48.31} \approx 6.95$$

Then, a 95% confidence interval would be:

$$\left((\bar{x}_1 - \bar{x}_2) - t_{\alpha/2} \cdot s_p \cdot \sqrt{\frac{1}{n_1} + \frac{1}{n_2}}, (\bar{x}_1 - \bar{x}_2) + t_{\alpha/2} \cdot s_p \cdot \sqrt{\frac{1}{n_1} + \frac{1}{n_2}} \right)$$

$$= \left((74.2 - 73.9) - 2.048(6.95)\sqrt{\frac{1}{15} + \frac{1}{15}}, (74.2 - 73.9) + 2.048(6.95)\sqrt{\frac{1}{15} + \frac{1}{15}} \right)$$

$$\approx (0.30 - 5.20, 0.30 + 5.20) \approx (-4.90, 5.50)$$

where $t_{\alpha/2} = t_{0.05/2} = t_{0.025} = 2.048$ for $n_1 + n_2 - 2 = 15 + 15 - 2 = 28$ degrees of freedom.

Interpreting this confidence interval within the context of our problem suggests that we are 95% confident that the unknown population difference in the mean lifetimes between these two brands of cell phone batteries is between −4.90 and 5.50 hours. Notice that the form of the confidence interval estimates the *difference* in the true population mean battery lifetimes between Brand 1 and Brand 2, namely, $\mu_1 - \mu_2$, as follows:

$$-4.90 < \mu_1 - \mu_2 < 5.50$$

By looking at the range of numbers given in this confidence interval, we can expect that if we subtract the population mean lifetime of Brand 2 from that of Brand 1, we could be 95% confident that the unknown difference in the population mean lifetimes between these two brands of batteries would fall somewhere between −4.90 and 5.50 hours. Notice that this confidence interval covers the three cases where the difference between the population means for Brand 2 subtracted from Brand 1 can be 0, positive, or negative.

This confidence interval provides a range of differences in the population mean lifetimes between the two brands of cell phone batteries. Since this interval contains both positive and negative values, and thus the value of 0, we cannot infer that either brand of cell phone battery lasts longer than the other.

Similarly, if we were to assume unequal variances, then:

$$s = \sqrt{\frac{s_1^2}{n_1} + \frac{s_2^2}{n_2}} = \sqrt{\frac{6.85^2}{15} + \frac{7.05^2}{15}} \approx 2.538$$

There would be 27 degrees of freedom, which can be obtained by rounding down the results of the following calculation to an integer:

$$df = \frac{\left(\dfrac{s_1^2}{n_1} + \dfrac{s_2^2}{n_2} \right)^2}{\dfrac{\left(s_1^2/n_1 \right)^2}{n_1 - 1} + \dfrac{\left(s_2^2/n_2 \right)^2}{n_2 - 1}} = \frac{\left(\dfrac{6.85^2}{15} + \dfrac{7.05^2}{15} \right)^2}{\dfrac{\left(6.85^2/15 \right)^2}{15 - 1} + \dfrac{\left(7.07^2/15 \right)^2}{15 - 1}} \approx 27.98$$

Therefore, a 95% confidence interval would be:

$$\left((\bar{x}_1 - \bar{x}_2) - t_{\alpha/2} \cdot s, \ (\bar{x}_1 - \bar{x}_2) + t_{\alpha/2} \cdot s \right)$$

$$\approx \left((74.2 - 73.9) - 2.052(2.538), \ (74.2 - 73.9) + 2.052(2.538) \right)$$

$$\approx \left(0.30 - 5.21, \, 0.30 + 5.21 \right) \approx (-4.91, 5.51)$$

where $t_{\alpha/2} = t_{0.025} = 2.052$ for 27 degrees of freedom.

Notice that the confidence interval assuming unequal variances is very similar to that obtained when we assumed equal variance. This is because the sample variances we obtained $\left(s_1^2 \approx 46.92, \ s_2^2 \approx 49.70 \right)$ are not much different from each other.

5.3 Using Minitab to Calculate a Confidence Interval for the Difference Between Two Means

Minitab can easily calculate a confidence interval for the difference between two independent population means by selecting **Stat > Basic Statistics > 2-Sample t**, as can be seen in Figure 5.1.

This brings up the dialog box for a two-sample *t*-test and a confidence interval where we can enter either the raw data or the summarized data, as can be seen in Figure 5.2.

Clicking on the **Options** tab, we can select the level of confidence and specify the alternative hypothesis, as can be seen in Figure 5.3. Recall that we always need to have the

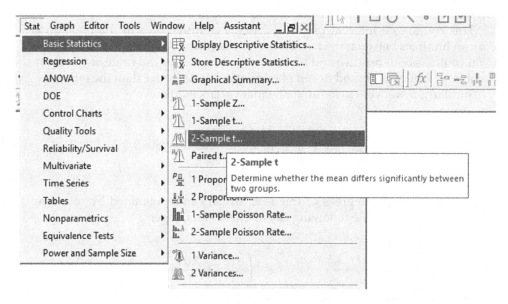

FIGURE 5.1
Minitab commands for a two-sample confidence interval for the difference between two population means.

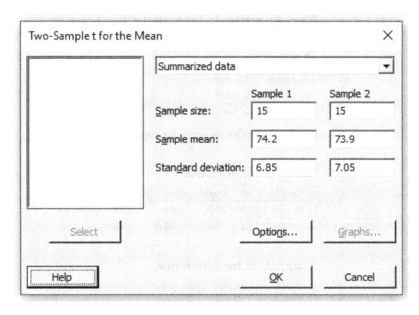

FIGURE 5.2
Minitab dialog box for an independent two-sample confidence interval for the difference between two population means.

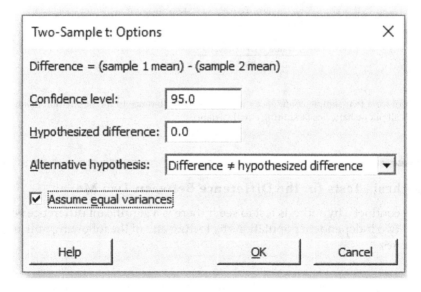

FIGURE 5.3
Options dialog box for a two-sample confidence interval to estimate the difference between two population means.

alternative hypothesis as *"not equal to"* when we are calculating confidence intervals. If we are assuming equal variances, then we need to check the box **Assume equal variances** (by not checking this box you are assuming unequal variances).

Figure 5.4 gives the Minitab printout if we assume equal variances, and Figure 5.5 gives the Minitab printout if we assume unequal variances.

Two-Sample T-Test and CI

Method

μ_1: mean of Sample 1
μ_2: mean of Sample 2
Difference: $\mu_1 - \mu_2$

Equal variances are assumed for this analysis.

Descriptive Statistics

Sample	N	Mean	StDev	SE Mean
Sample 1	15	74.20	6.85	1.8
Sample 2	15	73.90	7.05	1.8

Estimation for Difference

Difference	Pooled StDev	95% CI for Difference
0.30	6.95	(-4.90, 5.50)

Test

Null hypothesis $H_0: \mu_1 - \mu_2 = 0$
Alternative hypothesis $H_1: \mu_1 - \mu_2 \neq 0$

T-Value	DF	P-Value
0.12	28	0.907

FIGURE 5.4
Minitab printout for a two-sample confidence interval for the difference in the population mean lifetimes for two brands of cell phone batteries (assuming equal variances).

5.4 Hypothesis Tests for the Difference Between Two Means

We can also conduct a hypothesis test to see if there is a significant difference between the means from two independent populations by testing any of the following null and alternative hypotheses:

$$H_0 : \mu_1 - \mu_2 = 0$$
$$H_1 : \mu_1 - \mu_2 \neq 0$$

$$H_0 : \mu_1 - \mu_2 = 0$$
$$H_1 : \mu_1 - \mu_2 > 0$$

$$H_0 : \mu_1 - \mu_2 = 0$$
$$H_1 : \mu_1 - \mu_2 < 0$$

Two-Sample T-Test and CI

Method

μ_1: mean of Sample 1
μ_2: mean of Sample 2
Difference: $\mu_1 - \mu_2$

Equal variances are not assumed for this analysis.

Descriptive Statistics

Sample	N	Mean	StDev	SE Mean
Sample 1	15	74.20	6.85	1.8
Sample 2	15	73.90	7.05	1.8

Estimation for Difference

Difference	95% CI for Difference
0.30	(-4.91, 5.51)

Test

Null hypothesis	$H_0: \mu_1 - \mu_2 = 0$
Alternative hypothesis	$H_1: \mu_1 - \mu_2 \neq 0$

T-Value	DF	P-Value
0.12	27	0.907

FIGURE 5.5
Minitab printout for the two-sample confidence interval for the difference in the population mean lifetimes for two brands of cell phone batteries (assuming unequal variances).

Assuming equal variances, the test statistic for comparing two means follows a t-distribution with $n_1 + n_2 - 2$ degrees of freedom:

$$T = \frac{\bar{x}_1 - \bar{x}_2}{s_p \sqrt{\dfrac{1}{n_1} + \dfrac{1}{n_2}}}$$

where $s_p = \sqrt{\dfrac{(n_1 - 1)s_1^2 + (n_2 - 1)}{n_1 + n_2 - 2}}$ is the pooled standard deviation, which is used when we assume equal variances, and this value serves as an estimate of the population standard deviation. If we assume unequal variances, then the test statistic would be:

$$T = \frac{\bar{x}_1 - \bar{x}_2}{s}$$

where

$$s = \sqrt{\dfrac{s_1^2}{n_1} + \dfrac{s_2^2}{n_2}}$$

The number of degrees of freedom can be found by rounding down to an integer:

$$df = \frac{\left(\dfrac{s_1^2}{n_1} + \dfrac{s_2^2}{n_2}\right)^2}{\dfrac{\left(s_1^2/n_1\right)^2}{n_1 - 1} + \dfrac{\left(s_2^2/n_2\right)^2}{n_2 - 1}}$$

The three different null and alternative hypotheses can also be represented as follows:

$$H_0 : \mu_1 - \mu_2 = 0 \qquad H_0 : \mu_1 = \mu_2$$
$$H_1 : \mu_1 - \mu_2 \neq 0 \qquad H_1 : \mu_1 \neq \mu_2$$

$$H_0 : \mu_1 - \mu_2 = 0 \qquad H_0 : \mu_1 = \mu_2$$
$$H_1 : \mu_1 - \mu_2 > 0 \qquad H_1 : \mu_1 > \mu_2$$

$$H_0 : \mu_1 - \mu_2 = 0 \qquad H_0 : \mu_1 = \mu_2$$
$$H_1 : \mu_1 - \mu_2 < 0 \qquad H_1 : \mu_1 < \mu_2$$

Example 5.2

Two types of drugs are used to treat migraine headaches, Drug 1 and Drug 2. A researcher wants to know if Drug 2 provides a *faster* relief time for migraine headaches as compared to Drug 1. The researcher collected data as given in Table 5.1 which consists of the time taken (in minutes) for a random sample of 15

TABLE 5.1

Time (in Minutes) for Patients to Feel Pain Relief Using Either Drug 1 or Drug 2

Drug 1	Drug 2
22	26
18	28
31	29
26	24
38	23
25	25
29	28
31	28
31	27
28	29
26	31
19	30
45	22
31	27
27	28
16	
24	

patients using Drug 1 and a random sample of 17 patients using Drug 2 to begin feeling pain relief from their migraine headaches.

The descriptive statistics for Drug 1 and Drug 2 obtained from Table 5.1 are as follows:

For Drug 1:

$$n_1 = 17$$

$$\bar{x}_1 \approx 27.47 \text{ minutes}$$

$$s_1 \approx 7.14 \text{ minutes}$$

For Drug 2:

$$n_2 = 15$$

$$\bar{x}_2 = 27.00 \text{ minutes}$$

$$s_2 \approx 2.56 \text{ minutes}$$

Notice that it is not necessary to have equal sample sizes from each population for a two-sample *t*-test.

To test whether the mean time taken for Drug 2 to provide relief is *faster* than the mean time taken for Drug 1 ($\alpha = 0.01$), the appropriate null and alternative hypotheses would be:

$$H_0 : \mu_1 = \mu_2$$

$$H_1 : \mu_1 > \mu_2$$

The rejection region, which is based on the direction of the alternative hypothesis, would be in the right tail, and the value of $t = 2.457$ that defines this region would be for $\alpha = 0.01$ and $17 + 15 = 30$ degrees of freedom, as illustrated in Figure 5.6.

If we are assuming equal variances, then the pooled standard deviation is calculated as follows:

$$s_p = \sqrt{\frac{(n_1 - 1)s_1^2 + (n_2 - 1)}{n_1 + n_2 - 2}}$$

$$= \sqrt{\frac{(17 - 1)(7.14)^2 + (15 - 1)(2.56)^2}{17 + 15 - 2}}$$

$$\approx \sqrt{30.25} \approx 5.50$$

Thus, the test statistic for the two-sample *t*-test would be:

$$T = \frac{\bar{x}_1 - \bar{x}_2}{s_p\sqrt{\frac{1}{n_1} + \frac{1}{n_2}}} = \frac{27.47 - 27.00}{5.50\sqrt{\frac{1}{17} + \frac{1}{15}}} \approx 0.24$$

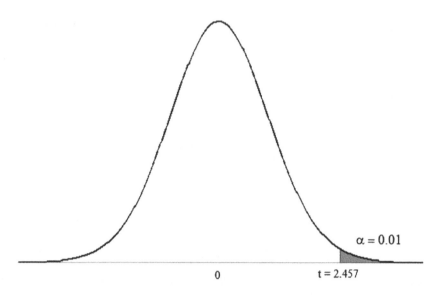

FIGURE 5.6
Rejection region defined by $\alpha = 0.01$ and $17 + 15 - 2 = 30$ degrees of freedom for a right-tailed, two-sample t-test.

The value of the test statistic T does not fall into the defined rejection region ($t = 2.457$, as in Figure 5.6; this is because we are performing a right-tailed test and the value of the test statistic T is less than t). This suggests that the population mean time taken for Drug 2 to provide relief for migraine headaches cannot be claimed to be significantly faster (i.e., less than) than the population mean time taken for Drug 1.

5.5 Using Minitab to Test the Difference Between Two Means

Once the sample data is entered in two separate columns or the summarized data from each of the two populations is entered in Minitab, a two-sample t-test can be carried out by selecting **Stat > Basic Statistics** and then **2-Sample t**, similar to when we were finding confidence intervals.

Note that similar to when deriving confidence intervals for the difference between two means, either the raw data or the summarized data from each of the two samples can also be entered. Also, the box for **Assume equal variances** needs to be checked if we are assuming equal variances. The level of significance and the appropriate direction of the alternative hypothesis need to be specified and this is done through the **Options** tab. We can also test whether there is a specific difference between the two population means by specifying such a difference in the **Options** box (see Exercise 15).

The Minitab printout for the two-sample t-test assuming equal variances is presented in Figure 5.7, and the Minitab printout for the two-sample t-test assuming unequal variances is presented in Figure 5.8. Note that the value of the test statistic $T \approx 0.24$ in Figure 5.7 is the same as that calculated manually. Similarly, the p-value of 0.405 supports the same

Two-Sample T-Test and CI: Drug 1, Drug 2

Method

μ_1: mean of Drug 1
μ_2: mean of Drug 2
Difference: $\mu_1 - \mu_2$

Equal variances are assumed for this analysis.

Descriptive Statistics

Sample	N	Mean	StDev	SE Mean
Drug 1	17	27.47	7.14	1.7
Drug 2	15	27.00	2.56	0.66

Estimation for Difference

Difference	Pooled StDev	99% Lower Bound for Difference
0.47	5.50	-4.32

Test

Null hypothesis H_0: $\mu_1 - \mu_2 = 0$
Alternative hypothesis H_1: $\mu_1 - \mu_2 > 0$

T-Value	DF	P-Value
0.24	30	0.405

FIGURE 5.7
Minitab printout for a two-sample *t*-test for whether Drug 2 provides a faster relief time for migraine headaches than Drug 1 (assuming equal variances).

conclusion—we cannot claim that Drug 2 provides a faster relief time for migraine headaches than Drug 1. This is because the *p*-value is greater than our predetermined level of significance $\alpha = 0.01$.

5.6 Using Minitab to Create an Interval Plot

Another technique that can be used to compare the difference between two population means is to plot separate confidence intervals for each population mean on a single graph and visually inspect whether the two confidence intervals overlap each other. To do this, we can create what is called an *interval plot* with Minitab. An interval plot is a plot of the confidence intervals for one or more population means on the same graph. Using the pain relief data from Table 5.1, the Minitab commands to draw such an interval plot first requires selecting **Interval Plot** under the **Graphs** menu.

The dialog box for creating an interval plot is presented in Figure 5.9, which depicts how to select if a simple interval plot for one variable or a multiple interval plot for two or more variables are required to be drawn.

Two-Sample T-Test and CI: Drug 1, Drug 2

Method

μ_1: mean of Drug 1
μ_2: mean of Drug 2
Difference: $\mu_1 - \mu_2$

Equal variances are not assumed for this analysis.

Descriptive Statistics

Sample	N	Mean	StDev	SE Mean
Drug 1	17	27.47	7.14	1.7
Drug 2	15	27.00	2.56	0.66

Estimation for Difference

Difference	99% Lower Bound for Difference
0.47	-4.22

Test

Null hypothesis	$H_0: \mu_1 - \mu_2 = 0$
Alternative hypothesis	$H_1: \mu_1 - \mu_2 > 0$

T-Value	DF	P-Value
0.25	20	0.401

FIGURE 5.8
Minitab printout for a two-sample *t*-test for whether Drug 2 provides a faster relief time for migraine headaches than Drug 1 (assuming unequal variances).

We can then select a simple plot of the **Multiple Ys** so as to compare the confidence intervals for two different population means (namely, Drug 1 and Drug 2), as illustrated in Figure 5.10.

The interval plot for comparing the true population mean time to pain relief from migraines for Drugs 1 and 2 is given in Figure 5.11.

The graph in Figure 5.11 illustrates that there is no difference in the mean time to pain relief between the two different brands of drugs. This is because these two confidence intervals overlap each other. If a significant difference in the mean times were to exist, the confidence intervals would not show much overlap.

Another way of using the interval plot is as an empirical check of the assumption of equal variances. For example, the interval plot in Figure 5.11 shows that the assumption of equal variances may not likely hold true. This is because there appears to be more variability in the time to relief for Drug 1 as compared to Drug 2, as is illustrated by the length of the confidence interval for Drug 1 as compared to the length of the confidence interval for Drug 2.

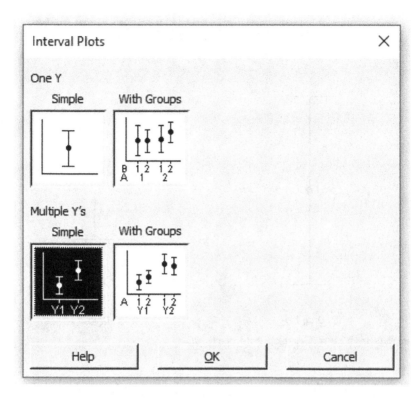

FIGURE 5.9
Minitab dialog box for selecting the type of interval plot.

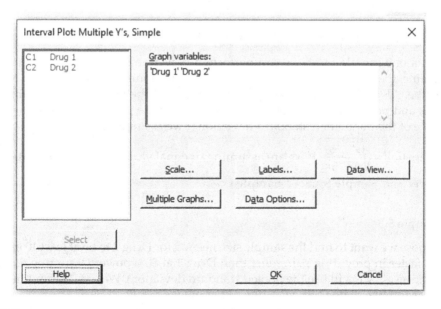

FIGURE 5.10
Minitab dialog box for an interval plot of multiple populations.

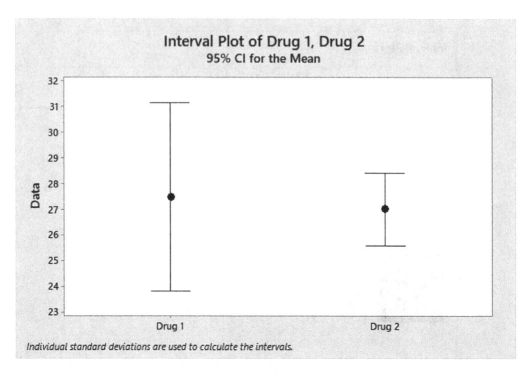

FIGURE 5.11
Interval plot illustrating the confidence intervals of the population mean time to pain relief for Drug 1 and Drug 2.

5.7 Using Minitab for a Power Analysis for a Two-Sample *t*-Test

Similar to a one-sample *t*-test, we can perform a power analysis for a two-sample *t*-test to determine the sample size needed to find a minimum difference for a specified power value. Also similar to a one-sample *t*-test, two of the three values of sample size, difference, and power need to be specified. We can use the pooled standard deviation as an estimate of the population standard deviation if we assume equal variances, or we can use the statistic $s = \sqrt{\dfrac{s_1^2}{n_1} + \dfrac{s_2^2}{n_2}}$ if we are assuming unequal variances. Under the **Stat** tab, we select **Power and Sample Size > 2-Sample t.**

Example 5.3

Suppose we want to find the sample size needed for Drug 2 to be at least 10 minutes faster in providing pain relief than Drug 1 at 80% power (assuming equal variances, so we will use the pooled standard deviation). We would input these values into the dialog box, as is illustrated in Figure 5.12.

By selecting the **Options** tab, as is illustrated in Figure 5.13, this allows us to specify a given alternative hypothesis and also a level of significance.

This would give the Minitab printout as presented in Figure 5.14.

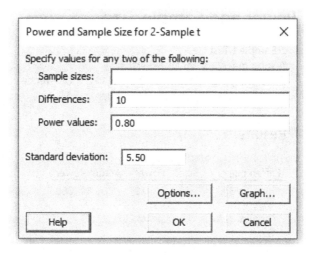

FIGURE 5.12
Minitab dialog box for a power analysis for a two-sample *t*-test for a difference of 10 minutes and a power value of 80%.

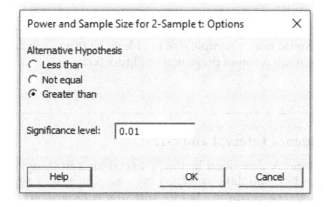

FIGURE 5.13
Minitab options box for a power analysis for a two-sample *t*-test for a right-tailed test and a significance level of $\alpha = 0.01$.

Power and Sample Size

2-Sample t Test
Testing mean 1 = mean 2 (versus >)
Calculating power for mean 1 = mean 2 + difference
$\alpha = 0.01$ Assumed standard deviation = 5.5

Results

Difference	Sample Size	Target Power	Actual Power
10	8	0.8	0.828960

The sample size is for each group.

FIGURE 5.14
Minitab power analysis for a two-sample *t*-test to detect a difference of 10 minutes at 80% power.

Power and Sample Size

2-Sample t Test
Testing mean 1 = mean 2 (versus >)
Calculating power for mean 1 = mean 2 + difference
α = 0.01 Assumed standard deviation = 5.5

Results

Difference	Sample Size	Target Power	Actual Power
3	69	0.8	0.801089

The sample size is for each group.

FIGURE 5.15
Minitab power analysis for a two-sample *t*-test to detect a difference of 3 minutes at 80% power.

As the Minitab printout in Figure 5.14 suggests, to find a difference of at least 10 minutes between the mean time to relief for the two different brands of pain relief drugs, a minimum sample size of 8 would be needed from each population.

If we wanted to detect a smaller difference, say a difference of at least 3 minutes, then we would need a sample size of at least 69 from each group, as illustrated in the Minitab printout presented in Figure 5.15.

5.8 Paired Confidence Interval and *t*-Test

The methods of inference described in the last sections can compare two population means from independent populations. When the assumption of independence does not hold true, such as when performing a before and after comparison on the same subject, we can calculate a confidence interval for *paired data* or a conduct a hypothesis test that is called a *paired t-test*. These techniques allow us to calculate confidence intervals and perform hypothesis tests of the population mean difference between paired observations, in other words, when we are sampling from *dependent* populations.

To calculate a confidence interval for paired data, we use the following formula:

$$\left(\bar{d} - t_{\alpha/2} \cdot \frac{s_d}{\sqrt{n}}, \bar{d} + t_{\alpha/2} \cdot \frac{s_d}{\sqrt{n}}\right)$$

Where $\bar{d}$ is the mean difference of the paired sample, s_d is the standard deviation of the difference, $t_{\alpha/2}$ is the value of the *T*-distribution for $n - 1$ degrees of freedom for an area of $\alpha/2$ in the upper right tail, and n is the number of paired observations.

A paired *t*-test is the same as testing whether the difference between paired comparisons is different from 0. The test statistic is quite similar to a one-sample *t*-test:

$$T = \frac{\bar{d}}{\frac{s_d}{\sqrt{n}}}$$

where $\bar{d} = \bar{x}_1 - \bar{x}_2$ is the difference of the sample means, s_d is the standard deviation of the difference, and n is the sample size.

We can test any of the following null and alternative hypotheses that describe whether the difference in the dependent population means is different from 0 as follows:

$$H_0 : \mu_1 - \mu_2 = 0 \quad H_0 : \mu_1 = \mu_2$$
$$H_1 : \mu_1 - \mu_2 \neq 0 \quad H_1 : \mu_1 \neq \mu_2$$

$$H_0 : \mu_1 - \mu_2 = 0 \quad H_0 : \mu_1 = \mu_2$$
$$H_1 : \mu_1 - \mu_2 > 0 \quad H_1 : \mu_1 > \mu_2$$

$$H_0 : \mu_1 - \mu_2 = 0 \quad H_0 : \mu_1 = \mu_2$$
$$H_1 : \mu_1 - \mu_2 < 0 \quad H_1 : \mu_1 < \mu_2$$

Example 5.4

The data set presented in Table 5.2 gives two measures of weight for a sample of 24 individuals—the individual's weight before beginning a weight loss

TABLE 5.2

Before and After Weights for 24 Individuals Participating in a 4-Week Weight Loss Program

Weight Before (lbs)	Weight After (lbs)
184	184
191	187
207	209
176	174
155	147
189	183
254	238
218	210
170	168
154	151
148	145
137	135
167	158
129	125
174	170
225	219
218	219
175	171
182	175
194	193
177	177
209	207
176	170
164	163

program and the individual's weight after participating in the weight loss program for 4 weeks.

Suppose we want to estimate whether the program helps people lose weight. Clearly, we can see that the sample weights before and after participating in the program are dependent samples because the weights were taken on the same individual.

We will begin by calculating a 95% confidence interval for the difference in the before and after weights. In order to do this, we first need to create a column that represents the difference between the weights before and after participating in the program. Table 5.3 gives the original data set along with a column with the before- and after-differences.

We can then find the mean and standard deviation of the difference column:

$$\bar{d} = \frac{\sum x_i}{n} = \frac{95}{24} \approx 3.958 \qquad s_d \approx 3.906$$

TABLE 5.3

Before and After Weights and Differences (in Pounds) for a Sample of 24 Individuals

Weight Before (lbs)	Weight After (lbs)	Difference (lbs) (Weight Before – Weight After)
184	184	$184 - 184 = 0$
191	187	$191 - 187 = 4$
207	209	$207 - 209 = -2$
176	174	$176 - 174 = 2$
155	147	$155 - 147 = 8$
189	183	$189 - 183 = 6$
254	238	$254 - 238 = 16$
218	210	$218 - 210 = 8$
170	168	$170 - 168 = 2$
154	151	$154 - 151 = 3$
148	145	$148 - 145 = 3$
137	135	$137 - 135 = 2$
167	158	$167 - 158 = 9$
129	125	$129 - 125 = 4$
174	170	$174 - 170 = 4$
225	219	$225 - 219 = 6$
218	219	$218 - 219 = -1$
175	171	$175 - 171 = 4$
182	175	$182 - 175 = 7$
194	193	$194 - 193 = 1$
177	177	$177 - 177 = 0$
209	207	$209 - 207 = 2$
176	170	$176 - 170 = 6$
164	163	$164 - 163 = 1$
		$\sum_{i=1}^{n} x_i = 95$

The value of $t_{\alpha/2} = 2.069$ for a 95% confidence interval with $n - 1 = 24 - 1 = 23$ degrees of freedom is illustrated in Figure 5.16.

The confidence interval is calculated as follows:

$$\left(\bar{d} - t_{\alpha/2} \cdot \frac{s_d}{\sqrt{n}}, \ \bar{d} + t_{\alpha/2} \cdot \frac{s_d}{\sqrt{n}} \right)$$

$$= \left(3.958 - 2.069 \cdot \frac{3.906}{\sqrt{24}}, \ 3.958 + 2.069 \cdot \frac{3.906}{\sqrt{24}} \right)$$

$$= \left(3.958 - 1.650, \ 3.958 + 1.650 \right) \approx \left(2.308, \ 5.608 \right)$$

To interpret this finding, we can claim to be 95% confident that the mean difference in weight loss for the population of individuals participating in the weight loss program is between 2.30 and 5.61 pounds.

We can also do a hypothesis test to see if the mean weight after participating in the program is significantly less than the mean weight at the beginning of the program. In this case, we would consider the following null and alternative hypothesis:

$$H_0 : \mu_{Before} - \mu_{After} = 0$$

$$H_1 : \mu_{Before} - \mu_{After} > 0$$

Another way of approaching the null and alternative hypothesis is as follows:

$$H_0 : \mu_{Before} = \mu_{After}$$

$$H_1 : \mu_{Before} > \mu_{After}$$

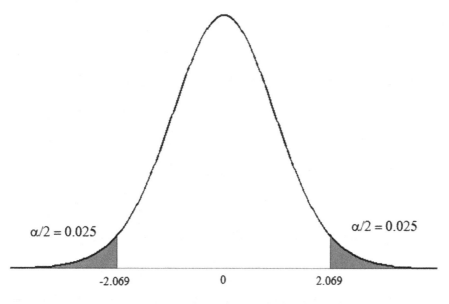

$\alpha/2 = 0.025$ $\alpha/2 = 0.025$

-2.069   0   2.069

FIGURE 5.16
T-distribution with the value of $t = 2.069$ that defines the upper boundary point for a 95% confidence interval with 23 degrees of freedom.

The test statistic would be calculated as follows:

$$T = \frac{\bar{d}}{\frac{s_d}{\sqrt{n}}} = \frac{3.958}{\frac{3.906}{\sqrt{24}}} \approx 4.96$$

We then compare this value to $t = 1.714$, as is illustrated in Figure 5.17.

Since the value of the test statistic $T = 4.96$ falls in the rejection region, we can accept the alternative hypothesis and reject the null, and claim that the mean weight for the population of program participants after the 4-week weight loss program is significantly less than the weight for the population before beginning the program.

5.9 Using Minitab for a Paired Confidence Interval and *t*-Test

To calculate a confidence interval for paired samples, we need to select **Stat > Basic Statistics > Paired t**. This brings up the dialog box that is given in Figure 5.18.

We can specify each sample in a column or we can enter the summarized data. To calculate a confidence interval, we need to click the **Options** tab to bring up the dialog box that is given in Figure 5.19.

We need to specify the confidence level, the hypothesized difference (which in this case is 0), and the alternative hypothesis. Recall that to calculate a confidence interval, the alternative hypothesis always needs to be specified as "not equal to." Then entering **OK** gives the printout that is presented in Figure 5.20.

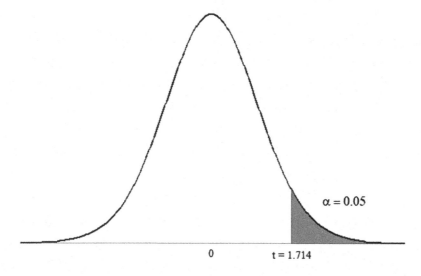

FIGURE 5.17
T-distribution where $t = 1.714$ that defines the rejection region for a right-tailed test with 23 degrees of freedom ($\alpha = 0.01$).

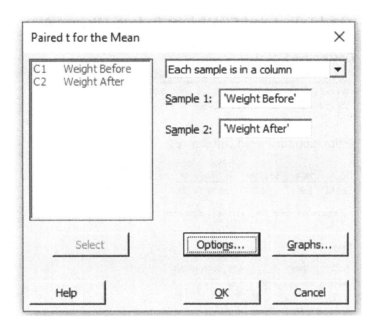

FIGURE 5.18
Minitab dialog box to calculate a confidence interval and perform a hypothesis test for two dependent populations.

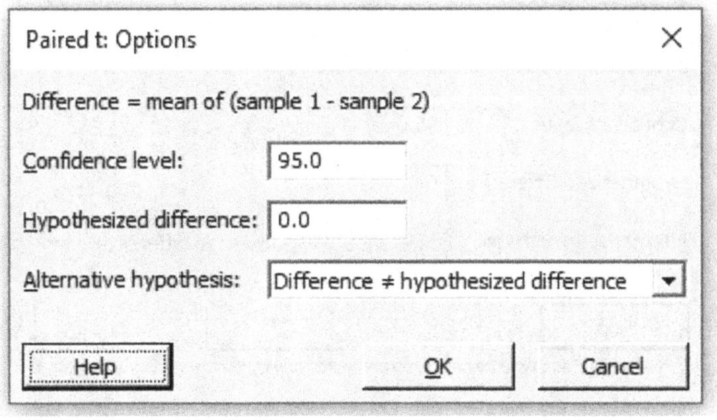

FIGURE 5.19
Minitab dialog box for two dependent samples.

If we were to conduct a hypothesis test, we would need to specify the direction of the alternative hypothesis as can be seen in Figure 5.21, the results of which are given in Figure 5.22.

Note that in Figure 5.22, the direction of the null and alternative hypothesis is given along with the value of the test statistic $T = 4.96$ and the corresponding p-value.

Paired T-Test and CI: Weight Before, Weight After

Descriptive Statistics

Sample	N	Mean	StDev	SE Mean
Weight Before	24	182.21	29.24	5.97
Weight After	24	178.25	28.38	5.79

Estimation for Paired Difference

Mean	StDev	SE Mean	95% CI for μ_difference
3.958	3.906	0.797	(2.309, 5.608)

μ_difference: mean of (Weight Before - Weight After)

Test

Null hypothesis	H_0: μ_difference = 0
Alternative hypothesis	H_1: μ_difference ≠ 0

T-Value	P-Value
4.96	0.000

FIGURE 5.20
Minitab printout for a 95% confidence interval for the paired data given in Table 5.2.

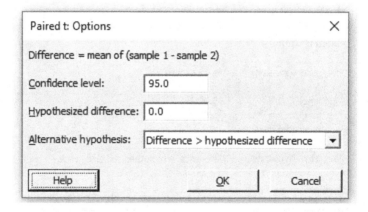

FIGURE 5.21
Minitab dialog box to perform a right-tailed paired *t*-test.

5.10 Differences Between Two Proportions

We can also calculate confidence intervals and perform hypothesis tests of whether there is a difference in the proportions between two *independent* populations. Similar to a one-sample proportion for large samples ($n \geq 30$), the sampling distribution of the difference between two proportions is normally distributed and is centered about the true population difference between the two proportions.

Paired T-Test and CI: Weight Before, Weight After

Descriptive Statistics

Sample	N	Mean	StDev	SE Mean
Weight Before	24	182.21	29.24	5.97
Weight After	24	178.25	28.38	5.79

Estimation for Paired Difference

Mean	StDev	SE Mean	95% Lower Bound for μ_difference
3.958	3.906	0.797	2.592

μ_difference: mean of (Weight Before - Weight After)

Test

Null hypothesis	H₀: μ_difference = 0
Alternative hypothesis	H₁: μ_difference > 0

T-Value	P-Value
4.96	0.000

FIGURE 5.22
Minitab printout for a two-sample paired *t*-test of whether the mean weight for the population of participants before participating in the weight loss program is significantly greater than the mean weight for the population after participating in the program.

The formula for a confidence interval for the difference between two independent population proportions is as follows:

$$\left((\hat{p}_1 - \hat{p}_2) - z_{\alpha/2} \cdot \sqrt{\frac{\hat{p}_1(1-\hat{p}_1)}{n_1} + \frac{\hat{p}_2(1-\hat{p}_2)}{n_2}}, (\hat{p}_1 - \hat{p}_2) + z_{\alpha/2} \cdot \sqrt{\frac{\hat{p}_1(1-\hat{p}_1)}{n_1} + \frac{\hat{p}_2(1-\hat{p}_2)}{n_2}} \right)$$

Where $\hat{p}_1$ is the sample proportion for the first sample, $\hat{p}_2$ is the sample proportion for the second sample, n_1 is the sample size for the first sample, n_2 is the sample size for the second sample, and $z_{\alpha/2}$ is the z-value that defines the upper $\alpha/2$ portion of the standard normal distribution.

Example 5.5

The data set in Table 5.4 gives a comparison of the number of students who planned to graduate in the spring semester versus those who actually graduated in the spring semester at two competing colleges.

We want to generate a 95% confidence interval for the true population difference in the actual graduation rates between the two colleges.

From Table 5.4, the sample proportions would be calculated as follows:

$$\hat{p}_1 = \frac{218}{275} \approx 0.7927, n_1 = 275$$

$$\hat{p}_2 = \frac{207}{289} \approx 0.7163, n_1 = 289$$

TABLE 5.4

Comparison of the Number of Students Who Plan to Graduate versus the
Number of Students Who Actually Graduate for Two Competing Colleges

	College A	College B
Number of students who planned to graduate in spring	275	289
Number of students who actually graduated in spring	218	207

Thus, a 95% confidence interval for the difference between the actual graduation rates for the population of students at College A and the population of
students at College B would be as follows:

$$\left((\hat{p}_1 - \hat{p}_2) - z_{\alpha/2} \cdot \sqrt{\frac{\hat{p}_1(1-\hat{p}_1)}{n_1} + \frac{\hat{p}_2(1-\hat{p}_2)}{n_2}}, (\hat{p}_1 - \hat{p}_2) + z_{\alpha/2} \cdot \sqrt{\frac{\hat{p}_1(1-\hat{p}_1)}{n_1} + \frac{\hat{p}_2(1-\hat{p}_2)}{n_2}} \right)$$

$$= \left((0.7927 - 0.7163) - 1.96 \cdot \sqrt{\frac{0.7927(0.2073)}{275} + \frac{0.7163(0.2837)}{289}}, \right.$$

$$\left. (0.7927 - 0.7163) + 1.96 \cdot \sqrt{\frac{0.7927(0.2073)}{275} + \frac{0.7163(0.2837)}{289}} \right)$$

$$\approx (0.0764 - 0.0707, 0.0764 + 0.0707) \approx (0.0057, 0.1471) \approx (5.7\%, 14.71\%)$$

This confidence interval suggests that we are 95% confident that the difference
in the population percentage of students who actually graduated from the two
competing schools is between 0.57% and 14.71%. Because the confidence interval is testing the difference in the graduation rates of College B subtracted from
College A, this confidence interval would also suggest that College A has a
graduation rate that is 0.57%–14.71% *higher* than that of College B as follows:

$$0.57\% < p_A - p_B < 14.71\%$$

Similarly, we could also perform a hypothesis test to see if the proportion of
students who graduate from College A is significantly different from the proportion of students who graduate from College B. The test statistic is as follows:

$$Z = \frac{\hat{p}_1 - \hat{p}_2}{\sqrt{\frac{\hat{p}_1(1-\hat{p}_1)}{n_1} + \frac{\hat{p}_2(1-\hat{p}_2)}{n_2}}}$$

where $\hat{p}_1$ is the proportion of students that graduated from College A, $\hat{p}_2$ is the
proportion of students that graduated from College B, n_1 is the sample size from
College A, and n_2 is the sample size from College B.

Typically, when testing for a difference of 0, the null and alternative hypotheses can be written as follows:

$$H_0 : p_1 - p_2 = 0 \qquad H_0 : p_1 = p_2$$
$$H_1 : p_1 - p_2 \neq 0 \qquad H_1 : p_1 \neq p_2$$

$$H_0 : p_1 - p_2 = 0 \qquad H_0 : p_1 = p_2$$
$$H_1 : p_1 - p_2 > 0 \qquad H_1 : p_1 > p_2$$

$$H_0 : p_1 - p_2 = 0 \qquad H_0 : p_1 = p_2$$
$$H_1 : p_1 - p_2 < 0 \qquad H_1 : p_1 < p_2$$

Example 5.12

Using the data in Table 5.4, suppose we want to estimate if there is a difference in the proportion of students that College A graduates compared to that of College B ($\alpha = 0.05$). Therefore, the appropriate null and alternative hypotheses would be:

$$H_0 : p_1 = p_2$$

$$H_1 : p_1 \neq p_2$$

The rejection region for the test statistic would be established by the level of significance and the direction of the alternative hypothesis, as shown in Figure 5.23. The value of the test statistic is then calculated as follows:

$$Z = \frac{\hat{p}_1 - \hat{p}_2}{\sqrt{\dfrac{\hat{p}_1(1-\hat{p}_1)}{n_1} + \dfrac{\hat{p}_2(1-\hat{p}_2)}{n_2}}} = \frac{0.7927 - 0.7163}{\sqrt{\dfrac{0.7937(1-0.7937)}{275} + \dfrac{0.7163(1-0.7163)}{289}}} \approx 2.12$$

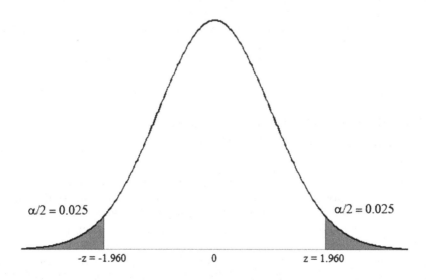

FIGURE 5.23
Rejection region for testing whether there is a difference in the proportion of students that College A graduates compared to College B ($\alpha = 0.05$).

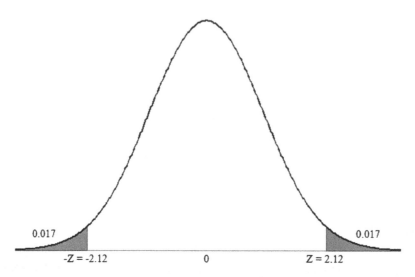

FIGURE 5.24
p-Value representing twice the area to the right of the corresponding value of the test statistic $Z = 2.12$.

By comparing this test statistic to the value of z that defines the rejection region, as illustrated in Figure 5.23, we can see that the test statistic does fall in the rejection region because it is greater than $z = 1.96$. Therefore, we can claim that there is a significant difference in the population proportion of students who graduate from College A compared to the population proportion of students who graduate from College B.

Because the sampling distribution for a two-sample proportion follows a normal distribution, we can also calculate the *p*-value by finding the area that is defined by the value of the test statistic, as presented in Figure 5.24. Because this is a two-sided test, the *p*-value is found by taking twice the area that is to the right of the test statistic Z. Therefore, since the area to the right of $Z = 2.12$ is 0.017, the *p*-value would be $2(0.017) = 0.034$.

5.11 Using Minitab for Two-Sample Proportion Confidence Intervals and Hypothesis Tests

To use Minitab for a two-sample proportion using the data in Table 5.4, under the **Stat** menu, select **Basic Statistics** and then **2 Proportions**, as illustrated in Figure 5.25.

This gives the dialog box presented in Figure 5.26, where the raw data or the summarized data can be entered to represent the number of events and trials for each of the two samples.

To calculate a confidence interval, we select the **Options** tab to bring up the dialog box in Figure 5.27.

In order to match our manual calculations, we need to select **Estimate the proportions separately** for the test method under the options tab (typically, the pooled estimate of the proportions is less accurate as compared to estimating them separately). Selecting **OK** then gives the Minitab printout that is presented in Figure 5.28.

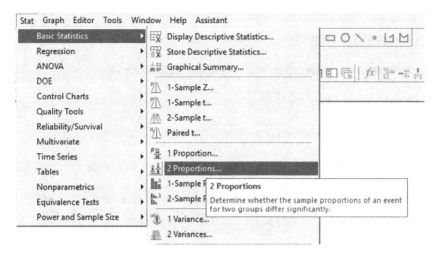

FIGURE 5.25
Minitab commands for comparing two proportions.

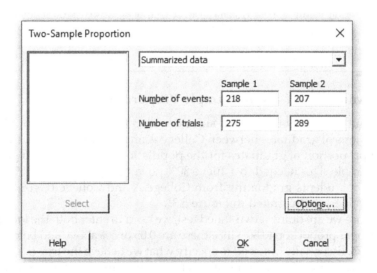

FIGURE 5.26
Minitab dialog box for a two-sample proportion using the summarized data.

Note that Figure 5.28 provides a 95% confidence interval of the difference in the graduation proportions for College A and College B, similar to what we have calculated manually.

If we were interested in testing whether College A graduated a *larger* proportion of students than College B ($\alpha = 0.01$), we could test the following null and alternative hypotheses:

$$H_0 : p_1 = p_2$$

$$H_1 : p_1 > p_2$$

To adjust for these factors in Minitab, we would simply have to change the direction of the alternative hypothesis and the confidence level as illustrated in Figure 5.29, gives the Minitab printout in Figure 5.30.

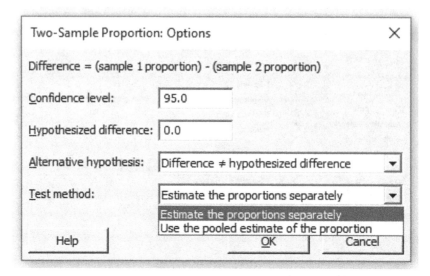

FIGURE 5.27
Minitab dialog box for calculating a confidence interval for the difference between two proportions.

5.12 Power Analysis for a Two-Sample Proportion

Suppose we want to find out the sample size needed to find a difference of 5% in the population proportions of graduates between College A and College B, and also suppose that the estimated proportion of graduates for the population of all colleges in general is 75%.

To find the sample sizes needed to achieve 80% power that can detect a 5% difference in the proportion of students graduating from College A and College B, you could enter in the respective values, as illustrated in Figure 5.31.

Notice that since we are doing a two-tailed test, we need to enter both the lower proportion (0.70) and the upper proportion (0.80) since these are 0.05 or 5% above and below the baseline proportion of 0.75. Essentially, we need to specify what we believe the true (i.e., baseline) population proportion is and then add and subtract the difference we were interested in testing.

Figure 5.32 gives the Minitab printout for the given power analysis. We can see that we need a sample size of at least 1,251 from each population to detect a 5% difference in the mean graduation rate between the two colleges for a proportion of 0.70, and a sample size of 1,094 from each population for a proportion of 0.80.

5.13 Confidence Intervals and Hypothesis Tests for Two Variances

Similar to finding confidence intervals and performing hypothesis tests for the difference between two means or the differences between two proportions, we can also calculate confidence intervals and perform hypothesis tests on whether or not there is a difference in the *variances* of two independent populations.

Test and CI for Two Proportions

Method

p_1: proportion where Sample 1 = Event
p_2: proportion where Sample 2 = Event
Difference: $p_1 - p_2$

Descriptive Statistics

Sample	N	Event	Sample p
Sample 1	275	218	0.792727
Sample 2	289	207	0.716263

Estimation for Difference

Difference	95% CI for Difference
0.0764643	(0.005777, 0.147151)

CI based on normal approximation

Test

Null hypothesis	H_0: $p_1 - p_2 = 0$
Alternative hypothesis	H_1: $p_1 - p_2 \neq 0$

Method	Z-Value	P-Value
Normal approximation	2.12	0.034
Fisher's exact		0.040

FIGURE 5.28
Minitab printout for the 95% confidence interval for the difference between the population proportion of students who graduate from College A and College B.

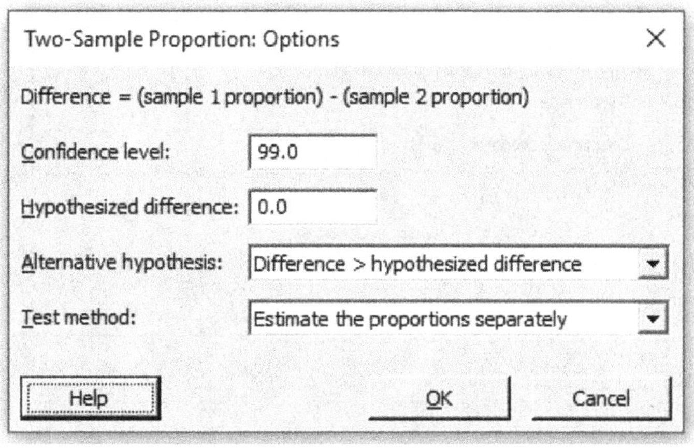

Two-Sample Proportion: Options ✕

Difference = (sample 1 proportion) - (sample 2 proportion)

Confidence level: 99.0

Hypothesized difference: 0.0

Alternative hypothesis: Difference > hypothesized difference ▼

Test method: Estimate the proportions separately ▼

Help   OK   Cancel

FIGURE 5.29
Minitab dialog box to test whether College A graduated a larger proportion of students than College B ($\alpha = 0.01$).

Test and CI for Two Proportions

Method

p₁: proportion where Sample 1 = Event
p₂: proportion where Sample 2 = Event
Difference: p₁ - p₂

Descriptive Statistics

Sample	N	Event	Sample p
Sample 1	275	218	0.792727
Sample 2	289	207	0.716263

Estimation for Difference

Difference	99% Lower Bound for Difference
0.0764643	-0.007436

CI based on normal approximation

Test

Null hypothesis	H₀: p₁ - p₂ = 0
Alternative hypothesis	H₁: p₁ - p₂ > 0

Method	Z-Value	P-Value
Normal approximation	2.12	0.017
Fisher's exact		0.022

FIGURE 5.30
Minitab printout of the hypothesis test whether College A graduated a larger proportion of students than College B ($\alpha = 0.01$).

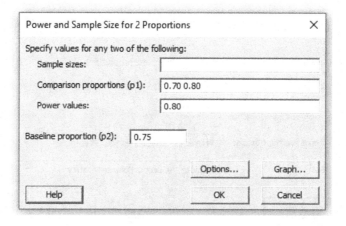

FIGURE 5.31
Minitab dialog box for a power analysis to detect a 5% difference in the proportion of students graduating from College A and College B.

Power and Sample Size

Test for Two Proportions
Testing comparison p = baseline p (versus ≠)
Calculating power for baseline p = 0.75
$\alpha = 0.05$

Results

	Comparison p	Sample Size	Target Power	Actual Power
	0.7	1251	0.8	0.800090
	0.8	1094	0.8	0.800095

The sample size is for each group.

FIGURE 5.32
Minitab output for the power analysis to find a 5% difference from a hypothesized population proportion of 0.75 with 80% power.

One difference in calculating confidence intervals and performing hypothesis tests with two variances is that the confidence interval and hypothesis test are expressed as a *ratio of the two variances*.

A $100(1 - \alpha)\%$ confidence interval for the ratio of two population variances would be as follows:

$$\frac{\frac{s_1^2}{s_2^2}}{F_{\alpha/2}} < \frac{\sigma_1^2}{\sigma_2^2} < \frac{\frac{s_1^2}{s_2^2}}{F_{1-(\alpha/2)}}$$

Where s_1^2 is the variance of the first sample, s_2^2 is the variance of the second sample, $F_{\alpha/2}$ is the value of the F-distribution for an area of $\alpha/2$ in the right tail, $F_1 - (\alpha/2)$ is the value obtained from the F-distribution for an area of $1 - (\alpha/2)$ in the right tail with $n_1 - 1$ degrees of freedom for the numerator and $n_2 - 1$ degrees of freedom for the denominator, where n_1 is the sample size collected from the first population, n_2 is the sample size collected from the second population, σ_1^2 is the unknown variance from the first population, and σ_2^2 is the unknown variance from the second population.

This confidence interval gives a range of values that represents the ratio of two variances from two different populations. If this interval contains the value of 1, it suggests that there is not a significant difference in the variability between the two different populations. If both the lower bound and the upper bound of the confidence interval are numbers that are greater than 1, this suggests that the variability of the first population is greater than that of the second population. Similarly, if both the lower bound and the upper bound of the confidence interval are less than 1, then this would suggest that the first population does not vary as much as the second population.

One thing to remember when finding confidence intervals and testing two population variances is that it is usual practice for the larger sample variance to be in the numerator. This is because F-tables typically have the rejection region only in the right tail. By having the larger variance in the numerator of the test statistic, this allows us to use the F-tables as provided in Appendix A, since these tables have the rejection region only in the right tail. However, if Minitab is used instead of the tables to find the F values that define the rejection regions, then it does not matter which variance is in the numerator.

Example 5.6

One common way of measuring the risk profile of a stock is to see how much the price varies (volatility). Stocks whose prices tend to fluctuate are often seen as being more risky. We can estimate the risk profiles of two different stocks by calculating a confidence interval for the ratio of the variances.

Suppose that we have collected a sample of prices for stocks from two different mutual fund companies. This data is provided in Table 5.5.

Suppose we want to find a 95% confidence interval for the ratio of the variances for these two different stocks. Let's begin by first finding the descriptive statistics for the two different stock funds, as is provided in Table 5.6.

Note that since the variance for Stock B is greater than the variance for Stock A, the variance for Stock B will be placed in the numerator and the variance for Stock A in the denominator as follows:

$$\frac{\frac{s_B^2}{s_A^2}}{F_{\alpha/2}} < \frac{\sigma_B^2}{\sigma_A^2} < \frac{\frac{s_B^2}{s_A^2}}{F_{1-(\alpha/2)}}$$

TABLE 5.5

Random Sample of Stock Prices (in Dollars)
for Two Different Stock Funds

Stock A	Stock B
27.45	29.85
26.85	30.45
27.46	28.54
23.54	30.55
24.68	31.47
29.24	32.66
27.55	38.15
28.99	28.73
26.55	27.44
29.58	24.88
27.22	29.68
26.58	27.55
25.32	29.49
28.31	26.54
29.47	29.99
24.50	
28.99	
27.44	

TABLE 5.6

Descriptive Statistics for the Two Stock
Funds from Table 5.5

Statistic	Stock A	Stock B
Sample size	18	15
Mean	27.207	29.731
Standard deviation	1.789	3.047
Variance	3.202	9.282

We can now find the *f*-values that represent a 95% confidence interval for the
F-distribution with the degrees of freedom for the numerator as $15 - 1 = 14$, and
the degrees of freedom for the denominator as $18 - 1 = 17$. This distribution is
presented in Figure 5.33.

The confidence interval for the ratio of the two variances can be calculated as
follows:

$$\frac{\frac{s_B^2}{s_A^2}}{F_{\alpha/2}} < \frac{\sigma_1^2}{\sigma_2^2} < \frac{\frac{s_B^2}{s_A^2}}{F_{1-(\alpha/2)}} \approx \frac{\frac{9.282}{3.202}}{2.753} < \frac{\sigma_B^2}{\sigma_A^2} < \frac{\frac{9.282}{3.202}}{0.3448}$$

$$1.05 < \frac{\sigma_B^2}{\sigma_A^2} < 8.41$$

This confidence interval suggests that the population variance for Stock B is
more than 1.05 times the population variance for Stock A and less than 8.41
times the population variance for Stock A. You can see this by multiplying
through the inequality by σ_A^2 to get the following:

$$(1.05) \cdot \sigma_A^2 < \sigma_B^2 < (8.41) \cdot \sigma_A^2$$

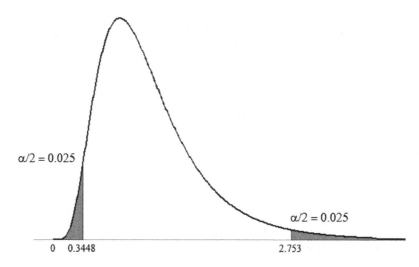

FIGURE 5.33
Shaded area of the rejection region for an F-distribution with 14 degrees of freedom for the numerator and 17 degrees of freedom for the denominator ($\alpha = 0.01$).

As for hypothesis testing, similar to what we have seen with two means and two proportions, we can formulate any of the following null and alternative hypotheses for the ratio of the variances:

$$H_0 : \frac{\sigma_1^2}{\sigma_2^2} = 1$$

$$H_1 : \frac{\sigma_1^2}{\sigma_2^2} \neq 1$$

$$H_0 : \frac{\sigma_1^2}{\sigma_2^2} = 1$$

$$H_1 : \frac{\sigma_1^2}{\sigma_2^2} > 1$$

$$H_0 : \frac{\sigma_1^2}{\sigma_2^2} = 1$$

$$H_1 : \frac{\sigma_1^2}{\sigma_2^2} < 1$$

where the parameter σ_1^2 is the unknown population variance for the first population, and the parameter σ_2^2 is the unknown population variance for the second population.

We can also express the aforementioned null and alternative hypotheses as the relationship of one variance to the other:

$$H_0 : \sigma_1^2 = \sigma_2^2$$

$$H_1 : \sigma_1^2 \neq \sigma_2^2$$

$$H_0 : \sigma_1^2 = \sigma_2^2$$

$$H_1 : \sigma_1^2 > \sigma_2^2$$

$$H_0 : \sigma_1^2 = \sigma_2^2$$

$$H_1 : \sigma_1^2 < \sigma_2^2$$

The sampling distribution for the difference between two variances follows an F-distribution with the test statistic:

$$F = \frac{s_1^2}{s_2^2}$$

Where the statistic s_1^2 is the sample variance obtained from the first sample and s_2^2 is the sample variance obtained from the second sample.

The degrees of freedom for the numerator is $n_1 - 1$, where n_1 is the sample size for the first sample, and the degrees of freedom for the denominator is $n_2 - 1$, where n_2 is the sample size for the second sample.

Just like with confidence intervals, one thing to remember when testing two population variances is that it is usual practice for the larger sample variance to be in the numerator. Notice that since the variance for Stock B is greater than the variance for Stock A, we will let the variance for Stock B be in the numerator and the variance for Stock A be in the denominator.

The null and alternative hypothesis for testing whether the population variances are different from each other would be:

$$H_0 : \sigma_B^2 = \sigma_A^2 \quad \text{or} \quad H_0 : \frac{\sigma_B^2}{\sigma_A^2} = 1$$

$$H_1 : \sigma_B^2 \neq \sigma_A^2 \qquad H_1 : \frac{\sigma_B^2}{\sigma_A^2} \neq 1$$

$$F = \frac{s_B^2}{s_A^2} = \frac{9.282}{3.202} \approx 2.90$$

For a level of significance of 0.05, this would mean that we would put 0.025 in each tail of the F-distribution. The degrees of freedom for the numerator would be $15 - 1 = 14$, and the degrees of freedom for the denominator would be $18 - 1 = 17$. This corresponds to the same shaded region found with $f = 0.3448$ and $f = 2.753$ on the F-distribution that was presented in Figure 5.33.

The test statistic $F = 2.90$ falls in the rejection region because 2.90 is greater than 2.753. Therefore, we can accept the alternative hypothesis and reject the null hypothesis. We claim that there is a significant difference in the variability between the two stocks. More specifically, we can go one step further and claim that Stock B has a higher degree of volatility when compared to Stock A.

5.14 Using Minitab for Testing Two Sample Variances

To use Minitab for testing two variances, we need to go to **Stat > Basic Statistics > 2 Variances**. This brings up the dialog box that is given in Figure 5.34.

Similar to what you have seen before, we can enter each sample in its own column or enter summarized data. Since we want the variance of Stock B to be in the numerator and the variance of Stock A to be in the denominator, we need to specify **Sample 1** as Stock B and **Sample 2** as Stock A. Then, clicking the **Options** tab brings up the dialog box that is given in Figure 5.35.

We need to select the ratio of sample variances, specify the direction of the alternative hypothesis, and check the box that bases the calculations on the normal distribution so that our manual calculations will match those obtained from Minitab. The

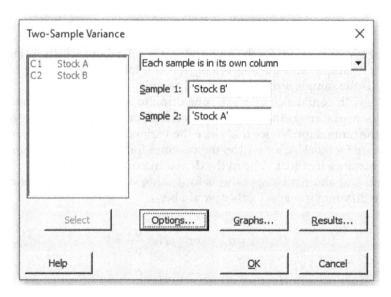

FIGURE 5.34
Minitab dialog box for testing two variances.

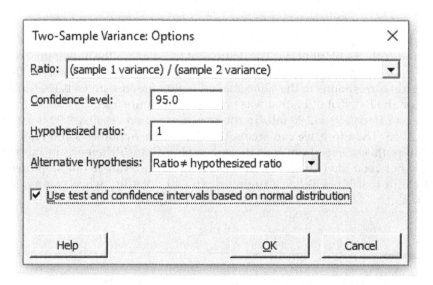

FIGURE 5.35
Minitab options dialog box for testing two variances.

hypothesized ratio of 1 corresponds to how we specified our null and alternative hypothesis as $H_0 : \dfrac{\sigma_B^2}{\sigma_A^2} = 1, \quad H_A : \dfrac{\sigma_B^2}{\sigma_A^2} \neq 1.$

This gives the Minitab output in Figure 5.36.

Notice from the Minitab output in Figure 5.36, that the null and alternative hypotheses are expressed as a ratio. And since the p-value = 0.040 is less than our specified level of significance of 0.05, we can reject the null hypothesis and accept the alternative hypothesis.

Test and CI for Two Variances: Stock B, Stock A

Method

σ_1: standard deviation of Stock B
σ_2: standard deviation of Stock A
Ratio: σ_1/σ_2
F method was used. This method is accurate for normal data only.

Descriptive Statistics

Variable	N	StDev	Variance	95% CI for σ^2
Stock B	15	3.047	9.282	(4.975, 23.087)
Stock A	18	1.789	3.202	(1.803, 7.196)

Ratio of Variances

Estimated Ratio	95% CI for Ratio using F
2.89920	(1.053, 8.408)

Test

Null hypothesis	H_0: $\sigma_1^2 / \sigma_2^2 = 1$
Alternative hypothesis	H_1: $\sigma_1^2 / \sigma_2^2 \neq 1$
Significance level	$\alpha = 0.05$

Method	Test Statistic	DF1	DF2	P-Value
F	2.90	14	17	0.040

FIGURE 5.36
Minitab output for testing the difference in the variability of the prices between the two stocks in Table 5.5 ($\alpha = 0.05$).

Thus, there is a significant difference in the variability of the prices between the two different stocks.

5.15 Power Analysis for Two-Sample Variances

Just as with other power analyses we have seen thus far, we need to specify two of the three values of sample size, ratio, and power level. Suppose we want to know the sample size needed to find a variance ratio of 1.25 at 80% power. In other words, we want to know how big of a sample would be needed to detect a variance ratio of at least 1.25 (80% of the time). In order to do this, select **Stat > Power and Sample Size > 2 Variances** to bring up the dialog box that is illustrated in Figure 5.37.

Notice in Figure 5.37 that we need to enter the ratio of the variances. The other option is to enter the ratio of the standard deviations.

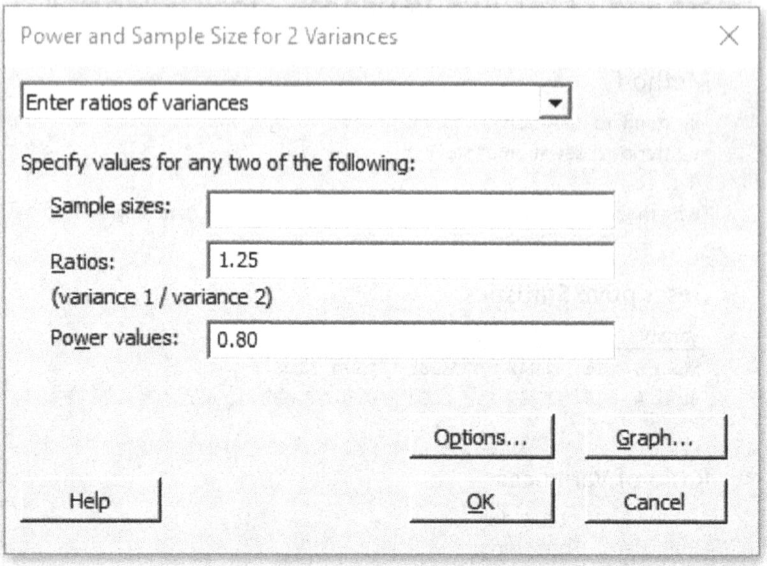

FIGURE 5.37
Minitab dialog box to find the sample size needed to find a variance ratio of at least 1.25 at 80% power.

We also need to click on the **Options** tab to select the appropriate alternative hypothesis and use the method that is based on the *F*-test, as illustrated in Figure 5.38.

The results for this power analysis are provided in Figure 5.39.

Notice from Figure 5.39 that we would need to obtain a sample of at least 633 observations from each population to be able to detect a variance ratio of at least 1.25 at 80% power.

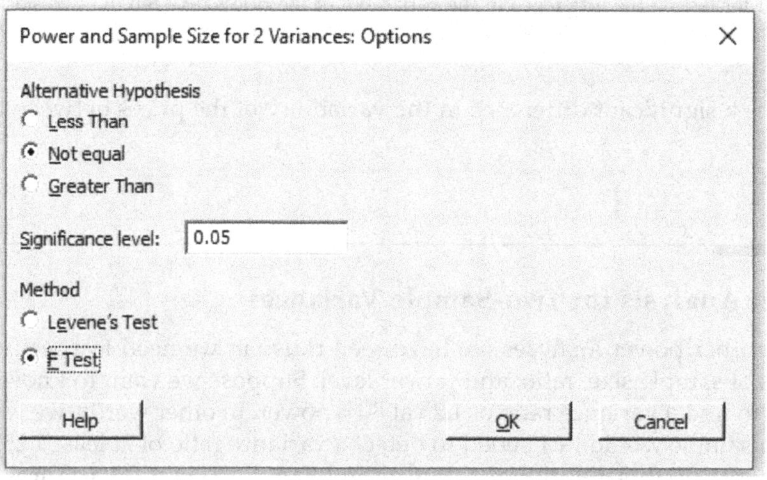

FIGURE 5.38
Minitab dialog box to select the alternative hypothesis, significance level, and method for a power analysis for two variances.

Power and Sample Size

Test for Two Variances
Testing (variance 1 / variance 2) = 1 (versus ≠)
Calculating power for (variance 1 / variance 2) = ratio
$\alpha = 0.05$
Method: F Test

Results

Ratio	Sample Size	Target Power	Actual Power
1.25	633	0.8	0.800310

The sample size is for each group.

FIGURE 5.39
Minitab output for the power analysis to find the sample size needed to find a variance ratio of 1.25 at 80% power.

5.16 Confidence Intervals and Hypothesis Tests for Two-Count Variables

We may be interested in calculating confidence intervals and performing hypothesis tests using count data obtained from two independent populations. To calculate a confidence interval for the difference in the mean counts from two independent populations, we use the following formula:

$$\left(\left(\hat{\lambda}_1 - \hat{\lambda}_2 \right) - z_{\alpha/2} \sqrt{ \frac{\hat{\lambda}_1}{n_1 \cdot t_1} + \frac{\hat{\lambda}_2}{n_2 \cdot t_2} }, \left(\hat{\lambda}_1 - \hat{\lambda}_2 \right) + z_{\alpha/2} \sqrt{ \frac{\hat{\lambda}_1}{n_1 \cdot t_1} + \frac{\hat{\lambda}_2}{n_2 \cdot t_2} } \right)$$

where $\hat{\lambda}_1$ is the mean count for the first sample, $\hat{\lambda}_2$ is the mean count for the second sample, n_1 is the sample size from the first sample, n_2 is the sample size from the second sample, t_1 is the observation length for the first sample, t_2 is the observation length for the second sample, and $z_{\alpha/2}$ is the z-value on the standard normal distribution for an area of $\alpha/2$ in the right tail.

Example 5.7

Suppose we want to calculate a confidence interval for the difference in the mean counts of defective products made at two different factories. Table 5.7 gives a random sample of defective products collected from the two factories.

The mean sample counts for Factory A and Factory B are 3.571 and 4.000, respectively; there are 7 observations from Factory A and 9 observations from Factory B; and the observation length for both Factory A and Factory B would be equal to 1. This is because the units of the data are in number of defects per day and we are interested in estimating if there is a difference in the mean number

TABLE 5.7

Number of Defective Products from
Factory A and Factory B

Factory A	Factory B
4	5
7	8
2	4
0	1
6	4
4	7
2	0
	6
	1

of defects by factory per day. If we were interested in estimating whether there was a difference in the mean number of defects per hour, we would set the observation length for Factory A and Factory B equal to 24. We could then calculate a 95% confidence interval for the difference in the population mean number of defective items per day as follows:

$$\left(\left(\hat{\lambda}_1 - \hat{\lambda}_2 \right) - z_{\alpha/2} \sqrt{ \frac{\hat{\lambda}_1}{n_1 \cdot t_1} + \frac{\hat{\lambda}_2}{n_2 \cdot t_2} }, \left(\hat{\lambda}_1 - \hat{\lambda}_2 \right) + z_{\alpha/2} \sqrt{ \frac{\hat{\lambda}_1}{n_1 \cdot t_1} + \frac{\hat{\lambda}_2}{n_2 \cdot t_2} } \right)$$

$$= \left((3.571 - 4.000) - 1.96 \sqrt{ \frac{3.571}{7 \cdot 1} + \frac{4.000}{9 \cdot 1} }, (3.571 - 4.000) + 1.96 \sqrt{ \frac{3.571}{7 \cdot 1} + \frac{4.000}{9 \cdot 1} } \right)$$

$$= \left(-0.429 - 1.915, -0.429 + 1.915 \right) \approx \left(-2.344, 1.486 \right)$$

We are 95% confident that the population mean difference in defective items from the two factories is between −2.344 and 1.486 items per day.

To conduct a hypothesis test on the difference between two mean counts from two independent populations, we could consider one of the following null and alternative hypotheses:

$$H_0 : \lambda_1 = \lambda_2 \qquad H_0 : \lambda_1 = \lambda_2 \qquad H_0 : \lambda_1 = \lambda_2$$
$$H_1 : \lambda_1 \neq \lambda_2 \qquad H_1 : \lambda_1 > \lambda_2 \qquad H_1 : \lambda_1 < \lambda_2$$

The sampling distribution for the difference in the population counts follows a standard normal distribution centered about $\lambda_1 - \lambda_2 = 0$ with the following test statistic:

$$Z = \frac{\hat{\lambda}_1 - \hat{\lambda}_2}{\sqrt{ \frac{\hat{\lambda}_1}{n_1 \cdot t_1} + \frac{\hat{\lambda}_2}{n_2 \cdot t_2} }}$$

where $\hat{\lambda}_1$ is the mean count for the first sample, $\hat{\lambda}_2$ is the mean count for the second sample, n_1 is the sample size from the first sample, n_2 is the sample size from the second sample, t_1 is the observation length for the first sample, and t_2 is the observation length for the second sample.

Example 5.8

Suppose we want to estimate if there is a difference in the population mean number of defective items from the two factories. Our null and alternative hypothesis would be:

$$H_0 : \lambda_1 = \lambda_2$$

$$H_1 : \lambda_1 \neq \lambda_2$$

The test statistic would be calculated as follows:

$$Z = \frac{\hat{\lambda}_1 - \hat{\lambda}_2}{\sqrt{\dfrac{\hat{\lambda}_1}{n_1 \cdot t_1} + \dfrac{\hat{\lambda}_2}{n_2 \cdot t_2}}} = \frac{3.571 - 4.000}{\sqrt{\dfrac{3.571}{7 \cdot 1} + \dfrac{4.000}{9 \cdot 1}}} \approx -0.439$$

Comparing the test statistic to the sampling distribution that is given in Figure 5.40 indicates that we do not have enough evidence to accept the alternative hypothesis and reject the null hypothesis.

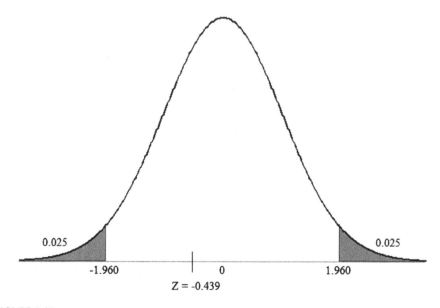

FIGURE 5.40
Sampling distribution for the difference in the population mean number of defective items from two different factories ($\alpha = 0.05$).

5.17 Using Minitab for a Two-Sample Poisson

To use Minitab to find confidence intervals and perform hypothesis tests for two sample count data, select **Stat < Basic Statistics < 2-Sample Poisson Rate**, as is shown in Figure 5.41.

This brings up the dialog box that is illustrated in Figure 5.42.

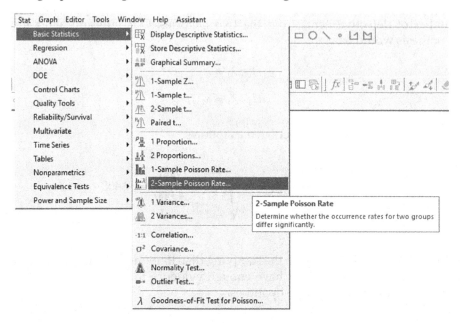

FIGURE 5.41
Minitab commands for two-sample count data.

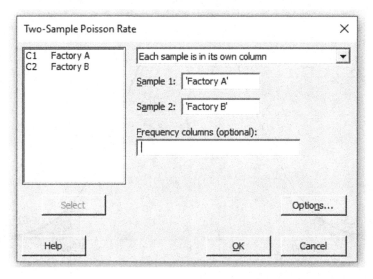

FIGURE 5.42
Minitab dialog box for a 2-Sample Poisson Rate.

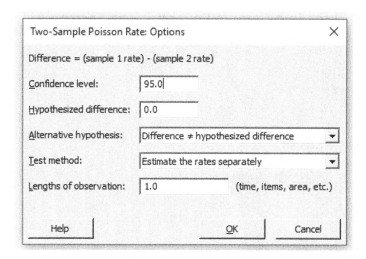

FIGURE 5.43
Minitab dialog box for a 2-Sample Poisson Rate for selecting the confidence level, direction of the alternative hypothesis, and the test method.

By clicking on the **Options** tab, we can enter the desired confidence level, the direction of the alternative hypothesis, and whether we want to estimate the rates separately or use a pooled estimate of the rate, as is shown in Figure 5.43. Estimating the rates separately is more accurate than using a pooled estimate.

This gives the Minitab printout as shown in Figure 5.44.

Just like with making inferences for a one-sample Poisson, the calculations we just described for a two-sample Poisson are based on the normal approximation and this is usually only appropriate when you have larger samples. If you have small samples, you may want to consider using an exact method. And while the calculations for exact methods for a two-sample Poisson are quite a bit more cumbersome to carry out manually, the inferences made remain the same.

5.18 Power Analysis for a Two-Sample Poisson Rate

Suppose we want to estimate the sample size needed to find a mean difference of two defective items at 80% power. The baseline rate (let's assume that we know this to be equal to 3), the comparison rate, and the power level need to be specified. Figure 5.45 gives the Minitab dialog box for such a power analysis.

The Minitab printout in Figure 5.46 gives the results.

The results given in Figure 5.46 suggest that a *sample size of at least 40 from each group* will be needed to detect a comparison rate of two defective items. Since the sample size of 7 taken from Factory A and the sample size of 9 taken from Factory B were not enough to be able to find a comparison rate of two items, our analysis is underpowered.

Test and CI for Two-Sample Poisson Rates: Factory A, Factory B

Method

λ_1: Poisson rate of Factory A
λ_2: Poisson rate of Factory B
Difference: $\lambda_1 - \lambda_2$

Descriptive Statistics

Sample	N	Total Occurrences	Sample Rate
Factory A	7	25	3.57143
Factory B	9	36	4.00000

Estimation for Difference

Estimated Difference	95% CI for Difference
-0.428571	(-2.34358, 1.48643)

Test

Null hypothesis $H_0: \lambda_1 - \lambda_2 = 0$
Alternative hypothesis $H_1: \lambda_1 - \lambda_2 \neq 0$

Method	Z-Value	P-Value
Exact		0.763
Normal approximation	-0.44	0.661

FIGURE 5.44
Minitab printout for a 2-Sample Poisson Rate for the population mean difference in defective items produced at the two factories.

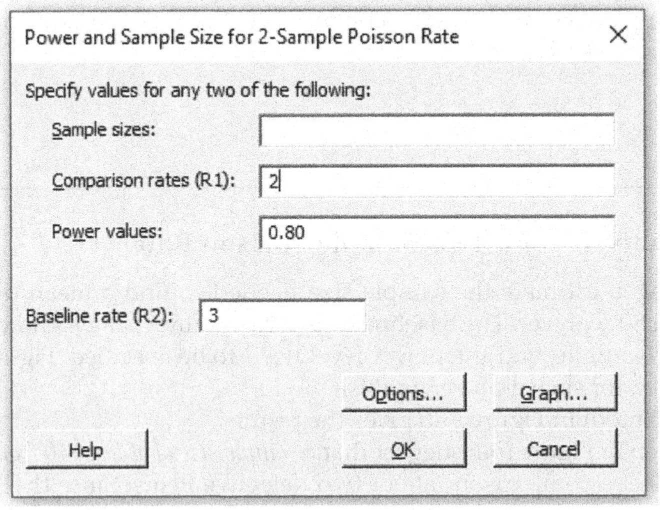

FIGURE 5.45
Minitab dialog box to find the sample size needed to detect a comparison rate of two defective items.

Power and Sample Size

Test for 2-Sample Poisson Rate
Testing comparison rate = baseline rate (versus ≠)
Calculating power for baseline rate = 3
$\alpha = 0.05$
"Lengths" of observation for sample 1, sample 2 = 1, 1

Results

Comparison Rate	Sample Size	Target Power	Actual Power
2	40	0.8	0.807430

The sample size is for each group.

FIGURE 5.46
Minitab output for a power analysis for a 2-Sample Poisson Rate for a comparison rate of two defective items at 80% power.

5.19 Best Practices

So your hypothesis test shows a significant effect, but is there a practical impact of such a difference? The following example gives a discussion about *statistical significance* and *practical significance*.

Example 5.9

Do men get paid more than women? A survey asked a random sample of 20,000 men and 20,000 women to provide their yearly salary and the descriptive statistics are presented in Table 5.8

Figure 5.47 gives the Minitab printout for the following hypothesis test (assuming equal variances and $\alpha = 0.05$):

$$H_0 : \mu_{Males} = \mu_{Females}$$

$$H_1 : \mu_{Males} > \mu_{Females}$$

From Figure 5.47 we can see that the p-value is 0.023, which is less than $\alpha = 0.05$, and so we can accept the alternative hypothesis and reject the null hypothesis. This means that the population mean salaries for men *is significantly higher* than the population mean salaries for women.

TABLE 5.8

Descriptive Statistics of the Yearly Salaries for a
Random Sample of 20,000 Males and 20,000 Females

Males	Females
$n = 20,000$	$n = 20,000$
$\bar{x} = \$50,000$	$\bar{x} = \$49,990$
$s = \$500$	$s = \$500$

Two-Sample T-Test and CI

Method

μ_1: mean of Sample 1
μ_2: mean of Sample 2
Difference: $\mu_1 - \mu_2$

Equal variances are assumed for this analysis.

Descriptive Statistics

Sample	N	Mean	StDev	SE Mean
Sample 1	20000	50000	500	3.5
Sample 2	20000	49990	500	3.5

Estimation for Difference

Difference	Pooled StDev	95% Lower Bound for Difference
10.00	500.00	1.78

Test

Null hypothesis	$H_0: \mu_1 - \mu_2 = 0$
Alternative hypothesis	$H_1: \mu_1 - \mu_2 > 0$

T-Value	DF	P-Value
2.00	39998	0.023

FIGURE 5.47
Minitab printout for testing if the population mean salaries for males is greater than the population mean salaries for females.

Often when we find a significant difference, one way to estimate the magnitude and direction of such a difference is to look at the difference between the sample means as follows:

$$\bar{x}_{Males} - \bar{x}_{Females} = \$50,000 - \$49,990 = \$10$$

The estimate of this very miniscule difference (\$10 per year) illustrates a finding that is *statistically significant* but does not carry much *practical significance* (this is sometimes referred to as *clinical significance*). *Practical significance* occurs when a finding has a meaningful impact on the study of interest. So clearly, a \$10 difference in yearly salaries between men and women does not have much of a practical impact at all. This can happen when the sample size is very large because bigger samples make it easier to detect small differences but such differences may not have a practical impact that is meaningful.

You may also run into a situation when you perform a hypothesis test that has not reached statistical significance, but the difference has a noticeable practical impact. For instance, consider the data provided in Table 5.9.

TABLE 5.9

Descriptive Statistics of the Yearly Salaries for a
Random Sample of Five Men and Five Women

Men	Female
$n = 5$	$n = 5$
$\bar{x} = \$50,000$	$\bar{x} = \$45,000$
$s = \$4,500$	$s = \$4,500$

The p-value for this test is 0.059 and this exceeds the desired significance level of $\alpha = 0.05$. Hence there is no significant difference in the population mean salaries between men and women. However, the estimated difference of \$5,000 per year does have a noticeable practical impact.

$$\bar{x}_{Males} - \bar{x}_{Females} = \$50,000 - \$45,000 = \$5,000$$

This can often occur when the sample size is small or if there is a large amount of sample variability.

Exercises

Unless otherwise specified, use a significance level of $\alpha = 0.05$ and a power level of 80%.

1. An organizational psychologist developed a 2-week training program in order to improve team behavior in an organization. The program consisted of working with teams to promote a more inclusive and satisfying work experience. To assess the program's effectiveness, a random sample of 15 employees was asked to complete a teamwork survey before they began the training program. This survey was scored on a scale of 10–150, where higher scores reflected a greater awareness of teamwork, communication, leadership, and support. After the training was complete, the same teamwork survey was given to the participants. The scores for each of these surveys are presented in Table 5.10.

   Is the training program effective in promoting teamwork? Clearly state the null and alternative hypotheses, calculate any relevant statistics, and clearly label the appropriate sampling distribution. If an effect is found, provide a clear and concise interpretation of the effect.

2. You may have heard that texting while driving can increase the time taken for you to react to avoid an accident. In order to test this, a university set up a driving simulator where 13 participants were asked to drive without texting and then asked to drive while texting. Three random accidents were programmed in the simulator and the time taken for the driver to react to each accident was recorded. The total reaction time (in seconds) that taken to potentially avoid the three different accidents are given in Table 5.11, where a lower score reflects a greater chance of avoiding an accident.

TABLE 5.10

Pre- and Post-Survey Responses on a Teamwork Survey

Survey Responses Before	Survey Responses After	Difference (Before – After)
69	76	–7
72	73	–1
39	58	–19
99	98	1
117	119	–2
115	120	–5
126	115	11
67	120	–53
49	69	–20
65	96	–31
62	87	–25
59	84	–25
86	96	–10
107	105	2
64	67	–3
69	76	–7

TABLE 5.11

Time (in Seconds) to React to a Set of Simulated Accidents Without Texting, While Texting, and the Difference

Without Texting	While Texting	Difference (Without Texting – While Texting)
15	17	–2
8	12	–4
5	7	–2
13	12	1
5	11	–6
11	17	–6
9	16	–7
11	15	–4
9	6	3
4	15	–11
14	10	4
13	16	–3
3	8	–5

Does texting while driving increase the time taken to react to an accident? Clearly state the null and alternative hypotheses, calculate any relevant statistics, and clearly label the appropriate sampling distribution. If an effect is found, provide a clear and concise interpretation of the effect.

3. Tire manufacturers are often interested in comparing the wear patterns of their tires with that of the other competitors. One way of comparing the wear patterns is to look at both the mean tread life and the variability in wear based on the brand

of the tire. Two different brands of tires were sampled. The number of miles until the tires were considered worn-out is given in Table 5.12.

a. Is there a difference in the mean tire life between the two different brands? Explain.

b. Find and interpret a 95% confidence interval for the ratio of the variances for the two different brands of tires.

c. Is there a difference in the variability in the tire life between the two different brands? Explain.

d. What is the minimum difference in the mean wear mileage between the two brands of tires that can be found with the given sample size?

e. How big of a sample would be needed to find at least a 5,000-mile difference between the mean wear patterns for the two brands of tires?

4. Which candidate will win the upcoming presidential election? A survey of 1,300 likely voters found that 652 planned to vote for Candidate A, 600 planned to vote for Candidate B, and the rest were undecided.

a. Find and interpret a 99% confidence interval for the difference in the population proportion of voters who will vote for either Candidate A or Candidate B.

b. Is there a significant difference between the proportion of voters who will vote for either Candidate A or Candidate B?

c. How big of a sample would be needed to find at least a 5% difference?

d. What is the minimum proportion difference that a sample of size 1,300 would be able to find (provided such a difference were to exist)?

TABLE 5.12

Number of Miles Until Worn-out for Two Different Brands of Tires

Brand 1	Brand 2
36,250	44,981
50,225	48,587
40,016	57,859
56,281	66,496
41,437	53,531
53,516	38,210
52,365	42,010
57,697	51,425
58,508	60,218
41,664	63,506
55,348	52,107
50,481	68,119
46,221	41,186
62,109	60,922
56,333	57,267
61,787	53,544
39,930	52,812
56,101	45,732

5. A new program was created to help people smoke less. Table 5.13 gives the average number of cigarettes smoked (per day) before participating in the program and the average number of cigarettes (per day) after participating in the program for a sample of nine program participants.

   Can you conclude whether the program was effective in helping people to smoke less?

6. Table 5.14 gives the yearly starting salaries for male and female bachelor's degree graduates who majored in the social sciences.

TABLE 5.13

Number of Cigarettes Smoked Per Day Before and After Participating in a Program Created to Help People Smoke Less

Number of Cigarettes (Per Day) Before	Number of Cigarettes (Per Day) After	Difference (Before − After)
15	13	2
12	10	2
10	10	0
22	20	2
16	15	1
11	12	−1
8	6	2
9	2	7
6	4	2

TABLE 5.14

Yearly Starting Salaries (in Dollars) for Male and Female Bachelor's Degree Graduates in the Social Sciences

Females	Males
46,500	52,260
53,240	58,390
49,685	61,480
53,555	53,280
58,675	49,350
42,500	48,200
53,550	56,570
54,650	62,150
48,290	33,860
50,500	58,250
55,650	50,150
51,250	42,700
52,650	61,640
53,000	53,640
51,220	51,600
56,690	49,500
47,230	53,850
45,700	55,100

a. Find and interpret a 99% confidence interval for the difference in the population mean salaries between males and females.

b. Do males with bachelor's degrees in the social sciences get paid more than females? Explain.

7. Can deep-breathing techniques help reduce mathematics anxiety? A researcher was interested in knowing if students who practice deep-breathing techniques tend to have less anxiety toward mathematics. An experiment was conducted where 14 students were first given a survey about their level of mathematics anxiety on a scale of 0–100, where higher scores reflected a higher level of mathematics anxiety (Pre Responses). The students were then asked to participate in a series of deep-breathing techniques and then given the exact same survey (Post Responses). These results are given in Table 5.15.

Can you conclude that deep-breathing techniques help to overcome mathematics anxiety?

8. A random sample of 2,750 students who majored in the social sciences found that 908 defaulted on their student loans, and a random sample of 3,560 students who majored in the physical sciences found that 1,245 defaulted on their student loans.

a. Is the loan default rate higher for students who major in the physical sciences as compared to the loan default rate for students who major in the social sciences?

b. How big of a sample would be needed to find a 10% difference in student loan default rates based on the major?

9. Two different presses are used to press shirts for a laundry service. A random sample of the number of shirts that can be pressed in 5 minutes for each type of press is given in Table 5.16.

a. Find and interpret a 99% confidence interval for the difference in the mean number of shirts pressed for the two different presses.

TABLE 5.15

Pre and Post Responses on a Mathematics Anxiety Survey

Observation	Pre Responses	Post Responses	Difference (Pre – Post)
1	61	52	9
2	83	71	12
3	73	66	7
4	80	60	20
5	65	51	14
6	87	61	26
7	59	60	−1
8	82	67	15
9	76	51	25
10	87	66	21
11	76	53	23
12	90	57	33
13	86	61	25
14	72	78	−6
Sample mean	76.93	61.00	15.93
Sample standard deviation	9.95	8.03	10.97

TABLE 5.16

Number of Shirts Pressed in 5 minutes for
Two Different Brands of Presses

Press A	Press B
1	3
2	5
3	2
4	3
1	4
5	2
2	5
2	1
1	3
3	5
5	1
4	2
1	5
5	2
2	–

    b. Is there a significant difference in the mean number of shirts pressed for the two different presses?

    c. How big of a sample would need to be collected in order to find a mean difference of at least two shirts?

10. A new small engine claims that it can get more miles per gallon (MPG) than the competitor's model. To verify this claim, a random sample of seven engines of each type are put through a series of tests and the fuel efficiency (in MPG) is given in Table 5.17.

    a. Does the new engine have better fuel efficiency?

    b. What is the estimated difference in the MPG between the two engines?

    c. Is the variance in the MPG of the new model less than the variance in the MPG of the competitors?

    d. How big of a sample would be needed to detect a variance ratio of at least 1.10?

TABLE 5.17

Fuel Efficiency (MPG) for a Random Sample of
Seven Engines from Each Brand

New Engine	Competitors
38	32
43	39
40	35
39	39
38	39
42	33
41	38

11. A heating and air-conditioning company wants to estimate the number of emergency service calls they receive during the week and on weekends so as to decide how many technicians to have on call. A random sample of the number of heating and/or air-conditioning calls taken during a 15-week period is presented in Table 5.18.

    a. Are there more emergency service calls on the weekends? Explain.

    b. Should the company staff have more on-call technicians during the week or on the weekend?

12. A survey was given to a random sample of 1,650 dog owners and a random sample of 1,300 cat owners. The survey asked if they used flea and tick prevention on their pets. Of the 1,650 dog owners, 1,485 responded that they did. Of the 1,300 cat owners, 1,079 responded that they did. Does a smaller percentage of cat owners use flea and tick prevention on their pets as compared to dog owners?

13. A new computer operating system was designed so that it does not need as many updates as the current operating system. A random sample of the number of updates for each operating system over the course of 12 months is given in Table 5.19.

    a. Does the current operating system require more updates than the new operating system?

    b. How big of a sample would be needed to detect a mean difference of at least three updates over the course of a 12-month period?

14. A landscaper wants to compare the variability in germination rates between two different brands of grass seed. In order to do this, a random sample of 15 plots of land was planted with each type of grass seed. The number of days it took for the two different brands of seeds to germinate on each of these plots is presented in Table 5.20.

    a. Does Brand 1 have more variability in the days it takes to germinate as compared to Brand 2?

    b. What is the minimum variance ratio that a sample size of 14 can find?

15. We can also test whether two populations' means differ by some constant, C. The procedure is similar to testing whether two population means are different from each other; however, the possible null and alternative hypotheses are

TABLE 5.18

Number of Emergency Service Calls During the Week and on the Weekend

Week	2	1	0	3	5	4	3	2	1	0	6	5	1	2	3
Weekend	2	6	5	4	8	4	5	6	3	5	2	4	5	3	6

TABLE 5.19

Random Sample of the Number of Updates of the New and Current Operating Systems Over the Course of a 12-Month Period

New operating system	1	0	0	2	1	0	1	0	1	3	1	0
Current operating system	1	1	0	2	1	0	1	2	3	3	2	2

TABLE 5.20

Germination Rates (in Days) for Two Different Brands of Grass Seed

Brand 1	7	8	13	12	9	6	14	9	7	10	13	9	7	7
Brand 2	7	8	10	7	6	8	9	10	8	7	9	10	11	10

framed somewhat differently in that we need to include the difference being tested for:

$$H_0 : \mu_1 - \mu_2 = C$$
$$H_1 : \mu_1 - \mu_2 \neq C$$

$$H_0 : \mu_1 - \mu_2 = C$$
$$H_1 : \mu_1 - \mu_2 > C$$

$$H_0 : \mu_1 - \mu_2 = C$$
$$H_1 : \mu_1 - \mu_2 < C$$

Assuming equal variances, the test statistic for comparing two means follows a t-distribution with $n_1 + n_2 - 2$ degrees of freedom:

$$T = \frac{(\bar{x}_1 - \bar{x}_2) - C}{s_p\sqrt{\frac{1}{n_1} + \frac{1}{n_2}}}$$

where $\bar{x}_1$ is the sample mean from the first sample, $\bar{x}_2$ is the sample mean from the second sample, n_1 is the sample size from the first sample, n_2 is the sample size from the second sample, and $s_p = \sqrt{\frac{(n_1-1)s_1^2 + (n_2-1)}{n_1 + n_2 - 2}}$ is the pooled standard deviation, which is used when we assume equal variances, where s_1^2 is the variance from the first sample, s_2^2 is the variance from the second sample, and C is the hypothesized difference.

If we assume unequal variances, then the test statistic would be:

$$T = \frac{(\bar{x}_1 - \bar{x}_2) - C}{s}$$

where

$$s = \sqrt{\frac{s_1^2}{n_1} + \frac{s_2^2}{n_2}}$$

The number of degrees of freedom can be found by rounding down to an integer:

$$df = \frac{\left(\frac{s_1^2}{n_1} + \frac{s_2^2}{n_2}\right)^2}{\frac{(s_1^2/n_1)^2}{n_1 - 1} + \frac{(s_2^2/n_2)^2}{n_2 - 1}}$$

TABLE 5.21

Wait Times (in Minutes) for Two Car Rental Agencies

Agency A	Agency B
25	19
18	20
20	21
12	18
19	19
12	17
19	20
22	21
21	22
20	17
21	15
25	8
28	12
24	19
28	20
24	22
35	23
15	25
8	27
16	
20	

The data set in Table 5.21 presents a sample of the wait times (in minutes) for renting a car from two competing car rental agencies at a local airport.

Suppose you want to determine if Car Agency B is at least 5 minutes faster in renting their cars than Car Agency A.

a. Estimate whether assuming equal variances would be appropriate in this case.

b. Set up the appropriate null and alternative hypotheses.

c. Find the value of the test statistic.

d. Interpret your findings. Justify if you can claim that Car Agency B is at least 5 minutes faster in renting a car than Car Agency A.

e. How large of a sample would be needed to find whether Car Agency B is at least 5 minutes faster in renting their cars than Car Agency A provided that such a difference exists?

6

Simple Linear Regression

6.1 Introduction

In the last two chapters, we described some basic techniques for making inferences about different population parameters such as the mean, proportion, variance, and counts using sample statistics. However, there may be occasions when we are interested in estimating whether two or more variables are related to each other. For instance, suppose we want to determine if there is a relationship between a student's high school ability in mathematics and how well he or she does during his or her first year of college. One way to quantify such a relationship could be to look at whether the score received by a student on the mathematics portion of the SAT examination (which students take when they are in high school) is related to his or her first-year college grade point average (GPA).

Consider the data set provided in Table 6.1 consisting of the scores on the mathematics portion of the SAT examination and the corresponding first-year GPAs for a random sample of ten university students.

The range of the possible scores on the SAT mathematics examination is from 200 to 800 points, and the range of the possible first-year GPAs is from 0.00 to 4.00 points.

We want to estimate if the score received on the SAT mathematics examination (SATM) is related to the first-year GPA. We would also want to further quantify this relationship by estimating how strong this relationship is, and we may want to develop some type of a model that can be used to describe the relationship between GPA and SATM that can also be used to predict GPA using SATM.

TABLE 6.1

SAT Mathematics Examination Scores (SATM) and Corresponding First-Year GPAs for a Random Sample of Ten College Students

Observation	SATM	GPA
1	750	3.67
2	460	1.28
3	580	2.65
4	600	3.25
5	500	3.14
6	430	2.82
7	590	2.75
8	480	2.00
9	380	1.87
10	620	3.46

We will be using a technique called *simple linear regression analysis* to study the relationship between two variables. The variable that we are interested in modeling or predicting (in this case, first-year GPA) is called the *dependent* or *response variable*. The variable that is used to predict the dependent or response variable (which in this case is the SATM) is called the *independent* or *predictor variable*. Simple linear regression analysis is a statistical technique that can be used to develop a model that estimates the linear relationship between a single continuous response variable (or *y*-variable) and a single continuous predictor variable (or *x*-variable).

Before we begin a detailed discussion of simple linear regression, it is important to understand that it is very difficult to infer a cause-and-effect relationship between two variables. In most situations, we can only infer an association or a relationship between two variables. It is extremely difficult to claim that any changes in *y* are *caused* by the variable *x* because there could be numerous factors influencing *y* other than the variable *x* alone.

For instance, in developing a model to predict first-year GPAs, although the score received on the mathematics portion of the SAT examination may have an influence on the first-year GPA, numerous other factors, such as motivation, study habits, number of credits attempted, and so on, could also impact a student's first-year GPA. Thus, it does not make sense to infer that only a single variable, such as the score received on the SAT mathematics examination, is the *only* variable that has an influence on first-year GPAs. And because it may not be possible to isolate every possible factor that influences the given response variable, simple linear regression is typically only used to describe an association or relationship between a response variable and a predictor variable.

6.2 The Simple Linear Regression Model

Recall in Chapter 2 that we described a scatterplot as a way to visualize the relationship between two variables. We can create a scatterplot to visualize the relationship between SATM and GPA, and we can also use a marginal plot to graph the scatterplot along with the histograms or boxplots for each of the two variables simultaneously. To create a scatterplot, we simply plot the ordered pairs (*x*, *y*) on the Cartesian plane, as illustrated in Figure 6.1.

By visually examining the trend in the scatterplot in Figure 6.1, we can decide whether we believe there is a relationship between SATM and GPA. We can also try to describe the pattern of this relationship. The scatterplot in Figure 6.1 appears to show that a positive trend between SATM and GPA, where an increase in SATM seems to suggest an increase in GPA (in other words, as *x* increases, *y* increases). We can also describe the relationship between SATM and GPA as one that could be reasonably modeled with a *straight line*. Using a straight line to approximate the relationship between two variables provides a simple way to model such a relationship because by using a linear model, the calculations are relatively simple and the interpretations are straightforward.

In order to use a straight line to model the relationship between two variables, we will begin by choosing an arbitrary line to model the relationship between our two variables and describe the assessment required to check how well this line fits the data. Consider Figure 6.2, which illustrates a line drawn on the scatterplot that connects the two most extreme data points— (380, 1.87) and (750, 3.67).

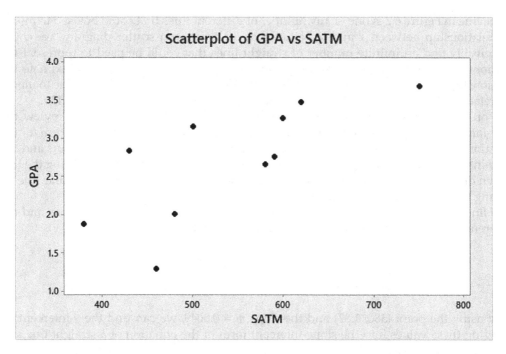

FIGURE 6.1
Scatterplot of first-year GPA versus the score received on the mathematics portion of the SAT examination (SATM) for the data in Table 6.1.

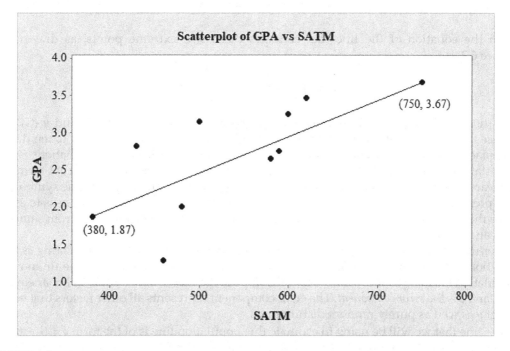

FIGURE 6.2
Scatterplot of GPA versus SATM that includes the line that connects the two most extreme points, namely, (380, 1.87) and (750, 3.67).

The line in Figure 6.2 is one of any number of different lines that could be used to model the relationship between x and y. Notice that for any given scatter diagram, we could conceivably find an infinite number of straight lines that could be used to represent the relationship between the two variables. Once we have found a line to approximate the relationship between our two variables, we can then use the equation of this line to model the relationship between SATM and GPA and also to predict GPA based on SATM.

In order to determine the equation of the line in Figure 6.2, we will first review the *slope-intercept form* of a linear equation. You may recall that the slope-intercept form for the equation of a straight line is of the form $y = mx + b$, where m is the slope of the line and b is the y-intercept (the y-intercept is the point where the line crosses the y-axis). Using the line drawn on the scatterplot in Figure 6.2, we can find the equation of this particular line by finding the slope and the y-intercept.

To find the slope, we simply find the quotient of the difference in the y-values and the difference in the x-values for the two selected points as follows:

$$m = \frac{y_2 - y_1}{x_2 - x_1} = \frac{3.67 - 1.87}{750 - 380} = \frac{1.80}{370} \approx 0.0048648649$$

Now using the point (380, 1.87) and the slope $m \approx 0.0049$, we can find the y-intercept by plugging these values into the slope-intercept form of the equation of a straight line and solving for b as follows:

$$1.87 = (0.0048648649) \cdot (380) + b$$

$$b = 0.0213513514$$

Then the equation of the line that connects these two extreme points, as drawn in Figure 6.2, is:

$$y = 0.0048648649x + 0.0213513514$$

This line can be used to approximate the relationship between x (SATM) and y (GPA). Notice that the calculations for the slope and the y-intercept were carried out to ten decimal places and this was done in order to avoid round-off error in future calculations.

In statistics, we use a slightly different notation to represent the slope-intercept form of a linear equation. We will use the symbol β_0 to represent the y-intercept and the symbol β_1 to represent the slope. The primary objective in simple linear regression analysis is to estimate the unknown population parameters β_0 and β_1 using the line obtained from sample data and determine how well this line fits the data.

The unknown population linear equation that we are interested in estimating is an equation of the form $y = \beta_0 + \beta_1 x + \varepsilon$, where β_0 represents the y-intercept of the unknown population linear equation, β_1 represents the slope of the unknown population linear equation, and ε is the *error component*. The error component represents all other factors that may affect y as well as purely random disturbances.

The line that we will be using to estimate this population line is of the form $\hat{y} = \hat{\beta}_0 + \hat{\beta}_1 x$, where the "hat" (^) symbol is used to denote the estimates of the slope and y-intercept of the unknown population linear equation. Thus, our estimate of the true population regression line, which connects the two most extreme points, is as follows:

$$\hat{y} = \hat{\beta}_0 + \hat{\beta}_1 \cdot x$$

$$\hat{y} \approx 0.0213513514 + 0.0048648649 \cdot x$$

One way to assess the usefulness of any line that describes the relationship between the predictor (or x) variable and the response (or y) variable is to look at the *vertical distances* between the y-values for each of the observations and the estimated line. We can measure the vertical distance between each of our observed (or sample) y-values and the line we are considering by simply taking the difference between the observed y-value for each observation in the sample and the estimated y-value for each given data point that is found using the estimated (i.e., sample) equation.

For every given data point in the sample (x_i, y_i), the *observed value* is y_i and the *estimated* (or *fitted*) *value* of y is denoted as $\hat{y}_i$. The estimated value $\hat{y}_i$ can be found by substituting the value of x_i into the equation of the estimated line that we are considering. The difference between the observed value y_i and the estimated value $\hat{y}_i$ is called the *i*th *residual*. The *i*th residual is found by taking the difference between the observed and estimated values for the *i*th observation as follows:

$$\hat{\varepsilon}_i = \left(y_i - \hat{y}_i \right) = y_i - \left(\hat{\beta}_0 + \hat{\beta}_1 \cdot x_i \right)$$

Table 6.2 presents the estimated values for the line drawn in Figure 6.2 along with the difference between the observed and estimated values for each of the observations in the sample. Notice that in doing the calculations in Table 6.2, the decimal values were carried out to ten decimal places in order to avoid round-off error.

For example, the residual for the tenth observation is the vertical distance between the observed y-value and the estimated y-value, which can be calculated as follows:

$$y_{10} - \hat{y}_{10}$$

TABLE 6.2

Estimated Values $\left(\hat{y}_i \right)$ and Residuals $\left(\hat{\varepsilon}_i = y_i - \hat{y}_i \right)$ for the Line Connecting the Points (380, 1.87) and (750, 3.67), as Illustrated in Figure 6.2.

Observation Number	Observed Response Value (y_i)	Predictor Value (x_i)	Estimated Value $\left(\hat{y}_i \right)$	Residual $\left(\hat{\varepsilon}_i = y_i - \hat{y}_i \right)$	$\hat{\varepsilon}_i^2$
1	3.67	750	3.6700000264	−0.0000000264	0.0000000000
2	1.28	460	2.2591892054	−0.9791892054	0.9588115000
3	2.65	580	2.8429729934	−0.1929729934	0.0372385762
4	3.25	600	2.9402702914	0.3097297086	0.0959324924
5	3.14	500	2.4537838014	0.6862161986	0.4708926712
6	2.82	430	2.1132432584	0.7067567416	0.4995050918
7	2.75	590	2.8916216424	−0.1416216424	0.0200566896
8	2.00	480	2.3564865034	−0.3564865034	0.1270826271
9	1.87	380	1.8700000134	−0.0000000134	0.0000000000
10	3.46	620	3.0375675894	0.4224324106	0.1784491415
			Total	$\sum_{i=1}^{10} \hat{\varepsilon}_i \approx 0.4548646750$	$\sum_{i=1}^{10} \varepsilon_i^2 \approx 2.3879687898$

We know that the observed y-value is $y_{10} = 3.46$. To find the estimated y-value $\hat{y}_{10}$, we need to substitute the value $x_{10} = 620$ into the estimated equation we are considering, as follows:

$$\hat{y}_{10} = 0.0213513514 + 0.0048648649(620) \approx 3.0375675894$$

$$y_{10} - \hat{y}_{10} = 3.46 - 3.037567589 \approx +0.4224324106$$

The residual term for the tenth observation, denoted as $\hat{\varepsilon}_{10}$, can also be illustrated as representing the vertical distance between the observed y-value of the data point $(x_{10} = 620, y_{10} = 3.46)$ and the estimated value $(\hat{y}_{10})$ as presented in Figure 6.3. The sign of this residual is positive because the observed y-value lies above the given line.

Now suppose that we find the residuals for each of the ten observations in the sample as given in Column 5 of Table 6.2 and then find the sum of all such residual terms as follows:

$$\sum_{i=1}^{10} \hat{\varepsilon}_i \approx 0.4548646750$$

And if we square each residual as in Column 6 of Table 6.2 and calculate the sum, the sum of the squares of all of the residual terms would be:

$$\sum_{i=1}^{10} \hat{\varepsilon}_i^2 \approx 2.3879687898$$

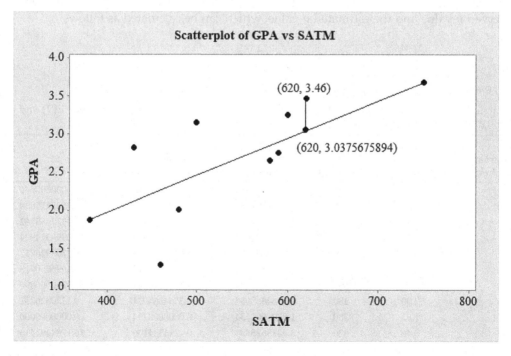

FIGURE 6.3
Vertical distance between the observed value of (620, 3.46) and the line that connects the two most extreme points.

These two summations tell us that the sum of the residuals is approximately 0.45, and the sum of the residuals squared is approximately 2.39. These summations can be used as a way to assess how well the line that we have chosen fits the underlying set of data.

But how do we know that the line we have chosen is the line that *best* represents the linear relationship between x and y? Clearly, there are many other lines that could be used to represent the relationship between SATM and GPA, and some of these lines may have a much better fit of the data than the line we have chosen (which is the line that connects the two most extreme points). There is one line in particular that is very interesting and it is called the *line of best fit*. This is the line that *best fits* a given set of data.

The line of best fit is *the unique* line such that the sum of the residuals is equal to 0 and the sum of the squares of the residuals is as small as possible. Essentially, this unique line is the line that *best fits* the data because for the sum of the residuals to equal 0, this implies that the vertical distance between the observations and the line of best fit is approximately the same for those observed values that lie above the line of best fit as for those observed values that lie below the line of best fit. Furthermore, the sum of the squares of the vertical distances between the observed values and the line of best fit (this is also called the sum of the squares of the errors and is denoted by SSE) is a minimum. This value being a minimum suggests that the observed values lie closer to the line of best fit than they do to any other line. In other words, for a line to be the line of best fit, the following two conditions need to be met:

1. The sum of the residuals should be 0. In symbols, $\sum_{i=1}^{n} \hat{\varepsilon}_i = 0$.

2. The sum of the squares of the residuals, in symbols, $\sum_{i=1}^{n} \hat{\varepsilon}_i^2$, needs to be as small as possiable.

The line of best fit is sometimes referred to as the *least squares line* or the *regression line*.

By minimizing the sum of the squares of the residuals from our sample data, we can calculate the estimates of the y-intercept $\left(\hat{\beta}_0\right)$ and slope of the line of best fit $\left(\hat{\beta}_1\right)$. These estimates can then be used to approximate the unknown population parameters β_0 and β_1. By calculating these values, we can estimate the unknown population linear equation by using the line of best fit as follows:

$$\hat{y} = \hat{\beta}_0 + \hat{\beta}_1 x$$

Notice that since we are using estimated values of the population parameters, we often use the "hat" ($\wedge$) symbol to denote an estimated parameter. The equation $\hat{y}$ gives the estimated (or predicted) values of the unknown population response variable and the values $\hat{\beta}_0$ and $\hat{\beta}_1$ are estimated values of the unknown population regression parameters β_0 and β_1.

Thus, by assuming that y is linearly related to x, we can use the equation of the line of best fit to estimate the *unknown population linear equation*:

$$y = \beta_0 + \beta_1 x + \varepsilon$$

where y denotes the population response (or dependent) variable, x corresponds to the population predictor (or independent) variable, ε represents all unknown variables as well as random disturbances that may impact y, and β_0 and β_1 are the unknown population y-intercept and slope parameters, respectively.

Similar to the discussions related to basic statistical inferences about population means or population proportions in Chapters 4 and 5, we can also make inferences about the unknown *population linear regression equation*. In other words, we can use the estimated line of best fit, which is based on sample data, to make inferences about the unknown population linear equation.

6.3 Model Assumptions for Simple Linear Regression

Before we can use the line of best fit to make inferences about the true population linear equation, we first need to consider several different model assumptions. These model assumptions are very important because they need to hold true in order to draw meaningful inferences about the unknown population linear equation using the estimated equation. If these assumptions do not hold reasonably true, then any inferences we make using the estimated regression equation about the unknown population equation may be suspect.

The model assumptions for a simple linear regression analysis rely on the unknown population error component, ε, being independent, normally distributed, and with constant variance, σ^2. Also, the specification of the regression model stipulates a linear relationship between x and y.

These assumptions are described in more detail below:

1. Any two observed values are independent of each other.
   This essentially means that for any two observations collected, one observation does not depend on the other.
   This assumption could be violated if we take repeated observations on the same individual over time. For instance, suppose we are interested in determining whether a relationship exists between the score received on the SAT examination and yearly GPAs. If the data were collected such that we are looking at the yearly GPAs for the same individuals over the course of 4 years, then these observations will not be independent of each other.

2. The error component is normally distributed.
   This assumption stipulates that the error component is essentially random and this allows us to make inferences about the population linear equation by creating confidence intervals and performing hypothesis tests using the estimated line of best fit.

3. The error component has constant variance.
   This assumption stipulates that the variance of the error component remains constant for all values of the independent variable x. This allows us to use simplified calculations to determine the estimated values of the population regression parameters (these will be described in the next section).

4. The functional form of the relationship between x and y can be established.
   If we assume that the population model, $y = \beta_0 + \beta_1 x + \varepsilon$, is linear with respect to the relationship between x and y, but if the unknown population model equation is not linear and we are using a linear equation to model the relationship between

x and y, then this assumption would be violated. This could be the case if x were quadratic or any other nonlinear power. For all of the models used throughout this text, we will always assume that the regression equations are linear with respect to the beta parameters. For instance, equations such as $y = \beta_0 + \beta_1 x^3 + \varepsilon$ can be considered and modeled appropriately but equations such as $y = \beta_0 + \beta_1^3 x + \varepsilon$ cannot.

6.4 Finding the Equation of the Line of Best Fit

Assuming that all of the model assumptions described in the last section do not appear to have been violated, we can now obtain the estimates of the population regression parameters β_0 and β_1 using the following formulas:

$$\hat{\beta}_1 = \frac{S_{xy}}{S_{xx}}$$

$$\hat{\beta}_0 = \bar{y} - \hat{\beta}_1 \cdot \bar{x}$$

These are the values of the estimated parameters that represent the slope and the intercept for the estimated line of best fit, where the estimated line of best fit is a unique line such that the sum of the residuals is 0 and the sum of the squared residuals is a minimum.

The following statistics are needed in order to calculate the estimated parameters for the line of best fit:

$$\bar{x} = \frac{\sum_{i=1}^{n} x_i}{n}, \qquad \bar{y} = \frac{\sum_{i=1}^{n} y_i}{n},$$

$$S_{xx} = \sum_{i=1}^{n} (x_i - \bar{x})^2,$$

$$S_{yy} = \sum_{i=1}^{n} (y_i - \bar{y})^2,$$

$$S_{xy} = \sum_{i=1}^{n} (x_i - \bar{x})(y_i - \bar{y})$$

Where the statistics $\bar{x}$ and $\bar{y}$ are the sample means for the x- and y-variables, respectively, from the sample of ordered pairs of size n; the values S_{xx} and S_{yy} are the sums of the squared differences between the individual observations from the means for x and y, respectively; and S_{xy} is the sum of the cross-products of the difference between each observation and the mean for both x and y. Even though S_{yy} is not used in the calculation for the line of best fit, it will be used in future calculations.

Example 6.1

Using the data given in Table 6.1, these statistics are calculated as follows (see Exercise 1):

$$\bar{x} = 539$$

$$\bar{y} = 2.689$$

$$S_{xx} = 107490$$

$$S_{yy} = 5.228$$

$$S_{xy} = 555.490$$

With these sample statistics, we can now obtain the estimates $\hat{\beta}_0$ and $\hat{\beta}_1$ for the unknown population parameters β_0 and β_1, as follows:

$$\hat{\beta}_1 = \frac{S_{xy}}{S_{xx}} = \frac{555.490}{107490} \approx 0.00517$$

$$\hat{\beta}_0 = \bar{y} - \hat{\beta}_1 \cdot \bar{x} = 2.689 - (0.00517) \cdot 539 \approx -0.0965$$

Once we find these estimates, the line of best fit is:

$$\hat{y} = \hat{\beta}_0 + \hat{\beta}_1 x = -0.0965 + 0.00517x$$

This is the line that best fits our data because it is the unique line such that the sum of the residuals is 0 and the sum of the squared residuals is a minimum. The residuals for each of the observations can be calculated as follows:

$$\hat{\varepsilon}_i = y_i - \hat{y}_i = y_i - \left(\hat{\beta}_0 + \hat{\beta}_1 \cdot x_i \right)$$

Table 6.3 gives the values of the observed and estimated values and the residuals (notice again that we did not round off in order to avoid errors in future calculations).

The values for the residuals in Column 5 of Table 6.3 are both positive and negative, and the sum of the residuals is approximately 0 (due to a slight rounding error):

$$\sum_{i=1}^{10} \hat{\varepsilon}_i \approx 0.000000000$$

Furthermore, the sum of the squares of the residuals (found by summing the values in Column 6 of Table 6.3) will be the minimum:

$$\text{SSE} = \sum_{i=1}^{10} \hat{\varepsilon}_i^2 \approx 2.3574123546$$

In comparing the sum of the residuals and the sum of the residuals squared for the line of best fit to those for the line that connected the two most extreme

TABLE 6.3

Estimated Values $(\hat{y}_i)$ and Residuals $(\hat{\varepsilon}_i = y_i - \hat{y}_i)$ for the Line of Best Fit

Observation Number	Observed Response Value (y_i)	Predictor Value (x_i)	Estimated Value $(\hat{y}_i)$	Residual $(\hat{\varepsilon}_i = y_i - \hat{y}_i)$	$\hat{\varepsilon}_i^2$
1	3.67	750	3.7794120383	−0.1094120383	0.0119709941
2	1.28	460	2.2807414643	−1.0007414643	1.0014834784
3	2.65	580	2.9008810122	−0.2508810122	0.0629412823
4	3.25	600	3.0042376035	0.2457623965	0.0603991555
5	3.14	500	2.4874546469	0.6525453531	0.4258154378
6	2.82	430	2.1257065774	0.6942934226	0.4820433567
7	2.75	590	2.9525593078	−0.2025593078	0.0410302732
8	2.00	480	2.3840980556	−0.3840980556	0.1475313163
9	1.87	380	1.8673150991	0.0026849009	0.0000072087
10	3.46	620	3.1075941948	0.3524058052	0.1241898515
			Total	$\sum_{i=1}^{10} \hat{\varepsilon}_i \approx 0.000000000$	$\sum_{i=1}^{10} \hat{\varepsilon}_i^2 \approx 2.3574123546$

points on the scatterplot (see Table 6.2), the line of best fit has the sum of the residuals approximately equal to 0, and the sum of squares of the residuals is a minimum, and notice that it is smaller than what was obtained using the line that connected the two most extreme data values.

In fact, if you compare the sum of the residuals and the sum of the squares of the residuals to any other line, you will find that the sum of the residuals will not equal 0 or the sum of the squares of the residuals will be greater than for the line of best fit.

The sum of the squares of the residuals, SSE, is also an interesting statistic because it can be used to estimate the variance of the unknown error component (σ^2):

$$\text{SSE} = \sum_{i=1}^{n} \hat{\varepsilon}_i^2$$

To obtain an estimate of σ^2, we first find the value of SSE and then divide it by $n - 2$ (we subtract 2 from the sample size because 2 degrees of freedom are used to estimate two parameters, β_0 and β_1).

Thus, an estimate of the variance for the unknown error component σ^2 can be found as follows:

$$S^2 = \frac{\text{SSE}}{n-2}$$

where S^2 is called the *mean square error.*

The *root mean square error* is the square root of the mean square error as follows:

$$S = \sqrt{S^2} = \sqrt{\frac{\text{SSE}}{n-2}}$$

Example 6.2

To find S^2 for the data presented in Table 6.3, since SSE ≈ 2.357412, then:

$$S^2 = \frac{\text{SSE}}{n-2} \approx \frac{2.357412}{10-2} \approx 0.294677$$

We can also find the root mean square error by taking the square root of the mean square error as follows:

$$S = \sqrt{S^2} = \sqrt{\frac{\text{SSE}}{n-2}} \approx \sqrt{0.294677} \approx 0.54284$$

6.5 Using Minitab for Simple Linear Regression

We can use Minitab to draw a scatterplot with the line of best fit and to perform all of the calculations that we have done thus far. In order to use Minitab for a simple linear regression analysis, the data have to be entered in two columns, one column for the x-variable and one column for the y-variable.

To draw the scatterplot with the line of best fit, select **Regression** from the **Stat** menu and then select **Fitted Line Plot** to get the dialog box that is presented in Figure 6.4.

Providing the appropriate response and predictor variables along with the type of relationship (which we are assuming is a linear relationship) and then hitting **OK** gives the fitted line plot, which includes the scatterplot along with the equation of the line of best fit, as illustrated in Figure 6.5.

To run a regression analysis to obtain the estimated regression parameters, click on **Stat** on the top menu bar, and select **Regression > Fit Regression Model** to open the regression dialog box that is presented in Figure 6.6.

By selecting GPA as the response (y) variable and SATM as one of the continuous predictor (or x) variables, selecting **OK** will give the Minitab printout presented in Figure 6.7.

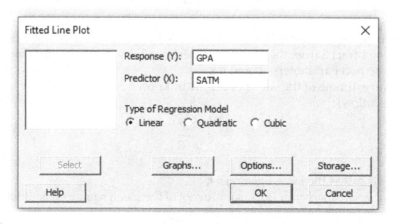

FIGURE 6.4
Dialog box for creating a fitted line plot.

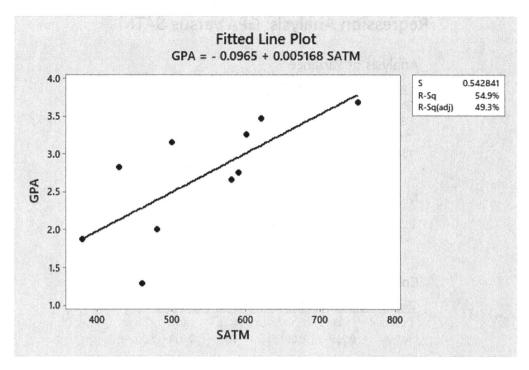

FIGURE 6.5
Fitted line plot describing the relationship between first-year GPA and the score received on the mathematics portion of the SAT examination (SAT).

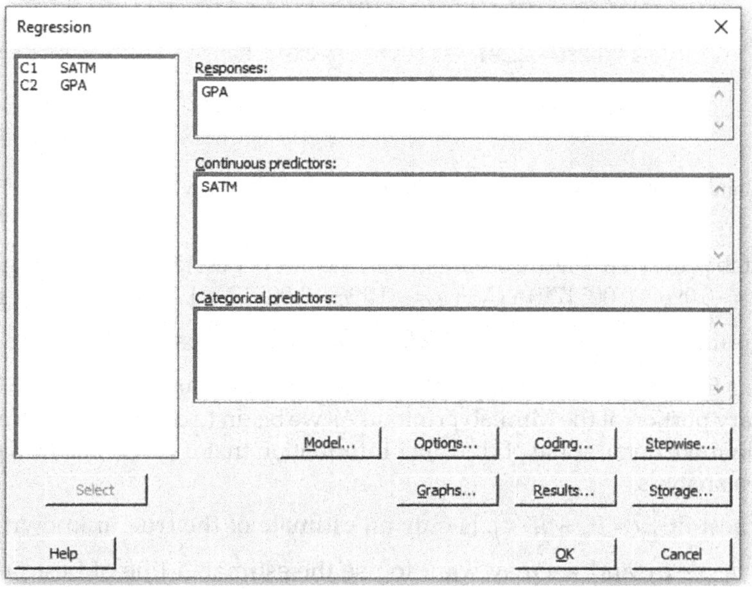

FIGURE 6.6
Regression dialog box.

Regression Analysis: GPA versus SATM

Analysis of Variance

Source	DF	Adj SS	Adj MS	F-Value	P-Value
Regression	1	2.871	2.8707	9.74	0.014
SATM	1	2.871	2.8707	9.74	0.014
Error	8	2.357	0.2947		
Total	9	5.228			

Model Summary

S	R-sq	R-sq(adj)	R-sq(pred)
0.542841	54.91%	49.27%	35.60%

Coefficients

Term	Coef	SE Coef	T-Value	P-Value	VIF
Constant	-0.096	0.909	-0.11	0.918	
SATM	0.00517	0.00166	3.12	0.014	1.00

Regression Equation

GPA =   -0.096 + 0.00517 SATM

Fits and Diagnostics for Unusual Observations

Obs	GPA	Fit	Resid	Std Resid	
2	1.280	2.281	-1.001	-2.01	R

R *Large residual*

FIGURE 6.7
Portion of the Minitab printout of a simple regression analysis for the data in Table 6.1.

As the Minitab printout in Figure 6.7 illustrates, the line of best fit (the regression equation) is given as GPA = −0.096 + 0.00517*SATM, $\left(\hat{y} = -0.096 + 0.00517 \cdot x\right)$. Figure 6.7 also provides the estimates of the intercept $\left(\hat{\beta}_0 \approx -0.096\right)$ and slope $\left(\hat{\beta}_1 \approx 0.00517\right)$ parameters in the Coefficients portion of the Minitab printout along with the root mean square error, $S = 0.542841$ in the Model Summary portion of the Minitab printout. As we begin to conduct further analyses, we will describe in more detail some of the other information that is given in a Minitab printout for a regression analysis.

The line of best fit, $\left(\hat{y} = \hat{\beta}_0 + \hat{\beta}_1 \cdot x\right)$, is only an estimate of the true unknown population line, $\left(y = \beta_0 + \beta_1 \cdot x + \varepsilon\right)$, and we may want to use the estimated line of best fit along with the residuals that represent an estimate of the error component to find confidence intervals and perform hypothesis tests about the unknown population parameters, β_0 and β_1.

6.6 Standard Errors for Estimated Regression Parameters

Similar to calculating confidence intervals and performing hypothesis tests about unknown population means and proportions, we can also calculate confidence intervals and perform hypothesis tests about the unknown population regression parameters β_0 and β_1. Just as using the sample mean to estimate the population mean, we will use the estimated (or sample) values $\hat{\beta}_0$ and $\hat{\beta}_1$ to estimate the unknown population parameters β_0 and β_1, and we will also use the residuals $\hat{\varepsilon}$ to estimate the unknown population error component ε.

In order to calculate confidence intervals and perform hypothesis tests on the population parameters, we first need to obtain estimates of the standard error for each of the estimated regression parameters. Such standard errors for $\hat{\beta}_0$ and $\hat{\beta}_1$ can be calculated as follows:

$$\text{SE}\left(\hat{\beta}_0\right) = S \cdot \sqrt{\frac{1}{n} + \frac{\bar{x}^2}{S_{xx}}}$$

$$\text{SE}\left(\hat{\beta}_1\right) = \frac{S}{\sqrt{S_{xx}}}$$

Where $S = \sqrt{\dfrac{\text{SSE}}{n-2}}$, $S_{xx} = \sum (x_i - \bar{x})^2$, $\bar{x}$ is the sample mean, and n is the sample size.

Example 6.3

We can use the above equations to calculate the standard error for each of the estimated regression parameters for the data from Table 6.3 as follows:

$$\text{SE}\left(\hat{\beta}_0\right) = (0.54284) \cdot \sqrt{\frac{1}{10} + \frac{(539)^2}{107490}} \approx 0.90879$$

$$\text{SE}\left(\hat{\beta}_1\right) = \frac{0.54284}{\sqrt{107490}} \approx 0.00166$$

Notice that the standard error for each of the estimated regression parameters appears in the Coefficients portion of the Minitab printout in Figure 6.7 as the standard error for the constant term (which is the intercept) and for the slope (SATM).

6.7 Inferences about the Population Regression Parameters

If the model assumptions for a regression analysis hold true, then we can make inferences about the slope of the unknown population line by calculating confidence intervals and testing whether or not the population slope parameter β_1 is significantly different from 0.

Similar to finding confidence intervals for population parameters such as the population mean or population variance, we can also find confidence intervals for the population parameter β_1 (the slope of the regression line) that can be estimated by using the sample statistic $\hat{\beta}_1$. The sampling distribution of the population slope parameter follows the t-distribution with $n - 2$ degrees of freedom and a $100(1 - \alpha)\%$ confidence interval can be calculated as follows:

$$\left(\hat{\beta}_1 - t_{\alpha/2} \cdot \frac{S}{\sqrt{S_{xx}}}, \ \hat{\beta}_1 + t_{\alpha/2} \cdot \frac{S}{\sqrt{S_{xx}}} \right)$$

where $\hat{\beta}_1$ is the estimate of the true population slope parameter β_1, which falls between $-t_{\alpha/2}$ and $t_{\alpha/2}$ for a given value of α, S is the root mean square error, and S_{xx} is the sum of the squared differences between the individual x-values and the mean of x $(\bar{x})$.

To test whether the true population slope parameter is significantly different from 0, we can use the following test statistic:

$$T = \frac{\hat{\beta}_1}{\dfrac{S}{\sqrt{S_{xx}}}}, \qquad df = n - 2$$

Where $\hat{\beta}_1$ is the estimated value of the slope parameter, S is the root mean square error, and S_{xx} is the sum of the squared differences between the observed values of x and the mean of x.

Provided that the regression model assumptions hold true, the sampling distribution of the sample statistic $\hat{\beta}_1$ has approximately the t-distribution centered about the true population slope parameter β_1, with $n - 2$ degrees of freedom, as is illustrated in Figure 6.8.

To determine if the predictor variable x has a significant impact on the dependent variable y, we need to look at the line of best fit and determine whether or not we can infer that the slope of the unknown population line is significantly different from 0, less than 0, or

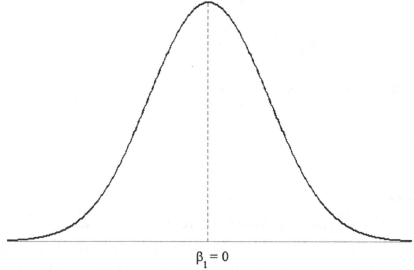

$\beta_1 = 0$

FIGURE 6.8 Distribution of the sample statistic $\hat{\beta}_1$ if H_0: $\beta_1 = 0$ were true.

greater than 0. In order to do this, we can test any of the following the null and alternative hypotheses:

$$H_0 : \beta_1 = 0$$

$$H_1 : \beta_1 \neq 0$$

$$H_0 : \beta_1 = 0$$

$$H_1 : \beta_1 < 0$$

$$H_0 : \beta_1 = 0$$

$$H_1 : \beta_1 > 0$$

Example 6.4

Suppose we are interested in testing whether or not our true population slope parameter β_1 is significantly different from 0 for the SAT–GPA data given in Table 6.1 ($\alpha = 0.05$).

The appropriate null and alternative hypotheses would be stated as follows:

$$H_0 : \beta_1 = 0$$

$$H_1 : \beta_1 \neq 0$$

Since we are testing whether the true population slope parameter is significantly different from 0, we could use a two-tailed test because to be different from 0 implies being either less than 0 or being greater than 0.

Then for $n - 2 = 10 - 2 = 8$ degrees of freedom, if the test statistic falls in the rejection region as defined by $t = 2.306$ and $t = -2.306$, we can reject the null hypothesis and claim that our unknown population parameter β_1 is significantly different from 0, as illustrated in Figure 6.9.

On the basis of our sample data, we have an estimate of the slope parameter $\hat{\beta}_1$, the root mean square error S, and S_{xx}, the sum of the squared difference between x and $\bar{x}$, which can be used to calculate the test statistic:

$$T = \frac{\hat{\beta}_1}{\frac{S}{\sqrt{S_{xx}}}} = \frac{0.00517}{\frac{0.54284}{\sqrt{107490}}} \approx 3.12$$

Since $T = 3.12 > t = 2.306$, the test statistic falls in the rejection region, as can be seen in Figure 6.9. Hence, the null hypothesis can be rejected and the alternative hypothesis can be accepted. Thus, we can infer that the unknown population slope parameter β_1 is significantly different from 0.

To interpret this finding within the context of our example, we claim to have strong evidence to suggest that higher SAT math examination scores from high school are associated with higher first-year GPAs in college. More specifically, for a 100-point increase in the score received on the mathematics portion of the SAT examination, the first-year GPA will increase by approximately 0.517 points. Notice that we did not claim that higher SAT math examination scores

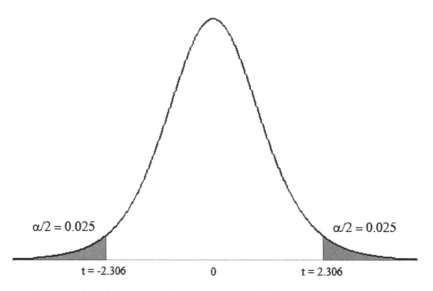

FIGURE 6.9
Rejection region for testing if the true population slope parameter β_1 is significantly different from 0 using the t-distribution with 8 degrees of freedom for $\alpha = 0.05$.

cause higher first-year GPAs in college; we only infer that higher SAT math scores are associated with higher first-year GPAs, and we can provide an estimate of the magnitude of such an effect.

We can also calculate a 95% confidence interval for the true population slope parameter β_1 with $n - 2$ degrees of freedom, as follows:

$$\left(\hat{\beta}_1 - t_{\alpha/2} \cdot \frac{S}{\sqrt{S_{xx}}}, \hat{\beta}_1 + t_{\alpha/2} \cdot \frac{S}{\sqrt{S_{xx}}} \right)$$

$$= \left(0.00517 - (2.306) \cdot \frac{0.54287}{\sqrt{107490}}, 0.00517 + (2.306) \cdot \frac{0.54287}{\sqrt{107490}} \right)$$

$$\approx \left(0.0014, 0.0090 \right)$$

Thus, if we take repeated samples of size ten and calculate confidence intervals in the same manner, then 95% of these intervals will contain the unknown population parameter β_1.

Another way to interpret this confidence interval would be to say that we are 95% confident that for a 100-point increase in the SAT math examination score, the unknown population first-year GPA will increase between 0.14 and 0.90 points.

6.8 Using Minitab to Test the Population Slope Parameter

Anytime we run a simple linear regression analysis, Minitab will automatically test whether the population slope parameter β_1 is significantly different from 0. In other words,

Coefficients

Term	Coef	SE Coef	T-Value	P-Value	VIF
Constant	-0.096	0.909	-0.11	0.918	
SATM	0.00517	0.00166	3.12	0.014	1.00

FIGURE 6.10
Portion of the Minitab printout showing the estimate of the true population regression slope parameter, $\hat{\beta}_1$, for the data given in Table 6.1.

the alternative hypothesis is H_1: $\beta_1 \neq 0$. Figure 6.10 gives a portion from the Minitab printout for the regression analysis of the coefficients and the p-value from the hypothesis test from Example 6.4.

Note that the standard error for $\hat{\beta}_1$ and the value of the test statistic T are what we obtained when we did the calculations manually. Also, notice the p-value of 0.014. Recall that the p-value gives the observed level of significance that is based on the test statistic, which in this case is less than our predetermined level of significance of $\alpha = 0.05$. So based on the p-value, we can reject the null hypothesis and accept the alternative hypothesis, which suggests that the true population slope parameter β_1 is significantly different from 0. Of course, this inference relies on the linear regression model assumptions having been reasonably met. Chapter 7 will describe in more detail some different techniques that can be used to check the regression model assumptions.

If we do not have enough evidence to reject the null hypothesis and accept the alternative hypothesis that the true population slope parameter is not significantly different and from 0 (in other words, if our test statistic does not fall in the rejection region and that happens when the p-value is greater than the predetermined level of significance), then we may infer that y is not related to x. This can be a misleading inference for two reasons. First, when determining the estimated model equation, we are only using a sample of x- and y-values. It could be the case that the given sample is not large enough for x to impact y. Second, we are assuming that all of the underlying model assumptions for the regression analysis hold true, and this includes the assumption that the unknown population model can be represented using a linear equation. However, if the true underlying population model is not reasonably represented by a linear equation, but yet the estimated model equation is linear, then it could be that x does have an influence on y, but such a relationship between x and y may not necessarily be linear.

Similar to testing whether the true population slope parameter β_1 is different from 0, we can also make inferences about whether the intercept of the true population equation β_0 is different from 0. Inferences about the intercept of the population equation are also based on the sampling distribution of the sample statistic following the t-distribution centered about the true population parameter $\beta_0 = 0$ with $n - 2$ degrees of freedom by using the following test statistic:

$$T = \frac{\hat{\beta}_0}{S\sqrt{\dfrac{1}{n} + \dfrac{\bar{x}^2}{S_{xx}}}}$$

where $\hat{\beta}_0$ is the estimate of the true population intercept parameter β_0 n is the sample size, S is the root mean square error, $\bar{x}$ is the sample mean, and S_{xx} is the sum of the squared differences between the observed values of x and the mean of x $(\bar{x})$.

We can also calculate confidence intervals for the value of the true population intercept value β_0.

A $100(1 - \alpha)\%$ confidence interval for β_0 with $n - 2$ degrees of freedom is as follows:

$$\left(\hat{\beta}_0 - t_{\alpha/2} \cdot S\sqrt{\frac{1}{n} + \frac{\bar{x}^2}{S_{xx}}}, \; \hat{\beta}_0 + t_{\alpha/2} \cdot S\sqrt{\frac{1}{n} + \frac{\bar{x}^2}{S_{xx}}} \right)$$

However, for many situations (such as with the last example), it may not make sense to cal-culate confidence intervals and perform hypothesis tests about the population intercept. This is because in many circumstances the x-values are not collected around 0. Recall that the y-intercept of a line is where the line crosses the y-axis, and this is where $x = 0$. One concern for our analysis is that a score of 0 is not even possible for the SAT mathematics examination because the scores on this examination range from 200 to 800. Thus, for data that is not collected around the value of 0, conducting a hypothesis test or finding a con-fidence interval about the population intercept parameter β_0 may not be of much interest.

6.9 Confidence Intervals for the Mean Response for a Specific Value of the Predictor Variable

We can use the line of best fit to estimate the *mean*, or average, response for *all* observa-tions that have a specific value of the predictor variable (in other words, a specific x-value). For example, suppose that we want to obtain an estimate of the *mean* first-year GPA for *all* of the students who have achieved a given score on the mathematics portion of the SAT examination. Using the line of best fit, the estimated response for a specific value of the predictor variable, x_s, can be calculated as follows:

$$\hat{y}_s = \hat{\beta}_0 + \hat{\beta}_1 \cdot x_s$$

Then, a $100(1 - \alpha)\%$ confidence interval with $n - 2$ degrees of freedom for the population mean response for a specific value of x (denoted as x_s) is given by:

$$\left(\left(\hat{\beta}_0 + \hat{\beta}_1 \cdot x_s \right) - t_{\alpha/2} \cdot S\sqrt{\frac{1}{n} + \frac{(x_s - \bar{x})}{S_{xx}}}, \; \left(\hat{\beta}_0 + \hat{\beta}_1 \cdot x_s \right) + t_{\alpha/2} \cdot S\sqrt{\frac{1}{n} + \frac{(x_s - \bar{x})}{S_{xx}}} \right)$$

Example 6.5

Suppose we want to calculate a 95% confidence interval for the population *mean* first-year GPA for *all* students who score 550 on the mathematics portion of the SAT examination. The estimated regression line is $\hat{y} = \hat{\beta}_0 + \hat{\beta}_1 x = -0.096 + 0.00517x$, so the predicted first-year GPA for the specific value of $x_s = 550$ would be:

$$y_s = \left(\hat{\beta}_0 + \hat{\beta}_1 \cdot x_s \right) \approx -0.096 + 0.00517(550) \approx 2.748$$

Then the estimated standard error for the confidence interval would be calcu-lated as follows:

$$S\sqrt{\frac{1}{n} + \frac{(x_s - \bar{x})}{S_{xx}}} \approx 0.54284 \cdot \sqrt{\frac{1}{10} + \frac{(550 - 539)^2}{107490}} \approx 0.1726$$

So a 95% confidence interval for the mean first-year GPA for all students who score 550 on the mathematics portion of the SAT exam would be:

$$\left(\left(\hat{\beta}_0 + \hat{\beta}_1 \cdot x_x\right) - t_{\alpha/2} \cdot S\sqrt{\frac{1}{n} + \frac{(x_s - \bar{x})}{S_{xx}}}, \left(\hat{\beta}_0 + \hat{\beta}_1 \cdot x_x\right) + t_{\alpha/2} \cdot S\sqrt{\frac{1}{n} + \frac{(x_s - \bar{x})}{S_{xx}}}\right)$$

$$\approx \left(2.748 - (2.306)(0.1726), 2.748 + (2.306)(0.1726)\right)$$

$$\approx \left(2.35, 3.15\right)$$

Therefore, if we take repeated samples of size $n = 10$ from the underlying population and calculate confidence intervals in the same manner, then 95% of these intervals will contain the population mean response value. In other words, we are 95% confident that for *all* students who score 550 on the SAT math examination, the *mean* first-year GPA for the entire population will fall between 2.35 and 3.15 points.

6.10 Prediction Intervals for a Response for a Specific Value of the Predictor Variable

We just illustrated how we can obtain confidence intervals that can be used for estimating the population *mean* first-year GPA for *all* students who scored 550 on the mathematics portion of the SAT examination.

However, there may be times when the *predicted* response for an *individual* with a specific value of the predictor variable needs to be estimated. For instance, we may want to estimate the *predicted* first-year GPA for an *individual* student who has achieved a specific score on the mathematics portion of the SAT examination.

A prediction interval is very similar to a confidence interval for a specific value of the predictor variable, except the standard error of the prediction interval will be larger than a confidence interval because using a single observation to predict a response generates more uncertainty than including all of the observations with that specific value.

The calculations needed to create a prediction interval are similar to those for finding a confidence interval to estimate the mean response for a specific predictor value (x_s), except the estimated standard error, is calculated as follows:

$$S\sqrt{1 + \frac{1}{n} + \frac{(x_s - \bar{x})}{S_{xx}}}$$

Therefore, a $100(1 - \alpha)\%$ prediction interval with $n - 2$ degrees of freedom for predicting a single response for a specified value of x (x_s) is given by:

$$\left(\left(\hat{\beta}_0 + \hat{\beta}_1 \cdot x_x\right) - t_{\alpha/2} \cdot S\sqrt{1 + \frac{1}{n} + \frac{(x_s - \bar{x})}{S_{xx}}}, \left(\hat{\beta}_0 + \hat{\beta}_1 \cdot x_x\right) + t_{\alpha/2} \cdot S\sqrt{1 + \frac{1}{n} + \frac{(x_s - \bar{x})}{S_{xx}}}\right)$$

Example 6.6

Suppose we want to *predict* (with 95% confidence) what the first-year GPA would be for an *individual* student who scores 550 on the SAT math examination.

The predicted first-year GPA for a SAT score of 550 would be:

$$\left(\hat{\beta}_0 + \hat{\beta}_1 \cdot x_s\right) \approx -0.096 + 0.00517(550) \approx 2.748$$

Then the prediction interval for the first-year GPA for a single student who scores a 550 on the SAT math examination would be:

$$\left(\left(\hat{\beta}_0 + \hat{\beta}_1 \cdot x_s\right) \pm t_{\alpha/2} \cdot S\sqrt{1 + \frac{1}{n} + \frac{(x_s - \bar{x})}{S_{xx}}}, \left(\hat{\beta}_0 + \hat{\beta}_1 \cdot x_s\right) \pm t_{\alpha/2} \cdot S\sqrt{1 + \frac{1}{n} + \frac{(x_s - \bar{x})}{S_{xx}}}\right)$$

$$\approx \left(2.748 - (2.306)(0.54284)\sqrt{1 + \frac{1}{10} + \frac{(550 - 539)^2}{107490}},\right.$$

$$\left. 2.748 + (2.306)(0.54284)\sqrt{1 + \frac{1}{10} + \frac{(550 - 539)^2}{107490}}\right)$$

$$\approx (1.43, 4.06)$$

If we take repeated samples of size $n = 10$ from the population and calculate prediction intervals in the same manner, then 95% of these intervals will contain the predicted GPA for an individual student who scored 550 on the SAT mathematics examination. In other words, we are 95% confident that an individual student who scores 550 on the SAT mathematics examination would achieve a first-year GPA that falls between 1.43 and 4.06 points.

Notice that the prediction interval is much wider than the confidence interval for the same specific value of the predictor variable (in this case, a SAT mathematics examination score of 550). Also, the upper limit of the prediction interval actually falls out of the range for the response variable (since GPA ranges from 0.00 to 4.00 points). This can happen when calculating prediction intervals because there is a large amount of uncertainty in estimating a single response using a single observation versus estimating the average, or mean, response by using the entire set of observations that have the specific value of the predictor variable, as is the case with confidence intervals. Such a large standard error could generate a prediction interval where some values in the interval may fall outside of the range of the possible values for the given variable.

One thing to keep in mind when making inferences with confidence and prediction intervals is that it is not usually a good idea to make inferences when the specific value of interest (x_s) is far removed from the range of the sample data that was used to obtain the line of best fit. Not only could a value out of range cause the confidence and prediction intervals to become very wide, but also the relationship between x and y may not be linear at more extreme values. The graph in Figure 6.11 shows a model with a different functional form outside the range where the data was collected.

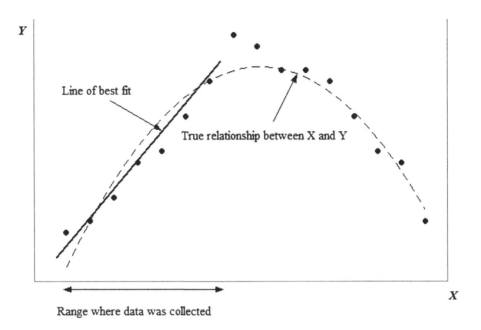

FIGURE 6.11
Graph of the true relationship being nonlinear outside the range of values used to obtain the line of best fit.

6.11 Using Minitab to Find Confidence and Prediction Intervals

To use Minitab to create confidence and prediction intervals for the data given in Table 6.1, use the **Regression** tab under the **Stat** menu and select **Fitted Line Plot** to get the dialog box that is illustrated in Figure 6.12.

Then selecting the **Options** tab on the dialog box and checking the boxes to display confidence and prediction intervals, as shown in Figure 6.13, gives the fitted line plot, including the confidence and prediction intervals illustrated in Figure 6.14.

The graph in Figure 6.14 provides the fitted line plot along with confidence and prediction intervals for the entire range of *x*-values that were used in estimating the line of best fit.

Exact confidence and prediction intervals can also be found by first running a regression analysis and then selecting **Stat > Regression > Regression > Predict**. This gives the dialog box as illustrated in Figure 6.15.

Then entering the specific *x*-value or a set of specific *x*-values and selecting **OK** gives the Minitab output that is in Figure 6.16, which provides the confidence and prediction intervals for a SAT mathematics score of 550. Notice that these match the exact confidence and prediction intervals we found manually.

Example 6.7

Does spending more time studying result in higher exam grades? To answer this question, a survey was given to a random sample of 32 students participating in

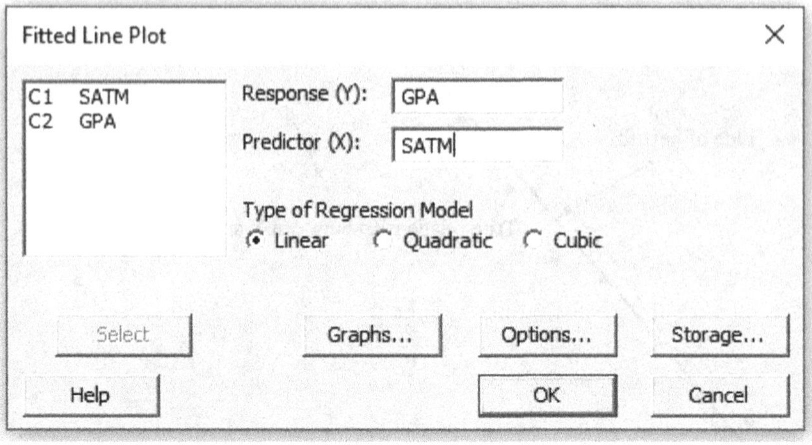

FIGURE 6.12
Minitab dialog box to draw the fitted line plot with confidence and prediction intervals.

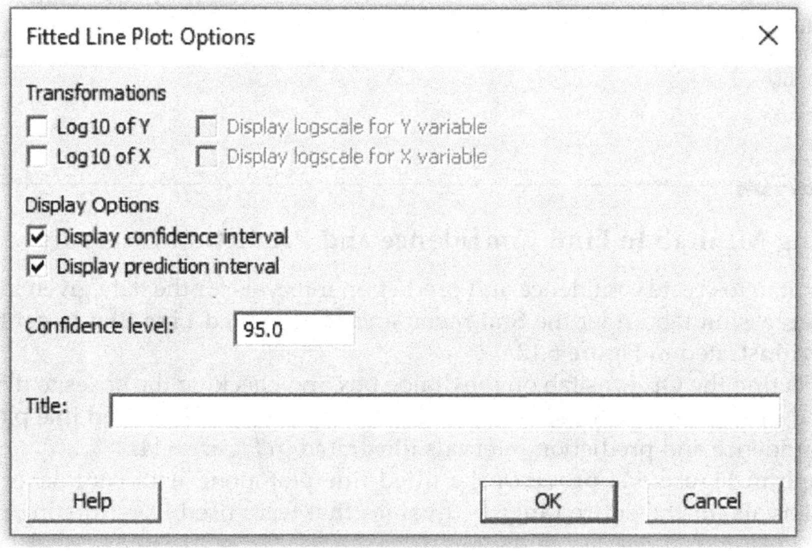

FIGURE 6.13
Options box to display confidence and prediction intervals on the fitted line plot.

an introductory statistics course. This survey asked this random sample of students how many hours they spent studying for a statistics examination during the week before they took the exam.

The data set provided in Table 6.4 gives the number of hours the students reported they spent studying for the examination along with the score they received on the examination.

Because we want to know if spending more time studying is related to higher examination grades, we could let the predictor (or x) variable be the number of

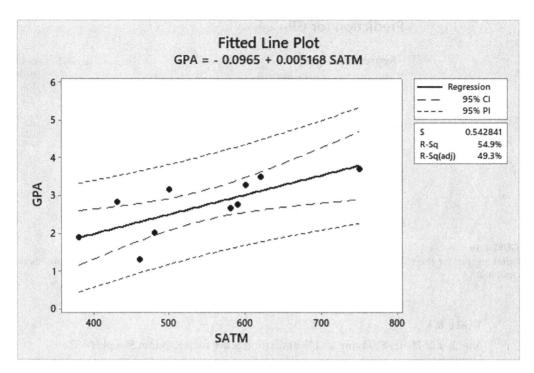

FIGURE 6.14
Fitted line plot that illustrates confidence and prediction intervals.

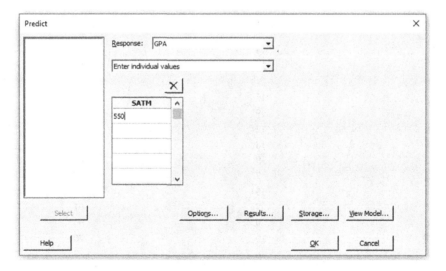

FIGURE 6.15
Regression options tab to find exact confidence and prediction intervals for a specific SAT mathematics score of 550.

hours studying and the response (or y) variable be the grade received on the examination.

We can first begin by drawing a scatterplot to see what the relationship between x and y looks like. This scatterplot is given in Figure 6.17.

Prediction for GPA

Regression Equation

GPA = -0.096 + 0.00517 SATM

Settings

Variable	Setting
SATM	550

Prediction

Fit	SE Fit	95% CI	95% PI
2.74585	0.172625	(2.34777, 3.14392)	(1.43228, 4.05941)

FIGURE 6.16
Minitab output for the 95% confidence and prediction intervals for a score of 550 on the SAT mathematics examination.

TABLE 6.4

Number of Hours Studying and Examination Score for a Random Sample of 32 Students

Observation	Hours Studying (x_i)	Score Received on Exam (y_i)
1	3	62
2	5	55
3	10	75
4	5	67
5	18	84
6	20	89
7	21	91
8	17	82
9	9	77
10	12	73
11	18	86
12	9	69
13	7	67
14	15	72
15	27	91
16	9	50
17	10	70
18	16	77
19	21	84
20	8	74
21	13	76
22	11	62
23	10	73
24	16	82
25	5	78

(Continued)

TABLE 6.4 (*Continued*)

Number of Hours Studying and Examination Score for a Random Sample of 32 Students

Observation	Hours Studying (x_i)	Score Received on Exam (y_i)
26	17	85
27	4	62
28	8	68
29	2	64
30	9	72
31	20	97
32	17	91

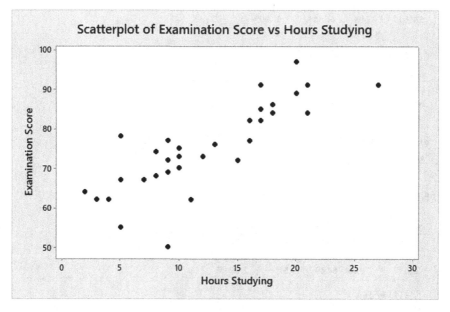

FIGURE 6.17

Scatterplot showing the relationship between the number of hours studying for an examination and the score received on the examination.

Notice that the scatterplot illustrates a positive trend between the number of hours studying and the score received on the examination. Also, it appears that this relationship is one that could be reasonably approximated by using a straight line.

For a significance level of $\alpha = 0.01$, we can use Minitab to run the regression analysis and generate the printout given in Figure 6.18.

From the Minitab printout in Figure 6.18, we can see that the line of best fit is given by the equation:

$$\hat{y} \approx 57.13 + 1.471 \cdot x$$

This equation can be used to estimate the true population equation:

$$y = \beta_0 + \beta_1 x + \varepsilon$$

Regression Analysis: Examination Score versus Hours Studying

Analysis of Variance

Source	DF	Adj SS	Adj MS	F-Value	P-Value
Regression	1	2598.2	2598.23	61.57	0.000
Hours Studying	1	2598.2	2598.23	61.57	0.000
Error	30	1266.0	42.20		
Lack-of-Fit	16	439.7	27.48	0.47	0.928
Pure Error	14	826.3	59.02		
Total	31	3864.2			

Model Summary

S	R-sq	R-sq(adj)	R-sq(pred)
6.49613	67.24%	66.15%	62.90%

Coefficients

Term	Coef	SE Coef	T-Value	P-Value	VIF
Constant	57.13	2.57	22.25	0.000	
Hours Studying	1.471	0.188	7.85	0.000	1.00

Regression Equation

Examination Score   =   57.13 + 1.471 Hours Studying

Fits and Diagnostics for Unusual Observations

Obs	Examination Score	Fit	Resid	Std Resid	
15	91.00	96.86	-5.86	-1.02	X
16	50.00	70.37	-20.37	-3.20	R
25	78.00	64.49	13.51	2.16	R

R *Large residual*
X *Unusual X*

FIGURE 6.18
Minitab printout of the regression analysis for the data in Table 6.4.

Thus, we can use $\hat{\beta}_0 = 57.13$ as an estimate of β_0 and $\hat{\beta}_1 = 1.471$ as an estimate of β_1.

We want to know if the unknown population slope parameter β_1 is significantly different from 0. By looking at the Minitab printout in Figure 6.18, we can see that the p-value for β_1 is less than our predetermined level of significance of 0.01. So can accept the alternative hypothesis and reject the null hypothesis, which suggests that our true population value of β_1 is significantly different from 0.

We can also interpret this finding within the context of our problem as follows:
For every additional 10 hours spent studying during the week prior to the examination, the grade on the examination will increase by approximately 14.71 points.

We can also illustrate confidence and prediction intervals and use Minitab to draw the fitted line plot with confidence and prediction intervals, as illustrated in Figure 6.19.

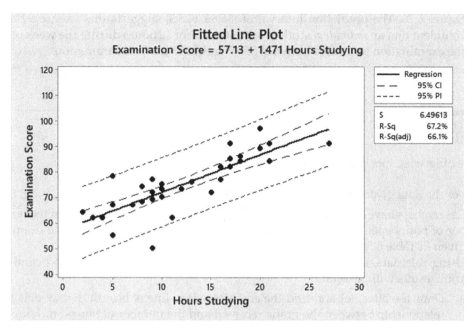

FIGURE 6.19
Minitab fitted line plot with 99% confidence and 99% prediction intervals for the data in Table 6.4.

Suppose that we wanted to know the *mean* examination score for *all* students who study for 12 hours the week before the examination. Using Minitab, we could get the value of the 99% confidence interval of (71.63, 77.95) for the unknown population mean examination score for all students who studied for 12 hours the week of the examination, as given in Figure 6.20.

Similarly, we could find the *predicted* examination score for an *individual* student who studied for 12 hours during the week of the examination in

Prediction for Examination Score

Regression Equation

Examination Score   =   57.13 + 1.471 Hours Studying

Settings

Variable	Setting
Hours Studying	12

Prediction

Fit	SE Fit	99% CI	99% PI
74.7884	1.14932	(71.6278, 77.9490)	(56.6466, 92.9302)

FIGURE 6.20
Confidence and prediction intervals for the specific value of the predictor of 12 hours.

Figure 6.20. The prediction interval of (56.65, 92.93) suggests that we are 99% confident that an *individual* student who studies for 12 hours during the week of the examination will score between 56.65 and 92.93 on the examination.

Exercises

Unless otherwise specified, use $\alpha = 0.05$.

1. For the data given in Table 6.1 calculate the sample statistics $\bar{x}$, $\bar{y}$, S_{xx}, S_{yy}, and S_{xy}.
2. Researchers have collected data from a random sample of six students on the number of hours spent studying for an exam and the grade received on the exam as given in Table 6.5.
   Using this data, answer each of the following questions manually and confirm your results with Minitab:
   a. Draw a scatterplot and find the equation of the line of best fit that models the relationship between the grade received and the number of hours studying.
   b. Find the residual for Observation #4.
   c. Interpret whether or not you believe there is a significant relationship between the grade received and the number of hours studying.
   d. Find and interpret a 90% confidence interval for the true population slope parameter.
   e. Find and interpret a 99% confidence interval for the predicted grade for an individual who spends 10 hours studying.
   f. Find and interpret a 99% confidence interval for the mean grade of all individuals who spend 10 hours studying.
3. Does using social media stress you out? A researcher is interested in whether college students who extensively use social media tend to have higher levels of stress. To test this, a sample of nine college students were asked to fill out a survey on their stress level and provide an estimate of the number of hours they spend per day on social media. The stress survey is on a scale of 0–100 where higher scores are indicative of a higher level of stress. The data is given in Table 6.6.
   a. Fill in the seven blank cells in Table 6.6 and describe which variable is the *response variable* and which variable is the *predictor variable*?

TABLE 6.5

Random Sample of the Number of Hours Studying and Grade

Observation	Grade	Number of Hours Studying
1	85	8
2	73	10
3	95	13
4	77	5
5	68	2
6	95	12

TABLE 6.6

Stress Level and Number of Hours a Random Sample of Nine Students Spend Per Day on Social Media

	Hours	Stress	$(x_i - \bar{x})$	$(x_i - \bar{x})^2$	$(y_i - \bar{y})$	$(y_i - \bar{y})^2$	$(x_i - \bar{x}) \cdot (y_i - \bar{y})$
	5	55	1.5556	2.4198	6.7778	45.9383	10.5432
	2	40	−1.4444		−8.2222	67.6049	
	3	46	−0.4444	0.1975	−2.2222	4.9383	0.9877
	4	46		0.3086	−2.2222	4.9383	−1.2346
	1	37		5.9753	−11.2222	125.9383	27.4321
	3	46		0.1975	−2.2222	4.9383	0.9877
	6	71	2.5556		22.7778	518.8272	
	2	38	−1.4444	2.0864	−10.2222	104.4938	14.7654
	5	55	1.5556	2.4198	6.7778	45.9383	10.5432
Total	31.0000	434.0000	0.0000	22.2222	0.0000	923.5556	134.1111

b. Using the data from Table 6.6, find the line of best fit.

c. Draw a scatterplot and test whether the population slope parameter is significantly different from 0.

d. Interpret whether you believe that spending more time on social media is related to higher levels of stress.

e. Find and interpret a 98% confidence interval for the true population slope parameter.

f. Find and interpret a 99% confidence interval for the mean level of stress for all individuals who spend 4 hours on social media per day.

g. Find and interpret a 99% confidence interval to predict the level of stress for an individual who spends 4 hours on social media per day.

4. Researchers have collected data from a sample of nine individuals on the number of hours of television watched in a day and the age of the individual. They are interested in estimating if age is related to the number of hours of television watched each day.

The scatterplot in Figure 6.21 shows the relationship between the number of hours of television watched in a day and the age of the individual.

a. Describe the relationship between age and the number of hours of television watched.

b. Which variable is the response variable and which variable is the predictor variable?

The equation of the *line of best fit*, $\hat{y} = 6.7859 - 0.1334x$ can be obtained from Figure 6.22:

c. Complete Table 6.7 (round to 4 decimal places).

d. Using Table 6.7, find $\sum_{i=1}^{n} \hat{\varepsilon}_i$ and $\text{SSE} = \sum_{i=1}^{n} \hat{\varepsilon}_i^2$ for the line of best fit.

Now consider the following linear equation $\hat{y} = 5.25 - 0.05x$ that is *not the line of best fit* that is illustrated in Figure 6.23.

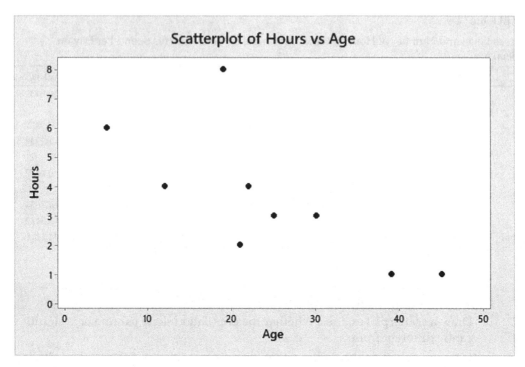

FIGURE 6.21
Scatterplot of hours of television watched versus age for a random sample of nine individuals.

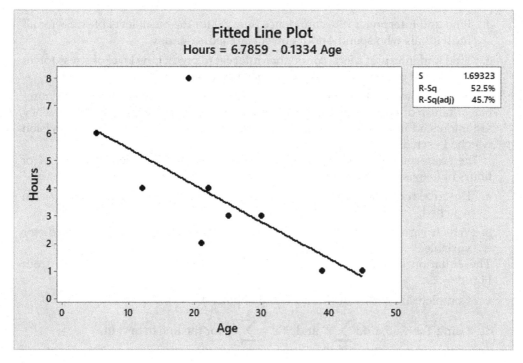

FIGURE 6.22
Fitted line plot of hours of television watched versus age.

TABLE 6.7

Hours of Television Watched Versus Age for a Random Sample of Nine Individuals Using the Line of Best Fit $\left(\hat{y} = 6.7859 - 0.1334x\right)$

Hours of Television	Age	$\hat{y}_i$	$\hat{\varepsilon}_i$	$\hat{\varepsilon}_i^2$
1	45			
3	30			
4	22			
3	25			
6	5			
4	12			
8	19			
2	21			
1	39			
—	—	—	$\sum_{i=1}^{n} \hat{\varepsilon}_i$	$\text{SSE} = \sum_{i=1}^{n} \hat{\varepsilon}_i^2$

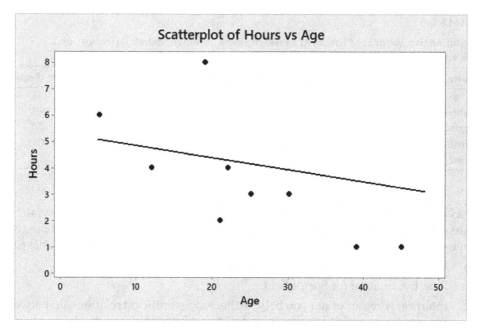

FIGURE 6.23

Scatterplot of hours of television watched versus age for a random sample of nine individuals including the new line $\hat{y} = 5.25 - 0.05x$.

    e. Complete Table 6.8 (round to 4 decimal places) using this new line.

    f. Find $\sum_{i=1}^{n} \hat{\varepsilon}_i$ and $\text{SSE} = \sum_{i=1}^{n} \hat{\varepsilon}_i^2$ for this new line.

    g. What can you say about the difference in the statistics $\sum_{i=1}^{n} \hat{\varepsilon}_i$ and $\text{SSE} = \sum_{i=1}^{n} \hat{\varepsilon}_i^2$ for the two different lines?

TABLE 6.8

Hours of Television Watched Versus Age for a Random Sample of Nine Individuals Using the New Line $\left(\hat{y} = 5.25 - 0.05x\right)$

Hours of Television	Age	$\hat{y}_i$	$\hat{\varepsilon}_i$	$\hat{\varepsilon}_i^2$
1	45			
3	30			
4	22			
3	25			
6	5			
4	12			
8	19			
2	21			
1	39			
—	—	—	$\sum_{i=1}^{n} \hat{\varepsilon}_i$	$\text{SSE} = \sum_{i=1}^{n} \hat{\varepsilon}_i^2$

TABLE 6.9

Random Sample of Five Towns in Massachusetts, Average Snowfall (Inches), and Elevation (Feet)

Town	Average Snowfall (Inches)	Elevation (Feet)
Boston	43.8	141
Amherst	37.0	295
Springfield	40.5	70
Pittsfield	62.1	1,039
Lowell	61.6	102

5. Do towns in Massachusetts with higher elevations tend to get more snowfall? To answer this question, a random sample of five towns in Massachusetts, their average yearly snowfall (in inches), and elevation (in feet) were recorded in Table 6.9.

   a. Draw a scatterplot and find the equation for the line of best fit.

   b. Find the residual for Springfield.

   c. Interpret whether or not you believe there is a significant relationship between the average snowfall and elevation in Massachusetts.

   d. Find and interpret a 95% confidence interval for the true population slope parameter.

7

More on Simple Linear Regression

7.1 Introduction

The focus of this chapter is to describe some different measures that can be used to assess model fit for a simple linear regression analysis. We will be introducing measures to assess how much variability in the response variable is due to the given model by using the coefficient of determination, and the coefficient of correlation will also be described as a way to measure the linear association between two variables. Finally, exploratory graphs and formal tests will be used to verify the regression model assumptions that were presented in the last chapter, in addition to assessing the impact of outliers.

7.2 The Coefficient of Determination

One simple strategy that can be used to estimate the population mean value of y for a simple linear regression analysis is to use the sample mean $\bar{y}$. In using the SAT–GPA data set provided in Table 6.1, the value of $\bar{y}$ is the mean first-year grade point average (GPA) for the given sample of ten students. The graph in Figure 7.1 illustrates a dashed line that represents the sample mean of the y-values, $\bar{y} = 2.689$.

If we were to use only the sample mean $\bar{y}$ to predict y, then we could represent the sum of the squared differences between $\bar{y}$ and the observed y-value for each observation as follows:

$$S_{yy} = \text{SST} = \sum_{i=1}^{n} \left(y_i - \bar{y} \right)^2$$

where $S_{yy} = \text{SST}$ represents the sum of the squared differences between each observed y-value and the mean of the y-values. However, by using only $\bar{y}$ to predict y, this clearly would not capture any impact that the variable x may have on y.

One way to determine whether the predictor variable x has an impact on the response variable y is to describe how well the regression equation predicts y. This can be done by comparing the effectiveness of the regression model that includes x to predict y to that using only the sample mean $\bar{y}$ to predict y. The measure that does this is called the *coefficient of determination*, which is denoted as R^2. The coefficient of determination is used to measure how well the regression model predicts y as compared to using only the sample mean $\bar{y}$ to predict y. The value of R^2 is found by using the following formula:

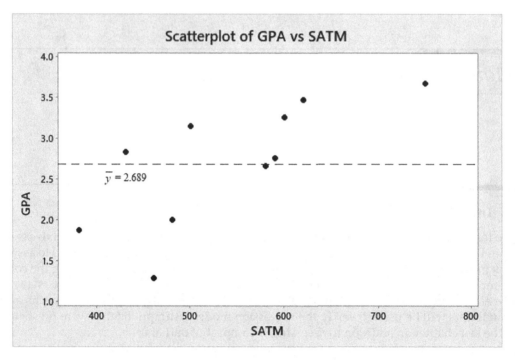

FIGURE 7.1
Scatterplot of GPA versus SATM where the dashed line is the mean GPA of $\bar{y} = 2.689$.

$$R^2 = \frac{\text{SST} - \text{SSE}}{\text{SST}}$$

where $\text{SST} = \sum_{i=1}^{n}(y_i - \bar{y})^2$ is the sum of the squared differences between each observed

y-value and the mean of y $(\bar{y})$, and $\text{SSE} = \sum_{i=1}^{n}(y_i - \hat{y})^2$ is the sum of the squared differences

between each observed y-value and the estimated y-value $(\hat{y})$.

If the variable x does not contribute any information about y, then SSE (the error sum of squares) will be equal to SST (the total sum of squares) and the value of R^2 would be equal to 0. Otherwise, if x perfectly predicts y, then the observed y-value will equal the fitted y-value and *SSE* will equal 0, so the value of R^2 would be equal to 1. Therefore, the R^2 statistic can be used to measure the proportion of the variability in y that is explained by the given regression model. Notice that the coefficient of determination will always be between 0 and 1, in other words, $0 \leq R^2 \leq 1$, and there are no units associated with R^2.

Example 7.1

Recall the data given in Table 6.1 that provides the score on the mathematics portion of the SAT examination and first-year grade point averages for a sample of ten students. To find the coefficient of determination R^2, we can use SST = 5.228 and SSE = 2.357412 as follows:

$$R^2 = \frac{\text{SST} - \text{SSE}}{\text{SST}} = \frac{5.228 - 2.357412}{5.228} \approx 0.549$$

Thus, the coefficient of determination $R^2 \approx 0.549 = 54.9\%$ suggests that 54.9% of the variability in first-year GPA (y) can be explained by using the score received on the mathematics portion of the SAT examination (x). The closer R^2 is to 100%, the more useful the model is in predicting y.

7.3 Using Minitab to Find the Coefficient of Determination

Every time you run a regression analysis, Minitab will automatically provide the value of R^2 just below the analysis of variance (ANOVA) table. Under the Model Summary heading in the Minitab printout in Figure 7.2, you can see the value of the coefficient of determination for the regression analysis $\left(R^2 = 54.91\%\right)$ as well as the root mean square error $\left(S = 0.542841\right)$. In the ANOVA table, you will find the sum of squares for the regression (SSR = 2.871), the sum of squares for the error (SSE = 2.357), and the total sum of squares (SST = 5.228).

Because R^2 measures how well the regression model predicts y, it can also be calculated by taking the sum of squares due to the regression analysis (SSR), which is found by taking the difference between the total sum of squares and the error sum of squares (SSR = SST – SSE), and then dividing this difference by the total sum of squares (SST), as follows:

$$R^2 = \frac{\text{SST} - \text{SSE}}{\text{SST}} = \frac{\text{SSR}}{\text{SST}}$$

Example 7.2

For the regression analysis of the data used to model the relationship between the score received on the SAT mathematics examination and first-year GPA, as given in Table 6.1, the "Analysis of Variance" table presented in Figure 7.2 shows that the total sum of squares is SST = 5.2281, and the regression sum of squares is SSR = 2.871. Using these values, we can calculate R^2 as follows:

$$R^2 = \frac{\text{SSR}}{\text{SST}} \approx \frac{2.871}{5.228} \approx 0.549 = 54.9\%$$

Analysis of Variance

Source	DF	Adj SS	Adj MS	F-Value	P-Value
Regression	1	2.871	2.8707	9.74	0.014
SATM	1	2.871	2.8707	9.74	0.014
Error	8	2.357	0.2947		
Total	9	5.228			

Model Summary

S	R-sq	R-sq(adj)	R-sq(pred)
0.542841	54.91%	49.27%	35.60%

FIGURE 7.2
Partial printout of the regression analysis for the data in Table 6.1.

7.4 The Coefficient of Correlation

There are sample statistics that we can use to measure the linear relationship between two variables. One such statistic is the *Pearson Product Moment Coefficient of Correlation*, typically denoted as r, that can be used as a measure to describe the linear trend between two continuous sample variables x and y. For a simple linear regression analysis, the value of r can be found by taking the square root of the coefficient of determination $\left(R^2\right)$ and attaching to it the sign that indicates whether the slope of the regression line is positive or negative.

The sign of r can be obtained by looking at the slope of the line of best fit. Recall that the slope describes the type of linear relationship between x and y. For instance, when r is negative, the x and y values are *negatively correlated* with each other; in other words, there is a negative linear relationship between x and y. This suggests that an increase in x implies a decrease in y, as is illustrated in Figure 7.3.

If $r = 0$, this suggests that there is *no apparent relationship* between the variables x and y, and thus a change in x does not bring about any change in y, as seen in Figure 7.4.

If r is positive, then there is a positive linear relationship and x and y are *positively correlated*. In other words, a positive increase in x brings about a positive increase in y, as is shown in Figure 7.5.

In addition to taking the square root of the coefficient of determination to find the coefficient of correlation, other formulas can be used to compute the value of r for two variables x and y, as follows:

$$r = \frac{S_{xy}}{\sqrt{S_{xx}} \cdot \sqrt{S_{yy}}}$$

The coefficient of correlation, r, is related to the estimated slope parameter and can be computed as follows:

$$r = \frac{s_x}{s_y} \cdot \hat{\beta}_1$$

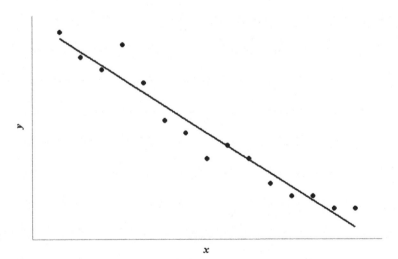

FIGURE 7.3
Negative linear relationship between x and y ($-1 \leq r < 0$).

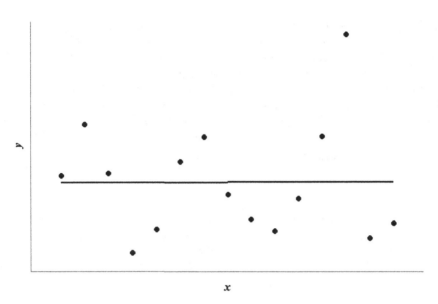

FIGURE 7.4
No apparent relationship between x and y $(r \approx 0)$.

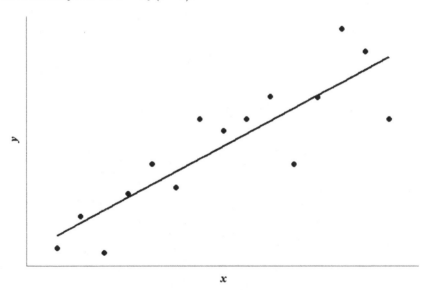

FIGURE 7.5
A positive linear relationship between x and y $(0 \le r \le 1)$.

where s_x is the sample standard deviation for the x-values, s_y is the sample standard deviation for the y-values, and $\hat{\beta}_1$ is the estimated slope of the regression line of y versus x.

Although two variables x and y can be highly correlated with each other, it is important to note that a strong correlation does not necessarily imply that a cause-and-effect relationship exists between x and y. Two variables could be highly correlated with each other because they are both associated with other unobserved variables. For instance, if the score on the mathematics portion of the SAT examination and the first-year GPA are both highly correlated with another variable, such as motivation, then a strong correlation between the scores received may not necessarily imply that a high SAT math score *causes* a high first-year GPA. A strong correlation between the SAT math score and the first-year GPA could also be due to their relationship with other variables.

Also notice that r is not equal to the slope of the regression line. Although positive values of r indicate that the slope of the simple linear regression line is positive (an increase in x implies an increase in y) and negative values of r indicate that the slope of the regression line is negative (an increase in x implies a decrease in y), the value of r is not the same as the estimate of the slope of the regression line $\left(\hat{\beta}_1\right)$.

Example 7.3

We can use the data in Table 6.1 to calculate the value of the sample coefficient of correlation. To find the value of r, we can take the square root of the value of the coefficient of determination $\left(R^2\right)$ as follows:

$$r = \sqrt{0.549} \approx 0.741$$

Because the slope of the estimated line of best fit is positive $\left(\hat{\beta}_1 \approx +0.00517\right)$, then $r = +0.741$.

We can also find the value of the sample coefficient of correlation by using other statistics such as S_{xx}, S_{yy}, and S_{xy}, as follows:

$$S_{xx} = 107490.000$$

$$S_{yy} = 5.228$$

$$S_{xy} = 555.290$$

Then:

$$r = \frac{S_{xy}}{\sqrt{S_{xx}} \cdot \sqrt{S_{yy}}} = \frac{555.290}{\sqrt{107490.000} \cdot \sqrt{5.228}} \approx 0.741$$

If we have the sample standard deviations for the x- and y-variables, we can also calculate the coefficient of correlation as follows:

$$r = \frac{s_x}{s_y} \cdot \hat{\beta}_1 = \frac{109.286}{0.7622} \cdot (0.00517) \approx 0.741$$

The coefficient of correlation is between -1 and $+1$ $(-1 \leq r \leq +1)$ and has no units.

Recall that the line of best fit from our regression analysis for the ten observations consisting of SAT math scores and first-year GPAs was

$$\hat{y} = -0.096 + 0.00517x,$$

where the estimate of the slope parameter is $\hat{\beta}_1 \approx 0.00517$. Notice that the sample coefficient of correlation r is not the same as the estimated slope parameter $\hat{\beta}_1$.

A strong correlation can only suggest whether a *linear trend* exists between two variables. Furthermore, a low value of r does not necessarily imply that there is no relationship between the two variables; it only suggests that no *linear relationship* exists. To illustrate what this means, consider the two scatterplots presented in Figures 7.6 and 7.7. Figure 7.6 illustrates a strong linear correlation between x and y. However, Figure 7.7 illustrates a weak correlation, but yet there appears to be a relationship between x and y, although it does not appear to be a linear relationship. Weak correlations do not suggest that no relationship exists between x and y, only that no *linear* relationship exists between x and y.

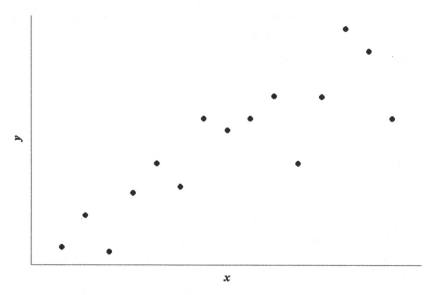

FIGURE 7.6
Scatterplot showing a strong linear trend of the relationship between x and y ($r \approx 0.96$).

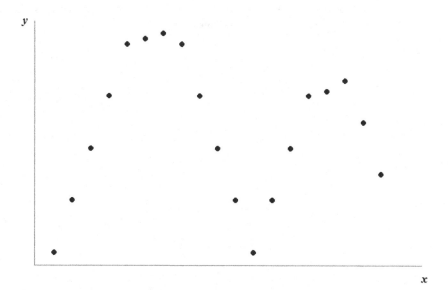

FIGURE 7.7
Scatterplot showing a weak coefficient of correlation but an apparent nonlinear relationship between x and y ($r \approx 0.016$).

7.5 Correlation Inference

Just like we used the estimated regression equation $\hat{y} = \hat{\beta}_0 + \hat{\beta}_1 \cdot x$ to estimate the unknown population equation $y = \beta_0 + \beta_1 \cdot x + \varepsilon$, we can perform a hypothesis test for the unknown population coefficient of correlation. In other words, we can use the value r, the sample coefficient of correlation, to estimate the unknown population coefficient of correlation ρ (pronounced "rho").

We can test whether the population coefficient of correlation is positive, negative, or different from 0 by running a hypothesis test using any one of the three different null and alternative hypotheses as follows:

$$H_0 : \rho = 0$$
$$H_1 : \rho \neq 0$$

$$H_0 : \rho = 0$$
$$H_1 : \rho < 0$$

$$H_0 : \rho = 0$$
$$H_1 : \rho > 0$$

Like many sampling distributions we have seen thus far, the sampling distribution of the sample coefficient of correlation follows the t-distribution with $n - 2$ degrees of freedom, and it is centered about the unknown population coefficient of correlation $\rho = 0$, as can be seen in Figure 7.8.

If we can show that $\rho > 0$ (for some given level of significance), then we can say that the population variables x and y are positively correlated with each other. Similarly, if we can show that $\rho < 0$, the population variables are negatively correlated with each other. And if ρ is not significantly different from 0, then the population variables are uncorrelated with each other and no linear relationship exists between x and y.

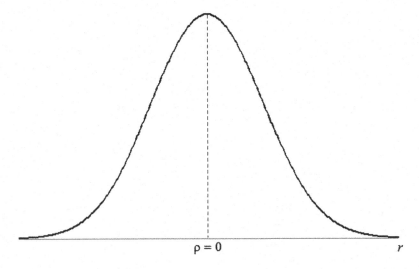

FIGURE 7.8
Distribution of the sample coefficient of correlation r if $H_0 : \rho = 0$ were true.

The following test statistic can be used to test the unknown population coefficient of correlation (ρ):

$$T = \frac{r}{\sqrt{\dfrac{1-r^2}{n-2}}}$$

where r is the sample coefficient of correlation and n is the sample size.

Example 7.4

Suppose that we want to test whether the correlation between SATM and GPA from Table 6.1 is significantly different from 0 ($\alpha = 0.05$). The null and alternative hypotheses would be as follows:

$$H_0 : \rho = 0$$

$$H_1 : \rho \neq 0$$

Using the value of r obtained from the sample of size $n = 10$ gives the value of the test statistic as follows:

$$T = \frac{r}{\sqrt{\dfrac{1-r^2}{n-2}}} = \frac{0.741}{\sqrt{\dfrac{1-(0.741)^2}{10-2}}} \approx 3.12$$

Comparing this value to the rejection region that is determined based on the direction of the alternative hypothesis and the level of significance, as illustrated in Figure 7.9, it can be seen that the null hypothesis can be rejected and the alternative hypothesis be accepted because the value of the test statistic T falls into the rejection region. Thus, the true population coefficient of correlation ρ is significantly different from 0 and we can infer that a linear trend exists between the score received on the SAT mathematics examination and the first-year GPA.

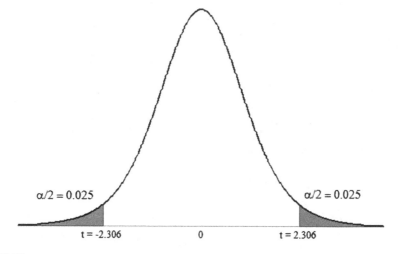

FIGURE 7.9
Rejection region for testing the population coefficient of correlation $H_1 : \rho \neq 0$ for $\alpha = 0.05$ with 8 degrees of freedom for a two-tailed test.

Since the two-sided test is significant, we can take this one step further and infer that the unknown population coefficient of correlation is significantly *greater* than 0 since the test statistic falls to the right of $t = 2.306$.

However, notice that the population coefficient of correlation would not be significantly different from 0 if we used a level of significance of $\alpha = 0.01$. This is because the value of t that defines the rejection region would be defined by $t = \pm 3.355$ and our test statistic would not fall into such a region.

Example 7.5

Suppose a sample of size $n = 18$ generates a sample coefficient of correlation of $r = 0.53$. We can test whether the true population coefficient of correlation (ρ) is significantly greater than 0. The appropriate null and alternative hypotheses would be as follows:

$$H_0 : \rho = 0$$

$$H_1 : \rho > 0$$

Given a level of significance of $\alpha = 0.01$, then:

$$T = \frac{0.53}{\sqrt{\dfrac{1-(0.53)^2}{18-2}}} \approx 2.50$$

for $18 - 2 = 16$ degrees of freedom.

We would then compare this value to $t = 2.921$, as illustrated in Figure 7.10. Since the value of the test statistic T does not fall in the rejection region defined by the alternative hypothesis, where $t = 2.921$, we do not have enough evidence to reject the null hypothesis and accept the alternative.

Therefore, we would infer that the true population coefficient of correlation is not significantly greater than 0.

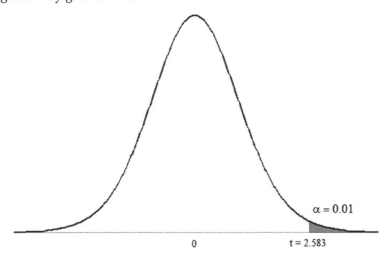

FIGURE 7.10
Rejection region for a right-tailed test of a population coefficient of correlation with a significance level of 0.01 and 16 degrees of freedom.

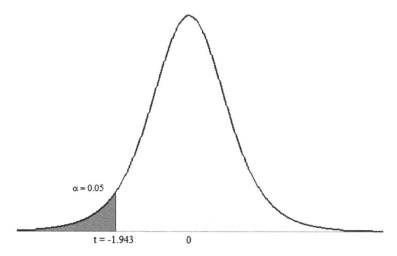

FIGURE 7.11
Rejection region for a left-tailed test of a population coefficient of correlation with a significance level of 0.05 and 6 degrees of freedom.

Example 7.6

Suppose we want to test whether a population coefficient of correlation is significantly less than 0 ($\alpha = 0.05$). For a sample of size $n = 8$ and a sample coefficient of correlation of $r = -0.76$, the appropriate null and alternative hypotheses would be:

$$H_0 : \rho = 0$$

$$H_1 : \rho < 0$$

And the test statistic would be:

$$T = \frac{r}{\sqrt{\dfrac{1-r^2}{n-2}}} = \frac{-0.76}{\sqrt{\dfrac{1-(-0.76)^2}{8-2}}} \approx -2.86$$

and $n - 2 = 8 - 2 = 6$ degrees of freedom. We then compare the value of this test statistic to $t = -1.943$, as illustrated in Figure 7.11.

Since $T < t$, the test statistic falls in the rejection region, we can infer that the true population coefficient of correlation is significantly less than 0.

7.6 Using Minitab for Correlation Analysis

To use Minitab to find the sample coefficient of correlation for the data given in Table 6.1, simply enter each variable in its own column, then under the **Stat** tab select **Basic Statistics < Correlation**. This gives the correlation dialog box that appears in Figure 7.12.

FIGURE 7.12
Minitab dialog box for calculating the sample coefficient of correlation.

By selecting the variables for which you want to find the correlation, simply highlight the variables and choose **Select**. The method is **Pearson correlation** and checking the box that says to **Display p-values** gives the Minitab printout that is presented in Figure 7.13.

Notice that there are two values reported on the Minitab printout, the value of the sample coefficient of correlation $r = 0.741$ and a p-value $= 0.014$. This p-value is the result of the hypothesis test of whether or not the true population coefficient of correlation is significantly different from 0; in other words,

$$H_0 : \rho = 0$$

$$H_1 : \rho \neq 0$$

Since the p-value is 0.014, this suggests that the true population coefficient of correlation is significantly different from 0. Recall that the p-value gives the observed level of significance, and this value is less than $\alpha = 0.05$ but not less than $\alpha = 0.01$, as was illustrated in Example 7.4.

Correlation: SATM, GPA

Correlations

Pearson correlation	0.741
P-value	0.014

FIGURE 7.13
Minitab printout out of sample correlation coefficient and the respective p-value.

7.7 Assessing Linear Regression Model Assumptions

Regression analysis involves so much more than simply entering data into a statistical program and performing a few hypothesis tests, constructing confidence intervals for population parameters, and interpreting p-values. In order to make meaningful inferences using the regression equation, we need to be reasonably sure that the following four assumptions have been met:

1. Any two observations are independent of each other.
2. The error component is normally distributed.
3. The error component has constant variance.
4. The functional relationship between x and y can be determined.

Exploratory data analysis and some formal tests can be used to assess whether or not these model assumptions have been reasonably met. One type of exploratory technique is to use *residual plots*, which are graphs of patterns of residual values, that is, the estimated errors. Recall that residuals represent the vertical distance between the observed y-value $\left(y_i\right)$ and the estimated value $\left(\hat{y}_i\right)$. In other words, $\hat{\varepsilon}_i = y_i - \hat{y}_i$. Using residual plots can give you some indication as to whether or not the model assumptions regarding the unknown error component may have been violated. For instance, an exploratory histogram of the residuals can be used to investigate whether the unknown error component is normally distributed. Formal tests of the assumptions can also be conducted if you are unsure of what you are seeing in the residual plots or if the sample size is too small to show any meaningful patterns.

7.8 Using Minitab to Create Exploratory Plots of Residuals

We can use Minitab to create a scatterplot, find the line of best fit, test the population parameters in our regression analysis, and check any regression assumptions. Using the data given in Table 6.1 and Minitab, the fitted line plot can be found by first going to the **Stat** tab; then selecting **Regression,** followed by **Line Plot** to generate the dialog box in Figure 7.14, which asks you to select the appropriate response and predictor variables along with the type of regression model that you are fitting.

Figure 7.15 gives the fitted line plot.

Notice that the box in the upper right corner of the fitted line plot shows the root mean square error (S) and the coefficient of determination $\left(R^2\right)$. We will describe the adjusted R-squared statistic in more detail in Chapter 9.

Notice that all of the points on the scatter diagram in Figure 7.15 do not lie perfectly on the line of best fit, and the fitted line represents the line where the residual values above and below the line sum to 0 and the squared distance between the line and all of the data points is a minimum.

Minitab will generate four plots that can be used to see if some of the model assumptions for a regression analysis have been met. To do this, select **Stat > Regression > Fit Regression Model** and then select the **Graphs** tab, as illustrated in Figure 7.16.

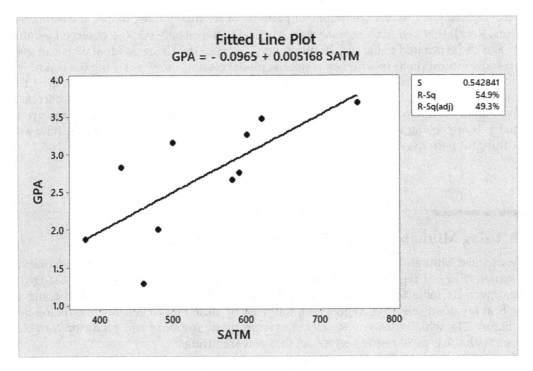

FIGURE 7.14
Minitab dialog box to create a fitted line plot.

FIGURE 7.15
Scatterplot with fitted line superimposed for the data given in Table 7.1.

Then select **Four in one** to get the plot in Figure 7.17.

The four-in-one graph consists of four panels. The bottom left panel is a histogram of the residuals. If the assumption that the error component is normally distributed holds true, then this histogram should resemble a bell-shaped curve. Remember that the residuals give us the unexplained or residual variation, and the estimated model equation is based on the assumption that the error component is normally distributed. By plotting a histogram of the residuals (the difference between the observed and the predicted value),

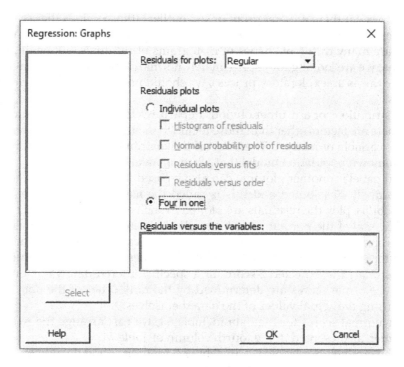

FIGURE 7.16
Minitab dialog box for checking the regression model assumptions.

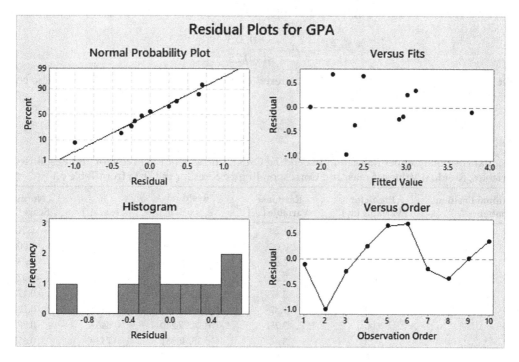

FIGURE 7.17
Four-in-one graph for the data in Table 6.1.

we can look to see if there are any pronounced outliers (those values that seem to be "far removed" from the rest of the data), and if the residuals appear roughly normally distributed. There are many different shapes of histograms of residuals, but again, we need to remember that we are looking for approximately normal distributions. We have to remember that since this is an exploratory process, we should not expect to see perfectly normal distributions.

Skewed distributions or any other obvious departures from normality could be an indication that there are factors other than those included in our model that may be influencing the response variable of interest. Therefore, if the distribution is skewed, any inferences about the unknown population model that we may make could be suspect.

The top left panel is another plot that can also be used to assess whether the error component is normally distributed and this is called the *normal probability plot*. To create a normal probability plot, the residuals are plotted in such a way that a straight-line pattern along the diagonal of the normal probability plot indicates normality. The normal probability plot is created by plotting the residuals in numerical order versus what is called the *normal score* for each index in the sample. Normal scores are theoretical values that can be used to estimate the expected z-score for a specific ordered data value on a standard normal curve. Normal scores are determined by the rank order of the data and are not determined using the actual values of the data themselves.

To find the normal scores for the data in Table 6.1, we first arrange the *residual* data in ascending order, as presented in the fourth column of Table 7.1.

To create the area column labeled A_i in Table 7.1, for each residual value we need to find the area in the lower tail of the standard normal distribution that is based on the rank of the residual observation. In other words, if the residual data comes from a population that is normally distributed, then for the ith residual, the expected area to the left of the observation can be found by using the following formula:

$$A_i = \frac{i - 0.375}{n + 0.25}$$

where i is the position number of the ordered residual and n is the sample size. The constants 0.375 and 0.25 are fixed.

TABLE 7.1

Residual Position Number Ranked in Ascending Order, Corresponding Predictor and Response Variables, Residual Values, Areas, and Corresponding z-Scores for the Data from Table 6.1

Residual Position Number	Predictor Variable (x_i)	Response Variable (y_i)	Residual $(\hat{\varepsilon}_i = y_i - \hat{y}_i)$	Area (A_i)	Normal Score (z_i)
1	460	1.28	−1.00074146	0.0610	−1.547
2	480	2.00	−0.38409806	0.1585	−1.000
3	580	2.65	−0.25088101	0.2561	−0.655
4	590	2.75	−0.20255931	0.3537	−0.375
5	750	3.67	−0.10941204	0.4512	−0.123
6	380	1.87	0.00268490	0.5488	0.123
7	600	3.25	0.24576240	0.6463	0.375
8	620	3.46	0.35240581	0.7439	0.655
9	500	3.14	0.65254535	0.8415	1.000
10	430	2.82	0.69429342	0.9390	1.547

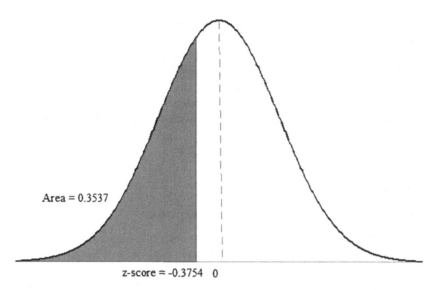

FIGURE 7.18
Area under the standard normal curve that corresponds to the fourth smallest residual value.

Then, z_i, the normal score for the position of the ith residual, is the z-score that corresponds to the area A_i under the standard normal curve.

For example, $A_4 = \dfrac{4 - 0.375}{10 + 0.25} \approx 0.3537$. This value corresponds to the area under the curve to the left of z_4, which corresponds to a z-score of approximately -0.3754, as illustrated in Figure 7.18.

If the residual data are normally distributed, then by plotting the value of the ith residual versus the z-score for the rank ordering of each observation in the sample, all of these points should lie along a straight diagonal line, as can be seen in the top left panel in Figure 7.17.

When Minitab generates the normal probability plot, notice that it does not plot the residual values versus the expected normal score, but instead plots the residual values versus the percentage of A_i. Also, notice that the normal probability plot generated directly from Minitab provides a reference line that can be used to assess if the residuals versus the percentile lie along a diagonal straight line. This line can be obtained by finding the z-score for any given percentage and multiplying the z-score by the standard deviation of the residuals. For instance, if we consider 95% of the normal distribution, and if the standard deviation of the residuals is 0.512, as it is for this example, then the theoretical value of the residual that corresponds to 95% of the normal distribution is:

$$1.645 \times 0.512 \approx 0.842$$

Therefore, one point on the diagonal reference line would be (0.842, 95).

The points in a normal probability plot should generally fall along a diagonal straight line if the residuals are normally distributed. If the points on the normal probability plot depart from the diagonal line, then the assumption that the errors are normally distributed may have been violated.

Minitab also provides a scatterplot of the residuals versus the fitted or estimated values that can be used to assess the assumption of constant variance. This plot should show a random pattern of residuals on both sides of the 0 line. There should not be any obvious patterns or clusters in the residual versus fitted plot. If the residual versus fitted plot shows any kind of pattern, this may suggest that the assumption of constant variance has been violated.

The bottom left panel of Figure 7.17 is a plot of the residuals versus the order of the observations. This plot can be used to check the assumption of independence. If your data was collected *sequentially* and there is not a random scatter around the 0 line, this could indicate that your data is dependent. If the data is independent, you would expect to see the residuals randomly scatter around the 0 line. One drawback of this plot is that it is only useful if your data is collected sequentially.

7.9 A Formal Test of the Normality Assumption

Assessing residual plots by eye to determine if the normality assumption of the error component has been violated can be a very difficult task, especially if the sample size is small. From the last example, it was difficult to determine whether the normality assumption appears to have been violated by simply looking at the histogram of the residuals and the normal probability plot in Figure 7.17.

The *Ryan–Joiner test* (Ryan and Joiner, 1976) can be used as a formal test of the normality assumption of the error component by comparing the distribution of the residuals to a normal distribution to see if they differ in shape. This is done by first computing a measure of correlation between the residuals and their respective normal scores and then using such a correlation as a test statistic. If we can infer that this correlation measure is close to 1.0, then the greater confidence we have that the unknown distribution of the error component is normally distributed. The Ryan–Joiner test statistic is calculated as follows:

$$RJ = \frac{\sum_{i=1}^{n} \hat{\varepsilon}_i \cdot z_i}{\sqrt{s^2 (n-1) \sum_{i=1}^{n} z_i^2}}$$

where $\hat{\varepsilon}_i$ is the ith residual, z_i is the normal score for the ith residual (see Section 7.8 for a review of how to calculate normal scores), n is the sample size, and s^2 is the variance of the residuals.

The null and alternative hypotheses for the Ryan–Joiner test are given as follows:

H_0 : The (population) error component follows a normal distribution.

H_1 : The (population) error component does not follow a normal distribution.

Because the Ryan–Joiner test statistic represents a measure of the correlation between the residuals and data that is obtained from a normal distribution, the distribution of the test statistic has a different shape from what we have already seen. The distribution of the test statistic RJ is skewed, where the rejection region is defined by the critical region rj, which is located in the left tail, as illustrated in Figure 7.19.

The value of the test statistic RJ is then compared to the value rj for a given level of significance in Table 7.2. The value of rj defines the smallest correlation between the residuals and data obtained from a normal distribution.

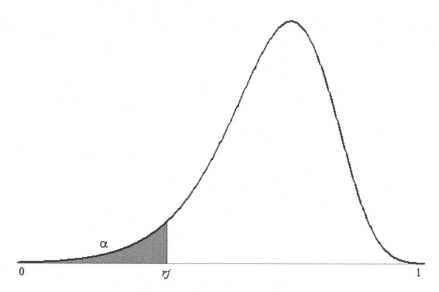

FIGURE 7.19
Distribution of the Ryan–Joiner test statistic.

TABLE 7.2

Approximate Rejection Regions for the Ryan–Joiner Test of Normality

n	α = 0.10	α = 0.05	α = 0.01
4	0.8951	0.8734	0.8318
5	0.9033	0.8804	0.8320
10	0.9347	0.9180	0.8804
15	0.9506	0.9383	0.9110
20	0.9600	0.9503	0.9290
25	0.9662	0.9582	0.9408
30	0.9707	0.9639	0.9490
40	0.9767	0.9715	0.9597
50	0.9807	0.9764	0.9644
60	0.9835	0.9799	0.9710
75	0.9865	0.9835	0.9757

Source: Minitab website (https://www.minitab.com/uploadedFiles/Content/News/Published_Articles/ normal_probability_plots.pdf).

If the test statistic *RJ* falls below any of these critical values, it is less than the smallest correlation between the residuals and a normal distribution that leads to accepting the alternative hypothesis that the error component does not follow a normal distribution. If the test statistic is greater than any of these critical values and falls closer to the value of 1, then there is a strong correlation between the residuals and a normal distribution. This leads to not accepting the alternative hypothesis, and thus we can infer that the assumption of normality of the error component does not appear to have been violated. Recall from Chapter 6 that two variables are positively correlated with each other if they have a correlation close to 1.

Example 7.7

Using the data given in Table 7.1, we find that:

$$\sum_{i=1}^{10} \hat{\varepsilon}_i \cdot z_i \approx 4.23592$$

$$\sum_{i=1}^{10} z_i^2 \approx 7.9560$$

$$s^2 \approx 0.2619$$

So the Ryan–Joiner test statistic would be calculated as follows:

$$RJ = \frac{\sum_{i=1}^{n} \hat{\varepsilon}_i \cdot z_i}{\sqrt{s^2 (n-1) \sum_{i=1}^{n} z_i^2}} = \frac{4.23592}{\sqrt{(0.2619)(10-1)(7.9560)}} \approx 0.9781$$

The value of this test statistic is then compared to the value rj, which defines the smallest correlation that can be used to infer normality for a given sample size. Thus, by comparing the value of the test statistic to the value of $rj = 0.9180$ obtained from Table 7.2 $(n = 10, \alpha = 0.05)$, the test statistic $RJ = 0.9780$ is greater than $rj = 0.9180$ (so it does not fall in the shaded region). This suggests that there is a strong correlation between the residuals and the data obtained from a normal distribution. Thus, the assumption that the unknown population error component is normally distributed does not appear to have been violated. In other words, because the test statistic is greater than the smallest correlation to infer normality defined by the corresponding value of rj for the given level of significance, we can infer that the assumption that the error component is normally distributed does not appear to have been violated.

The assumption of normality would likely be violated if the value of the Ryan–Joiner test statistic is *less than* the respective value that defines the rejection region for a given sample size and the desired level of significance. This implies that the measure of correlation between the residuals and the values obtained from a normal distribution is less than the smallest correlation that would suggest normality. This would lead to accepting the alternative hypothesis that the population error component *does not* follow a normal distribution.

7.10 Using Minitab for the Ryan–Joiner Test

To use Minitab to run the Ryan–Joiner test for the data in Table 6.1, we first need to store the residuals. This can be done by selecting the **Residuals** using the **Storage** box of the **Regression** menu, as illustrated in Figure 7.20.

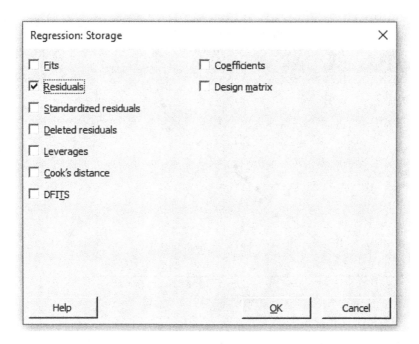

FIGURE 7.20
Minitab dialog box to store the residuals.

Then, once the residuals are found and stored in a column in the Minitab worksheet, the Ryan–Joiner test can be performed by selecting the **Stat** command, then **Basic Statistics**, and then **Normality Test**. The variable is selected as the residuals to run the Ryan–Joiner test in the Normality test dialog box, as is illustrated in Figure 7.21.

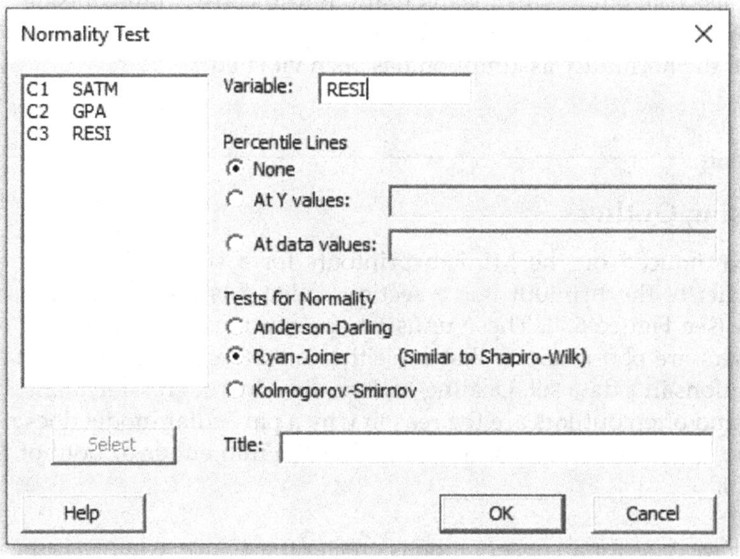

FIGURE 7.21
Minitab dialog box for conducting the Ryan–Joiner test on the residuals.

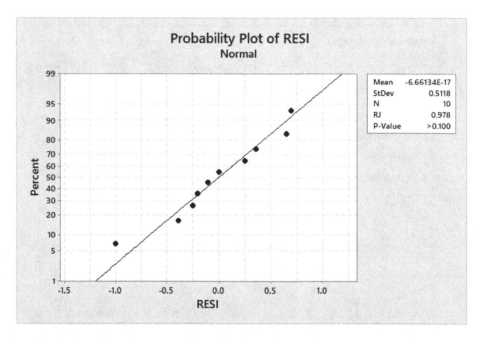

FIGURE 7.22
Normality plot and Ryan–Joiner test generated from Minitab.

Figure 7.22 gives the graph of the normality plot of the residuals along with the Ryan–Joiner test statistic and a corresponding *p*-value provided in the box on the upper right-hand side of the plot.

Notice in the box on the right-hand side of Figure 7.22 that we get the same value of the Ryan–Joiner test statistic as when calculated manually ($RJ = 0.978$). Also, notice that the *p*-value is *greater than* 0.100. This means that the test statistic falls outside of the shaded region, and thus we do not accept the alternative hypothesis and so we do not have reason to believe that the normality assumption has been violated.

7.11 Assessing Outliers

You may have noticed on the Minitab printouts for a simple regression analysis that the bottom part of the printout has a section called Fits and Diagnostics for Unusual Observations (see Figure 6.7). These unusual observations are sometimes referred to as *outliers*. Outliers are observations that are either much larger or much smaller than the other observations in a data set. Dealing with outliers in a regression analysis can be very challenging, and often outliers are the reason why a particular model does not fit well.

Outliers can be identified as observations that fall into either or both of the following two categories:

1. The *leverage value* for an observation is greater than $3 \cdot p/n$, where *p* is the number of beta parameters being estimated in the model and *n* is the sample size.
2. The *standardized residual* for an observation is less than −2 or greater than +2.

7.12 Assessing Outliers: Leverage Values

One way to see if a particular observation is having a strong impact on a regression analysis is to determine the *leverage value* for the given observation. The leverage value for a given observation x_i for a simple linear regression model can be found using the following formula:

$$h_i = \frac{1}{n} + \frac{(x_i - \bar{x})^2}{S_{xx}}$$

where x_i is the observation of interest, $\bar{x}$ is the mean of all the x-values, n is the sample size, and S_{xx} is the sum of the squared differences of the individual x-values from the mean of the x-values $(\bar{x})$.

Leverage values describe whether or not a given observation, x_i, is far removed from the mean of the x-values. Values of h_i will fall between 0 and 1. When h_i is close to 1, this suggests that the observation x_i is far removed from the mean, and if h_i is close to 0, this indicates that the observation is not far removed from the mean.

Example 7.8

The data in Table 6.1 present a sample of the SAT mathematics examination scores and the first-year GPAs for a sample of ten students. Suppose we want to find the leverage value for observation 2 (this is the observation that has an SATM score of 460 and a GPA of 1.28). The leverage value would be calculated as follows:

$$h_i = \frac{1}{n} + \frac{(x_i - \bar{x})^2}{S_{xx}} = \frac{1}{10} + \frac{(460 - 539)^2}{107490} \approx 0.15806$$

Table 7.3 gives the leverage values for all the observations of the data set in Table 6.1.

7.13 Using Minitab to Calculate Leverage Values

By clicking on the **Storage** tab in the regression dialog box, you can have Minitab calculate the leverage values for you, as illustrated in Figure 7.23.

TABLE 7.3

Table of Leverage Values for the Data in Table 7.1

Observation Number	SATM	Leverage Value (h_i)
1	750	0.51419
2	460	0.15806
3	580	0.11564
4	600	0.13462
5	500	0.11415
6	430	0.21053
7	590	0.12420
8	480	0.13238
9	380	0.33519
10	620	0.16104

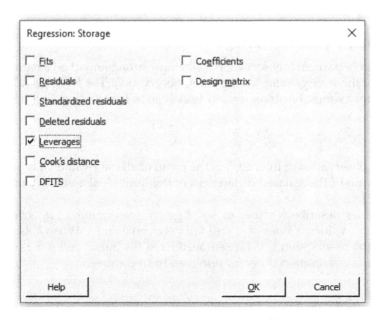

FIGURE 7.23
Minitab storage dialog box to calculate leverage values.

Minitab will then calculate the leverage values and store them in the current Minitab worksheet, as illustrated in Figure 7.24.

When running a regression analysis, Minitab will identify any observations that have a leverage value greater than $3 \cdot p/n$, where p is the total number of beta parameters being estimated in the model and n is the total number of observations. Observations that have

	C1	C2	C3
	SATM	GPA	HI
1	750	3.67	0.514187
2	460	1.28	0.158061
3	580	2.65	0.115639
4	600	3.25	0.134617
5	500	3.14	0.114150
6	430	2.82	0.210531
7	590	2.75	0.124198
8	480	2.00	0.132384
9	380	1.87	0.335194
10	620	3.46	0.161038

Worksheet 1 ***

FIGURE 7.24
Minitab storage of leverage values in column C3.

a leverage value greater than $3{\cdot}p/n$ will be identified with an "X" under the "Fits and Diagnostics for Unusual Observations" portion of the Minitab printout. It is important to note that the value of p is determined by counting all the beta parameters that are being estimated and this includes β_0.

Example 7.9

For the example using the data in Table 6.1, a leverage value would have to be greater than $\dfrac{(3\cdot p)}{n}=\dfrac{(3\cdot 2)}{10}=0.60$, in order for Minitab to identify such an observation as having a large leverage value. We let $p = 2$ because we are estimating two parameters, namely, β_0 and β_1. As can be seen in Table 7.3, since no observations have a leverage value greater than 0.60, there were no observations that were identified as having a large leverage value, as can be seen in the unusual observations portion of the Minitab printout illustrated in Figure 7.25.

If there were large leverage values, then Minitab would identify this observation with an "X" as will be seen in the next example.

Example 7.10

Using the data set provided in Table 7.4 and the Minitab printout of the regression analysis for this data set, as presented in Figure 7.26, we can see that observation 12 has a large leverage value.

Fits and Diagnostics for Unusual Observations

Obs	GPA	Fit	Resid	Std Resid	
2	1.280	2.281	-1.001	-2.01	R

R Large residual

FIGURE 7.25
Portion of the Minitab printout that identifies an unusual observation for the data presented in Table 6.1.

TABLE 7.4

Data Set to Illustrate Leverages

Observation	x	y	Leverage Value (h_i)
1	12	21	0.080660
2	18	19	0.121086
3	2	19	0.111976
4	5	19	0.089628
5	4	20	0.095844
6	6	24	0.084646
7	5	19	0.089628
8	2	19	0.111976
9	5	19	0.089628
10	4	20	0.095844
11	4	19	0.095844
12	45	18	0.852581
13	12	19	0.080660

Fits and Diagnostics for Unusual Observations

Obs	y	Fit	Resid	Std Resid	
6	24.000	19.745	4.255	2.96	R
12	18.000	18.318	-0.318	-0.55	X

R *Large residual*
X *Unusual X*

FIGURE 7.26
Fits and diagnostics for unusual observations portion of the Minitab printout for the data given in Table 7.4.

The reason Minitab identified observation 12 as an influential observation with a large leverage value is because the leverage value $h_{12} \approx 0.852581$ is greater than $3 \cdot \left(\dfrac{p}{n}\right) = 3 \cdot \left(\dfrac{2}{13}\right) \approx 0.46$. For this example, $p = 2$ because there are two parameters being estimated in the model, namely, β_0 and β_1, and the sample size is $n = 13$.

7.14 Assessing Outliers: Standardized Residuals

If a particular residual is more than two standard deviations away from the mean, then such an observation could have an impact on the regression analysis. We will be using what are called *standardized residuals* as a way to quantify whether a particular observation could be affecting the regression analysis. The formula to calculate a standardized residual for a given observation is as follows:

$$\varepsilon_i^s = \frac{\hat{\varepsilon}_i}{S\sqrt{1-h_i}}$$

where $\hat{\varepsilon}_i$ is the ith residual, S is the root mean square error from the regression analysis, and h_i is the leverage value for the ith residual.

Example 7.11

The standardized residual for observation 2 of the SATM-GPA data set that is provided in Table 6.1 can be calculated as follows:

$$\varepsilon_i^s = \frac{\hat{\varepsilon}_i}{S\sqrt{1-h_i}} = \frac{-1.0007}{(0.542841)\sqrt{1-0.15806}} \approx -2.0091$$

Table 7.5 presents the leverage values and the standardized residuals for each observation of the data given in Table 6.1.

TABLE 7.5

Table of Leverage Values and Standardized Residuals for the Data in Table 6.1

Observation Number	SATM	Leverage Value (h_i)	Standardized Residual (ε_i^s)
1	750	0.51419	−0.28917
2	460	0.15806	−2.00913
3	580	0.11564	−0.49145
4	600	0.13462	0.48667
5	500	0.11415	1.27720
6	430	0.21053	1.43947
7	590	0.12420	−0.39873
8	480	0.13238	−0.75964
9	380	0.33519	0.00607
10	620	0.16104	0.70876

7.15 Using Minitab to Calculate Standardized Residuals

Figure 7.27 shows the regression storage dialog box where you can select to have Minitab calculate the standardized residuals for each of the individual observations.

Using the data given in Table 6.1, Figure 7.28 shows that Minitab calculates these residuals and stores them in the Minitab worksheet.

Notice in the unusual observations portion of the Minitab printout, as in Figure 7.29, that observation 2 is identified with the value of "R." The designation of "R" indicates that the observation has a standardized residual that is either less than −2 or greater than +2.

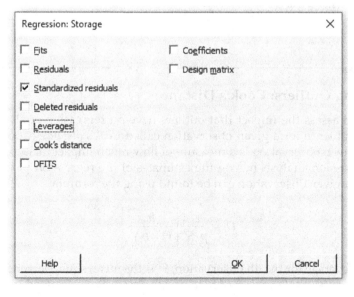

FIGURE 7.27
Regression dialog box to calculate the standardized residuals.

↓	C1	C2	C3 ☑	C4
	Observation	SATM	GPA	SRES
1	1	750	3.67	-0.28917
2	2	460	1.28	-2.00913
3	3	580	2.65	-0.49145
4	4	600	3.25	0.48667
5	5	500	3.14	1.27720
6	6	430	2.82	1.43947
7	7	590	2.75	-0.39873
8	8	480	2.00	-0.75964
9	9	380	1.87	0.00607
10	10	620	3.46	0.70876

Worksheet 1 ***

FIGURE 7.28
Minitab storage of standardized residuals.

Fits and Diagnostics for Unusual Observations

Obs	GPA	Fit	Resid	Std Resid	
2	1.280	2.281	-1.001	-2.01	R

R Large residual

FIGURE 7.29
Fits and diagnostics for unusual observations for a portion of a Minitab printout indicating that observation 2 has a standardized residual that is less than –2.00.

7.16 Assessing Outliers: Cook's Distances

Another way to assess the impact that outliers have on regression analysis is to look at a measure of influence for a given observation called *Cook's distance* (Cook, 1979). Cook's distance for a given observation is a measure of how much impact including the observation in the regression analysis has on the estimates of the regression parameters. Cook's distance for any given observation can be found using the formula:

$$D_i = \frac{\hat{\varepsilon}_i^2}{p \cdot S^2} \left[\frac{h_i}{(1-h_i)^2} \right]$$

where h_i is the leverage for the ith observation, S^2 is the mean square error from the regression analysis, p is the total number of beta parameters being estimated in the model, and $\hat{\varepsilon}_i^2$ is the squared residual for the ith observation.

Example 7.12

The value of Cook's distance for observation 2 from Table 6.1 can be calculated as follows:

$$D_i = \frac{\hat{\varepsilon}_i^2}{p \cdot S^2} \left[\frac{h_i}{(1-h_i)^2} \right] \approx \frac{(-1.0007)^2}{2 \cdot (0.5428)^2} \left[\frac{0.15806}{(1-0.15806)^2} \right] \approx 0.3789$$

7.17 Using Minitab to Find Cook's Distances

Similar to leverage values and standardized residuals, Minitab can calculate Cook's distance for all of the observations by using the **Storage** command on the regression dialog box. Figure 7.30 gives the Cook's distances for the data in Table 6.1.

One of the more interesting ways of using Cook's distances to assess the impact of outliers on a regression analysis is to draw a scatterplot of the observation number on the x-axis against the Cook's distance on the y-axis. Such a plot is illustrated in Figure 7.31.

This plot can be used to visually assess the impact that an outlier may have on a regression analysis by showing how far the Cook's distance for the observation falls away from 0. Notice in Figure 7.31 that observations 2 and 6 appear to be the most influential because the Cook's distances for these observations fall the farthest away from 0.

	C1	C2	C3 ☑	C4
	Observation	SATM	GPA	COOK
1	1	750	3.67	0.044253
2	2	460	1.28	0.378907
3	3	580	2.65	0.015791
4	4	600	3.25	0.018422
5	5	500	3.14	0.105100
6	6	430	2.82	0.276285
7	7	590	2.75	0.011273
8	8	480	2.00	0.044024
9	9	380	1.87	0.000009
10	10	620	3.46	0.048212

Worksheet 1 ***

FIGURE 7.30
Minitab storage of Cook's distances.

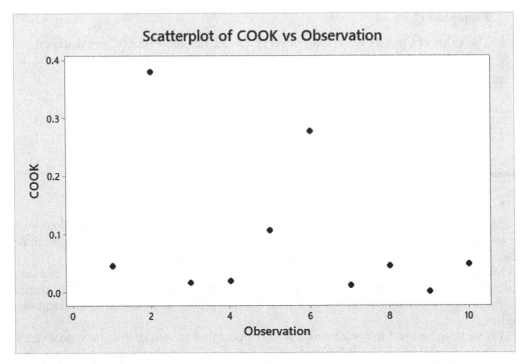

FIGURE 7.31
Scatterplot of Cook's distance versus the observation number.

7.18 How to Deal with Outliers

To determine the impact that an outlier has on a regression analysis, you can fit the model both with and without the outlier. Then you can to examine whether the values of the estimated coefficients, along with the associated p-values and the R^2 statistic, have changed between the two models. Whether to delete an outlier permanently is a matter of personal conviction and tends to differ by field of study. The convention in some fields is to never delete an outlier, whereas others may do so if they believe the outlier was obtained either erroneously or through measurement error.

Example 7.13

Figures 7.32 and 7.33 illustrate how the value of the estimated coefficients, p-values, and R^2 statistics change by deleting the influential observation 2.

By comparing the information given in Figures 7.32 and 7.33, notice that there is not a noticeable difference between the parameter estimates, p-values, and the value of the R^2 statistic for the two models. Since this unusual observation does not appear to be grossly affecting the analysis, it may be a good idea to keep this observation in the data set. Should there be a noticeable difference between the two models with respect to the R^2 statistic, parameter estimates, and p-values, you would want to investigate further why this particular observation is so influential.

Regression Analysis: GPA versus SATM

Analysis of Variance

Source	DF	Adj SS	Adj MS	F-Value	P-Value
Regression	1	2.871	2.8707	9.74	0.014
SATM	1	2.871	2.8707	9.74	0.014
Error	8	2.357	0.2947		
Total	9	5.228			

Model Summary

S	R-sq	R-sq(adj)	R-sq(pred)
0.542841	54.91%	49.27%	35.60%

Coefficients

Term	Coef	SE Coef	T-Value	P-Value	VIF
Constant	-0.096	0.909	-0.11	0.918	
SATM	0.00517	0.00166	3.12	0.014	1.00

Regression Equation

GPA = -0.096 + 0.00517 SATM

Fits and Diagnostics for Unusual Observations

Obs	GPA	Fit	Resid	Std Resid	
2	1.280	2.281	-1.001	-2.01	R

R Large residual

FIGURE 7.32
Portion of the Minitab printout of the regression analysis of the first-year GPA versus score received on the mathematics portion of the SAT examination including observation 2.

Exercises

Unless otherwise specified, use $\alpha = 0.05$.

1. An organizational psychologist is interested in determining what factors in the workplace are related to employee satisfaction. Survey data were collected from 20 employees regarding measures of physical working conditions, peer collaboration, salary and benefits, and leadership involvement. All of these variables are ranked on a scale from 0 to 100, where higher scores represent stronger agreement that the variable is important to employee satisfaction. Table 7.6 gives the correlations of working conditions, peer collaborations, salary and benefits, and leadership involvement with overall satisfaction.

   Find and interpret any significant correlations.

2. What a difference a single outlier can make! The data set in Table 7.7 presents a collection of ordered pairs (x, y).

   a. Using Minitab, draw a scatterplot and run a simple linear regression analysis to develop a model that describes the relationship between x and y.

Regression Analysis: GPA versus SATM

Analysis of Variance

Source	DF	Adj SS	Adj MS	F-Value	P-Value
Regression	1	1.854	1.8543	11.11	0.013
SATM	1	1.854	1.8543	11.11	0.013
Error	7	1.168	0.1668		
Total	8	3.022			

Model Summary

S	R-sq	R-sq(adj)	R-sq(pred)
0.408467	61.36%	55.84%	41.13%

Coefficients

Term	Coef	SE Coef	T-Value	P-Value	VIF
Constant	0.493	0.719	0.69	0.515	
SATM	0.00429	0.00129	3.33	0.013	1.00

Regression Equation

GPA = 0.493 + 0.00429 SATM

FIGURE 7.33
Portion of the Minitab printout of the regression analysis of the first-year GPA versus score received on the mathematics portion of the SAT examination deleting observation number 2.

TABLE 7.6

Correlations of Working Conditions, Peer Collaborations, Salary and Benefits, and Leadership Involvement with Overall Satisfaction

	Overall Satisfaction
Working conditions	0.41
Peer collaboration	0.45
Salary and benefits	0.43
Leadership involvement	−0.48

b. Check graphically to see if the regression model assumptions have been reasonably met.

c. Run the Ryan–Joiner test of normality and comment on the results.

d. Determine the leverage values, standardized residuals, and Cook's distances and draw a Cook's distance plot.

e. Notice the outlier that has the largest Cook's distance. Remove this outlier from the data set, and rerun steps (a) through (c). Comment on the impact that keeping this point in the data set has on your analysis.

3. Can the use of social media stress you out? A researcher is interested in finding out whether college students who extensively use social media tend to have higher levels of stress. To test this, a sample of nine college students were asked to fill

TABLE 7.7

Outlier Data Set

x	y
80	83
54	59
78	86
71	63
68	68
73	77
82	83
74	79
78	78
59	65
97	68
73	82
80	78
54	59
75	80
66	61
72	78
55	63
70	75
73	80
68	69
78	82
64	69
72	74
62	65
81	84
79	92
57	63
68	68
77	84

out a survey on their stress level and provide an estimate of the number of hours they spend per day on social media. The stress survey is on a scale of 0–100 where higher scores are indicative of a higher level of stress. The data is given in Table 7.8.

a. Fill in the five blank cells in the above table.

b. Use Table 7.8 to find and interpret the coefficient of determination.

c. Use Table 7.8 to find the coefficient of correlation.

d. Is there a significant correlation between the number of hours on social media and stress level?

4. A plant manager wants to estimate how the number of units produced at a plant is affected by the number of employees. The data in Table 7.9 gives a random sample of the number of units produced and the number of employees.

a. Which variable is the response variable and which is the predictor variable?

TABLE 7.8

Number of Hours a Random Sample of Nine Students Spend Per Day on Social Media and Their Stress Levels

	Hours	Stress	$\hat{y}_i$	$(y_i - \hat{y}_i)$	$(y_i - \hat{y}_i)^2$	$(y_i - \bar{y})$	$(y_i - \bar{y})^2$
	5	55	57.6100	−2.6100	6.8121	6.7778	45.9383
	2	40	39.5050	0.4950	0.2450		67.6049
	3	46	45.5400	0.4600	0.2116	−2.2222	4.9383
	4	46	51.5750	−5.5750	31.0806	−2.2222	4.9383
	1	37	33.4700	3.5300	12.4609		125.9383
	3	46	45.5400	0.4600		−2.2222	4.9383
	6	71	63.6450	7.3550	54.0960	22.7778	
	2	38	39.5050		2.2650	−10.2222	104.4938
	5	55	57.6100	−2.6100	6.8121	6.7778	45.9383
Total	31	434	434.0000	0.0000	114.1950	0.0000	923.5556

TABLE 7.9

Production Volume and Number of Employees

Number of Units Produced	Number of Employees
537	52
673	58
544	53
781	62
357	47
289	45
688	68
469	43
573	49
579	54
670	63
814	78

b. Draw a scatterplot and find the equation of the line of best fit that models the relationship between the number of units produced and the number of employees working.

c. Interpret whether or not you believe there is a significant relationship between the number of units produced and the number of employees working.

d. Find and interpret the coefficient of determination.

e. Check any relevant model assumptions and identify any outliers.

f. Find and interpret a 99% confidence interval for the true population slope parameter.

g. Find and interpret a 99% confidence interval for the predicted number of units produced when 50 employees are working.

h. Find and interpret a 99% confidence interval for the mean number of units produced when 50 employees are working.

5. Is there a relationship between buying an expensive car and the level of satisfaction with the car? To answer this question, a consumer researcher randomly sampled 15 individuals, asked them about the price of their car, and had them fill out a satisfaction survey. The survey was scored on a scale from 0 to 100, where higher scores indicated a higher degree of satisfaction with the vehicle. This data is provided in Table 7.10.

a. Which variable is the response variable and which is the predictor variable?

b. Draw a scatterplot and find the equation of the line of best fit that models the relationship between the price paid for a vehicle and the level of satisfaction.

c. Interpret whether or not you believe there is a significant relationship between the price paid for a vehicle and the level of satisfaction.

d. Find and interpret the coefficient of determination.

e. Check any relevant model assumptions and identify any outliers.

6. The correlation inference that was described in this chapter can only test whether a population coefficient of correlation is *significantly different from 0*. We may also be interested in testing hypotheses for population correlations other than 0. In order to do this, we need to use Fisher's Z-transformation so that we can obtain a test statistic that is approximately normally distributed. However, this transformation only gives an approximate normal distribution when the sample size is greater than 20.

TABLE 7.10

Price (in Dollars) and the Level of Satisfaction for a Random Sample of 15 Individuals

Price (in Dollars)	Satisfaction
38,650	89
45,210	75
18,690	83
23,340	76
99,850	90
63,920	40
19,830	45
56,674	79
19,250	93
23,674	69
34,250	47
16,584	39
26,984	85
37,684	85
98,684	37

The test statistic for testing correlations other than 0 is as follows:

$$Z = \frac{\frac{1}{2}\ln\left[\left(1+r\right)\Big/\left(1-r\right)\right] - \frac{1}{2}\ln\left[\left(1+\rho\right)\Big/\left(1-\rho\right)\right]}{\frac{1}{\sqrt{n-3}}}$$

where ln is the natural logarithm and ρ_0 is the population coefficient of correlation being tested under the null hypothesis $(\rho_0 \neq 0)$.

This test statistic is used in a manner similar to that of other test statistics that we have calculated throughout this text. However, because this test statistic follows a standard normal distribution, we will be using the standard normal distribution so we do not need to be concerned with the degrees of freedom.

Suppose we have a sample of size $n = 120$, and we want to test whether the true population correlation between two variables, x and y, is significantly different from 0.70. Suppose, based on sample data, we find that $r = 0.75$.

a. Set up the appropriate null and alternative hypotheses.

b. Calculate the test statistic Z.

c. Using the standard normal distribution, shade in the appropriate rejection region(s) and label the corresponding values of z.

d. Find the p-value and interpret your results.

7. If samples are taken from two *different populations*, we can also test whether the two population coefficients of correlation are significantly different from each other. To do this, we first need to find the appropriate Z-transformations for each of the different populations we are sampling from:

$$Z_1 = \frac{1}{2} \cdot \ln \frac{1+r_1}{1-r_1}$$

$$Z_2 = \frac{1}{2} \cdot \ln \frac{1+r_2}{1-r_2}$$

where r_1 is the sample coefficient of correlation from the first population, r_2 is the sample coefficient of correlation from the second population, and ln is the natural logarithm.

Given each of these transformations, the following test statistic can be used, which follows a standard normal distribution:

$$Z = \frac{Z_1 - Z_2}{\sqrt{\frac{1}{(n_1 - 3)} + \frac{1}{(n_2 - 3)}}}$$

Suppose that from two different samples of size $n_1 = 44$ and $n_2 = 88$, we have two sample correlations, $r_1 = 0.29$ and $r_2 = 0.31$. We want to determine if the population correlation for the first sample is significantly less than the population coefficient of correlation for the second sample.

a. State the appropriate null and alternative hypotheses.

b. Calculate the appropriate Z-transformation for each correlation.

c. Calculate the test statistic Z.

d. Using the standard normal curve, shade in the appropriate rejection region(s) and label the corresponding values of z that define the rejection region(s).

e. Find the *p*-value and interpret your results.

References

R. Cook (1979). Influential observations in linear regression, *Journal of the American Statistical Association*, 74: 169–174.

T. Ryan and B. Joiner. *Normal Probability Plots and Tests for Normality*. Technical report. University Park, PA: The Pennsylvania State University, 1976.

8

Multiple Regression Analysis

8.1 Introduction

It is often the case that more than one predictor, or x-variable, has an impact on a given response, or y-variable. For example, suppose we are interested in developing a model that can be used to predict executive salaries. An executive's salary can be determined not only by his or her level of education and the number of years of experience, but also by other factors, such as whether the business is profit or nonprofit, and the number of employees managed by the executive. You can probably even think of more factors that could also be used in determining an executive's salary beyond those just mentioned, such as the gender of the executive and the region of the country where he or she is employed. To account for multiple factors having an impact on a continuous outcome or response variable of interest, *multiple regression analysis* can be used.

This chapter begins by presenting the basics of a multiple regression analysis. We will work through some examples and provide an explanation of how to use Minitab to run a multiple regression analysis. We will also describe some measures that can be used to assess how well a multiple regression model fits the underlying set of data, how multicollinearity can impact the inferences made from a multiple regression analysis, and elaborate on how to use Minitab to assess some of the model assumptions.

8.2 Basics of Multiple Regression Analysis

In order to obtain a useful regression model that includes more than one independent or x-variable, it is important to consider *all* of the variables that may impact the outcome measure of interest. For instance, in modeling executive salaries, besides collecting data on the level of education and the number of years of experience, we also may want to collect data on any other factors that could impact executive salaries, such as the type of business (profit or nonprofit), the numbers of employees managed, gender, the region where employed, and so on.

The set of predictor variables, or x-variables, that may impact a given response variable, or y-variable, is also called *control variables*. A control variable is a variable included in a regression analysis because it may have a significant impact on the response variable of interest. *Multiple regression analysis* is a statistical technique that can be used to create a regression model that includes a collection of control variables, or x-variables, where such variables can be used to model a single continuous response, or y-variable.

The basic form of a linear multiple regression model is similar to that of a simple linear regression model, except that the additional predictor variables are included in the population model equation, as described below:

$$y = \beta_0 + \beta_1 x_1 + \beta_2 x_2 + \cdots + \beta_j x_j + \varepsilon$$

where y is the continuous response (or dependent) variable, $x_1, x_2, \ldots, x_j$ are the j-different predictor (or independent) variables, β_0 is the constant term, $\beta_1, \beta_2, \ldots, \beta_j$ are the unknown population coefficients that correspond to the effect that each of the different independent variables has on the response variable, and ε is the random error component. The assumptions underlying the error component in a multiple regression analysis are similar to the assumptions underlying the error component in a simple linear regression analysis. These assumptions are that the observations are independent and the error component is normally distributed and has constant variance.

The population linear multiple regression equation as just described stipulates that the response variable is linearly related to each of the independent variables when the other independent variables are held fixed. In other words, the effect of the ith independent variable, x_i, is found by estimating the population parameter β_i, and the estimate of β_i, denoted as $\hat{\beta}_i$, describes the estimated effect that the variable x_i has on the response variable when all of the other predictor variables are held fixed.

Just like with simple linear regression analysis, we will estimate the model equation that best fits the data, and confidence intervals can be found and hypothesis tests can be performed on the unknown population regression parameters by using the statistics obtained from the estimated model equation (i.e., the sample equation). Different strategies and techniques, such as exploratory analyses and formal tests, can be used to assess the model fit. However, by including more than one predictor variable in the model, the calculations used to find the equation of best fit become much more tedious and very complicated to carry out manually. This is where Minitab or any other statistical program can be very useful.

One way of getting more familiar with multiple regression analysis is by working through an example. In doing so, we will focus on describing the basic multiple regression equation, how to determine the estimates of the parameters, how to assess the model fit, and how to interpret the estimates of the parameters.

Example 8.1

Suppose we are interested in modeling the effect that the number of years of education and the number of years of experience have on executive salaries. The data provided in Table 8.1 consists of the number of years of education, the number of years of experience, and the yearly salary for a random sample of 15 executives. In this example, the continuous predictor variables are the number of years of education and the number of years of experience, and the continuous response variable is the yearly salary.

We want to develop a model that predicts the effect that the number of years of education and the number of years of experience have on executive salaries and because we have two independent variables (education and experience), the population linear model equation that we are interested in estimating can be written as follows:

$$y = \beta_0 + \beta_1 x_1 + \beta_2 x_2 + \varepsilon$$

TABLE 8.1

Education, Experience, and Yearly Salary for a Random
Sample of 15 Executives

Observation Number	Education in Years (x_1)	Experience in Years (x_2)	Yearly Salary in Thousands of Dollars (y)
1	12	12	95
2	13	19	145
3	16	20	164
4	14	24	186
5	15	30	197
6	12	16	139
7	13	19	163
8	15	15	125
9	16	25	122
10	13	26	173
11	14	25	152
12	12	5	75
13	13	19	165
14	13	22	167
15	14	28	187

We will use Minitab to find the estimated equation:

$$\hat{y} = \hat{\beta}_0 + \hat{\beta}_1 x_1 + \hat{\beta}_2 x_2$$

where $\hat{\beta}_1$ and $\hat{\beta}_2$ are the estimated effects of education and experience on executive salaries, respectively, and $\hat{\beta}_0$ is an estimate of the constant parameter β_0.

However, before we use Minitab to find the estimated model equation, we may first want to take a look at how each of the individual predictor variables is related to the response variable, y. We can do this by creating what is called a matrix plot.

8.3 Using Minitab to Create Matrix Plots

In simple linear regression when there is only one predictor variable, we used a scatterplot to assess the relationship between the predictor variable and the response variable. However, with multiple regression, there is more than one independent variable. Although a scatterplot can only assess the relationship between two variables, we can use a collection of scatterplots, called a *matrix plot*, to display the relationship between each of the individual predictor variables and the response variable. To create a matrix plot with Minitab (see Sections 2.14 and 2.15), select **Graph** and then **Matrix Plot**. In order to graph each predictor variable versus the response variable, we need to select a matrix plot for **Each Y versus each X** (and with a smoother), as presented in Figure 8.1.

To draw a matrix plot for all of the x- and y-variables, highlight and select the appropriate variables, as presented in Figure 8.2.

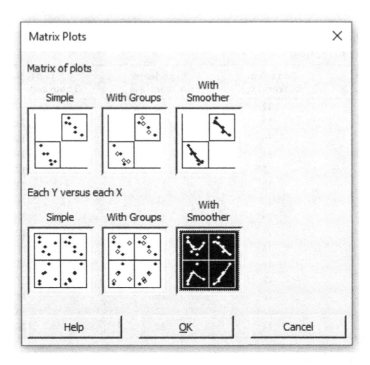

FIGURE 8.1
Minitab dialog box to select a plot of each *y* versus each *x* with a smoother.

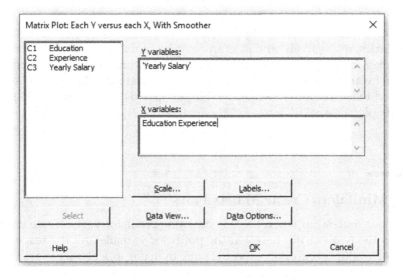

FIGURE 8.2
Matrix plot dialog box to select appropriate *y*- and *x*-variables.

Figure 8.3 presents the graph of the matrix plot, which consists of individual scatterplots for each of the predictor variables versus the response variable along with a smoothed curve that approximates the functional relationship between each *x* and *y*. The smoothed curve (the type of curve that Minitab draws is called a lowess smoother, which stands for

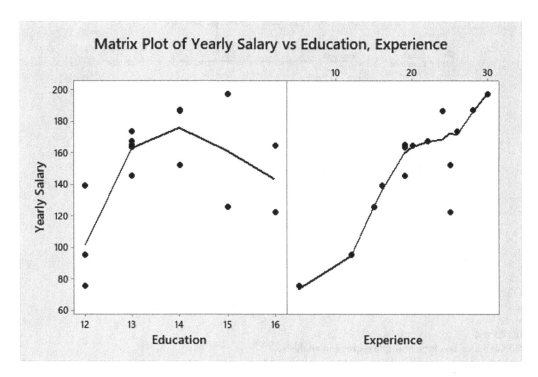

FIGURE 8.3
Matrix plot and lowess smoother describing the relationship between salary and education and salary and experience.

locally weighted scatterplot smoother) provides an empirical approximation of the functional form relationship between each predictor variable and the response variable.

Notice that the left panel of the matrix plot in Figure 8.3 describes the relationship between the years of education and yearly salary. The shape of the scatterplot and smoother suggest that the relationship between education and yearly salary may not be linear, but instead may be represented by a quadratic or some other nonlinear function. However, notice that the right panel of Figure 8.3 suggests more of a linear relationship between the number of years of experience and yearly salary.

Although a matrix plot will not tell you the best model that should be used in a regression analysis, it can be used in an exploratory manner to indicate whether the functional form of the relationship between each of the individual predictor variables and the response variable is linear.

8.4 Using Minitab for Multiple Regression

The same steps are required to use Minitab to run a multiple regression analysis as with a simple regression analysis. Under the **Stat** menu, select **Regression**, and **Fit Regression Model**. This gives the regression dialog box presented in Figure 8.4.

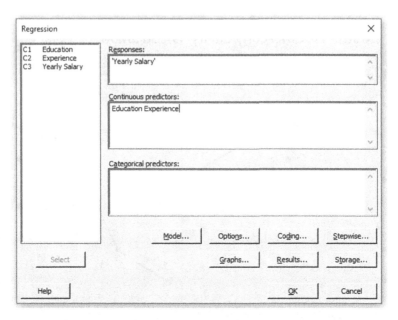

FIGURE 8.4
Minitab dialog box for a multiple regression analysis.

Then selecting the appropriate response and continuous predictor variables gives the Minitab output that is presented in Figure 8.5. The section in the lower portion of the Minitab printout in Figure 8.5 gives the estimated model equation (the regression equation):

$$\hat{y} \approx 108.2 - 4.18x_1 + 4.882x_2$$

Also given in Figure 8.5 are the parameter estimates for each of the predictor variables, standard errors for each of the estimated coefficients, values of the test statistic for each estimated coefficient, and the respective p-values. These can be found in the Coefficients section of the Minitab printout.

Before we describe in more detail how to make inferences using the information contained in the Minitab printout, we will first describe a way to assess the initial fit of a multiple regression model by using the coefficient of determination and conducting an overall test of the model.

8.5 The Coefficient of Determination for Multiple Regression

As with simple linear regression, the *coefficient of determination*, R^2, can be used to determine the proportion (or percentage) of the variability in y that is explained by the estimated regression model, and it can be calculated using the following formula:

$$R^2 = \frac{\text{SSR}}{\text{SST}} = \frac{\text{SST} - \text{SSE}}{\text{SST}}$$

Regression Analysis: Yearly Salary versus Education, Experience

Analysis of Variance

Source	DF	Adj SS	Adj MS	F-Value	P-Value
Regression	2	12096.0	6048.0	16.52	0.000
Education	1	317.6	317.6	0.87	0.370
Experience	1	10276.9	10276.9	28.07	0.000
Error	12	4393.4	366.1		
Lack-of-Fit	10	4150.7	415.1	3.42	0.247
Pure Error	2	242.7	121.3		
Total	14	16489.3			

Model Summary

S	R-sq	R-sq(adj)	R-sq(pred)
19.1341	73.36%	68.92%	52.42%

Coefficients

Term	Coef	SE Coef	T-Value	P-Value	VIF
Constant	108.2	54.0	2.00	0.068	
Education	-4.18	4.49	-0.93	0.370	1.39
Experience	4.882	0.922	5.30	0.000	1.39

Regression Equation

Yearly Salary = 108.2 - 4.18 Education + 4.882 Experience

Fits and Diagnostics for Unusual Observations

Obs	Yearly Salary	Fit	Resid	Std Resid	
9	122.00	163.36	-41.36	-2.56	R

R Large residual

FIGURE 8.5
Minitab printout for the multiple regression analysis that models the yearly salary based on years of education and years of experience.

where $\text{SST} = \sum_{i=1}^{n}(y_i - \bar{y})^2$ is the total sum of squares, $\text{SSE} = \sum_{i=1}^{n}(y_i - \bar{y}_i)^2$ is the error sum of squares, and $\text{SSR} = \text{SST} - \text{SSE}$ is the regression sum of squares.

Example 8.2

To calculate the coefficient of determination for the results of the multiple regression analysis presented in Figure 8.5, we can take the ratio of the regression sum of squares to the total sum of squares as follows:

$$R^2 = \frac{\text{SSR}}{\text{SST}} = \frac{\text{SST} - \text{SSE}}{\text{SST}} = \frac{16489.3 - 4393.4}{16489.3} \approx 0.734$$

Thus, 73.4% of the variability in executive salaries is attributed to the model that includes education and experience as predictor variables. Recall that for large values of the R^2 statistic, this suggests that a large amount of variability in the response variable is due to the given model, whereas an R^2 value near 0 suggests that the model does not account for a large amount of variability in the response variable.

8.6 The Analysis of Variance Table

When using Minitab to run a regression analysis, you may have noticed that an *analysis of variance* (ANOVA) table also appears on the top portion of the printout (see Figure 8.5). We can use the information in the ANOVA table to conduct an *F*-test to determine whether the multiple regression model that we have specified is a *useful model* for predicting the response variable. In other words, we want to know how well the overall model (which includes all of the given predictor variables) contributes to predicting the response variable.

Because we are interested in determining whether the overall model contributes to predicting the response, *y*, we need to test whether at least one of the predictor variables is significantly different from 0. To do this, we will consider the following null and alternative hypotheses:

$$H_0 : \beta_1 = \beta_2 = \cdots = \beta_j = 0$$

$$H_1 : \text{At least one } \beta_i \text{ is different from } 0$$

The test statistic that we will use consists of the ratio of the mean square due to the regression (MSR) and the mean square error (MSE), as follows:

$$F = \frac{\text{MSR}}{\text{MSE}}$$

In performing this *F*-test, we will be using the *F*-distribution, as is illustrated in Figure 8.6.

The *F*-distribution is a right-skewed distribution that begins at 0 and goes infinitely to the right, approaching but never touching the horizontal axis. Recall that the shape of the *F*-distribution relies on the numerator and the denominator degrees of freedom. Because

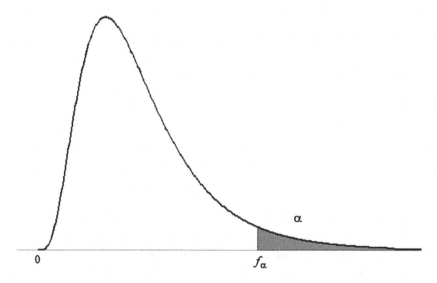

FIGURE 8.6
F-distribution.

our test statistic is the ratio of MSR and MSE, the number of degrees of freedom for the numerator would be $p-1$ and the number of degrees of freedom for the denominator would be $n-p$, where p is the total number of parameters being estimated in the model (the total number of parameters being estimated includes β_0) and n is the number of observations.

The value of f_α in Figure 8.6 defines the rejection region for a significance level of α. Notice that the rejection region for the F-distribution for this test will always be in the right tail. This is because we expect the MSR to be larger than the MSE if at least one predictor is significant.

We can use the following notation to describe the value of f for a specified level of significance and number of degrees of freedom for the numerator and the denominator as follows:

$$f_\alpha(p-1, n-p)$$

where α is the given level of significance.

Example 8.3

Using a significance level of $\alpha = 0.05$ for the executive salary data and the Minitab printout given in Figure 8.5, suppose we want to determine if the overall model contributes to predicting y. The null and alternative hypotheses we would be testing are as follows:

$$H_0 : \beta_1 = \beta_2 = 0$$

$$H_1 : \text{At least one } \beta_i \text{ is different from } 0$$

In order to do this, we need to find the ratio of MSR and MSE as follows:

$$F = \frac{\text{MSR}}{\text{MSE}}$$

The MSR can be found by taking the regression sum of squares (SSR) and dividing it by $p-1$, as follows:

$$\text{MSR} = \frac{\text{SSR}}{p-1}$$

where $\text{SSR} = \sum_{i=1}^{n}(\hat{y}_i - \bar{y})^2$ and p is the total number of beta parameters being estimated.

The MSE can be found by dividing SSE by $n-p$, as follows:

$$\text{MSE} = \frac{\text{SSE}}{n-p}$$

where $\text{SSE} = \sum_{i=1}^{n}(y_i - \hat{y}_i)^2$, p is the total number of beta parameters being estimated, and n is the sample size. Recall for a regression analysis, the sample size is the number of *observations* being considered and not the total number of data points collected.

Now using the data given in Table 8.2, we can find SSR and MSR as follows:

$$SSR = \sum_{i=1}^{15} (\hat{y}_i - \bar{y})^2 \approx 12095.97$$

$$MSR = \frac{SSR}{p-1} = \frac{12095.97}{3-1} \approx 6047.99$$

Note that $p = 3$ because there are three parameters that are being estimated, namely, β_0, β_1, and β_2.

Similarly, to find SSE and MSE:

$$SSE = \sum_{i=1}^{15} (y_i - \hat{y}_i)^2 \approx 4393.37$$

$$MSE = \frac{SSE}{n-p} = \frac{4393.37}{15-3} \approx 366.11$$

Therefore,

$$F = \frac{MSR}{MSE} = \frac{6047.99}{366.11} \approx 16.52$$

We can now compare the value of this test statistic to the F-distribution with $p - 1 = 3 - 1 = 2$ degrees of freedom for the numerator and $n - p = 15 - 3 = 12$ degrees of freedom for the denominator using the F-distribution in Table A.3.

The value of the test statistic $F = 16.52$ is greater than the value of $f = 3.885$, which defines the rejection region for $\alpha = 0.05$ with 2 degrees of freedom for the numerator and 12 degrees of freedom for the denominator, as is illustrated in Figure 8.7.

Since the test statistic $F = 16.52$ falls in the shaded region as defined by $f = 3.885$, we can accept the alternative hypothesis and reject the null hypothesis. Thus, we can infer that at least one of the population predictor parameters is significantly different from 0. In essence, this test suggests that the overall model that includes the independent variables of years of education and years of experience is useful in predicting yearly salary.

The ANOVA table obtained from the Minitab printout, given in Figure 8.8, provides approximately the same values for SSR, SSE, MSR, and MSE along with the corresponding degrees of freedom, and the value of the F-statistic (the slight differences are due to the round-off error).

Notice that the p-value for the F-test is 0.000. Since the p-value is less than our predetermined level of significance of 0.05, we can accept the alternative hypothesis that at least one β_i is different from 0, and reject the null hypothesis that all of the β_i's are equal to 0. Thus, at least one of our predictor variables is useful in predicting the response variable y.

Just because the model appears to be useful for predicting executive salaries, the F-test does not suggest that the specified model is the best, nor does it imply that the variables of years of education and years of experience are the *only* variables that are useful in predicting executive salaries. Using the F-test for a multiple regression analysis basically serves as a first step in determining if at least

TABLE 8.2

Estimated Salary $(\hat{y}_i)$, Difference Between the Observed and the Estimated Salary $(y_i - \hat{y}_i)$, and Difference Between the Estimated and the Mean Salary $(\hat{y}_i - \bar{y})$ for the Data in Table 8.1

Observation	Education in Years (x_1)	Experience in Years (x_2)	Yearly Salary (y)	Estimated Salary $(\hat{y}_i)$	$(y_i - \hat{y}_i)$	$(y_i - \hat{y}_i)^2$	$(\hat{y}_i - \bar{y})$	$(\hat{y}_i - \bar{y})^2$
1	12	12	95	116.6167	−21.6167	467.280	−33.7167	1136.814
2	13	19	145	146.6115	−1.6115	2.597	−3.7219	13.852
3	16	20	164	138.9478	25.0522	627.612	−11.3855	129.630
4	14	24	186	166.8415	19.1585	367.049	16.5081	272.519
5	15	30	197	191.9539	5.0461	25.463	41.6206	1732.270
6	12	16	139	136.1463	2.8537	8.144	−14.1870	201.272
7	13	19	163	146.6115	16.3885	268.584	−3.7219	13.852
8	15	15	125	118.7178	6.2822	39.466	−31.6155	999.541
9	16	25	122	163.3598	−41.3598	1710.636	13.0265	169.690
10	13	26	173	180.7883	−7.7883	60.658	30.4550	927.506
11	14	25	152	171.7239	−19.7239	389.032	21.3905	457.556
12	12	5	75	82.4398	−7.4398	55.351	−67.8935	4609.528
13	13	19	165	146.6115	18.3885	338.138	−3.7219	13.852
14	13	22	167	161.2587	5.74131	32.963	10.9254	119.363
15	14	28	187	186.3711	0.6289	0.396	36.0378	1298.720
					Total	4393.369		12095.965

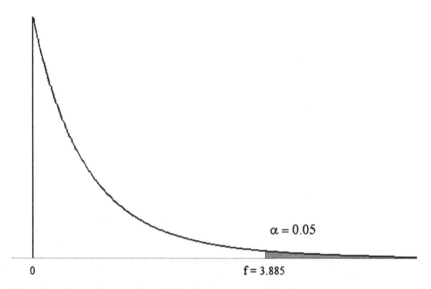

FIGURE 8.7
F-distribution for a significance level of 0.05 with 2 degrees of freedom for the numerator and 12 degrees of freedom for the denominator.

Analysis of Variance

Source	DF	Adj SS	Adj MS	F-Value	P-Value
Regression	2	12096.0	6048.0	16.52	0.000
Education	1	317.6	317.6	0.87	0.370
Experience	1	10276.9	10276.9	28.07	0.000
Error	12	4393.4	366.1		
Lack-of-Fit	10	4150.7	415.1	3.42	0.247
Pure Error	2	242.7	121.3		
Total	14	16489.3			

FIGURE 8.8
Analysis of variance table for the regression analysis predicting executive salaries based on experience and education.

one of the population predictor variables in the model is statistically significant in predicting the response variable. If the *p*-value for the *F*-test is less than the specified level of significance, this means that at least one of the predictor variables is significant in predicting the response and the analysis can be pursued further. If the *p*-value for the *F*-test is greater than the specified level of significance, this implies that none of the predictor variables are significant in predicting the response and you may want to consider formulating a different model.

8.7 Testing Individual Population Regression Parameters

After conducting an *F*-test and if the overall model is useful in predicting *y*, we may be interested in determining the effect that each of the *individual predictor variables* has on the response variable. For the last example, we might be interested in estimating the effect that the individual predictor variable of years of education or years of experience have on an executive's yearly salary.

If we can infer that the population parameter for a single independent variable is not different enough from 0, then we might expect that this variable does not have an impact on the response variable. If we can infer that the population parameter for a single independent variable is significantly different from 0, then we might expect that this variable does have an impact on the response variable. We can assess the effect of the ith predictor variable on the response variable by conducting the following hypothesis test:

$$H_0 : \beta_i = 0$$

$$H_1 : \beta_i \neq 0$$

where β_1 is the unknown population effect of the variable x_i.

In order to determine which of the individual population predictor variables has an impact on the response variable, we can test each of them individually by using the following test statistic:

$$T = \frac{\text{Parameter estimate for predictor } i}{\text{Estimated standard error for predictor } i} = \frac{\hat{\beta}_i}{\text{SE}\left(\hat{\beta}_i\right)}$$

Similar to many hypothesis tests we have done thus far, this test statistic is centered about the unknown population parameter β_i and follows the t-distribution with $n - p$ degrees of freedom, where p is the total number of beta parameters that are being estimated in the model and n is the number of observations.

Example 8.4

To test whether the amount of education has an impact on executive salaries (holding experience fixed), we can conduct a hypothesis test to see if the population parameter that represents the effect of the number of years of education on yearly salaries is significantly different from 0. Thus, our null and alternative hypotheses would be stated as follows:

$$H_0 : \beta_1 = 0$$

$$H_1 : \beta_1 \neq 0$$

Using the parameter estimate $\hat{\beta}_1$ for the number of years of education and the standard error for this coefficient that can be obtained from the Minitab printout given in Figure 8.5, the test statistic would be:

$$T = \frac{\hat{\beta}_1}{\text{SE}\left(\hat{\beta}_1\right)} = \frac{-4.182}{4.490} \approx -0.931$$

$$\text{df} = n - p = 15 - 3 = 12$$

We have that $n = 15$ because there are 15 observations and $p = 3$ because we are estimating three beta parameters, namely, $\beta_0, \beta_1,$ and β_2. By comparing the value of this test statistic $T = -0.931$ to the value $t = 2.179$ that defines the rejection region ($\alpha = 0.05$) as illustrated in Figure 8.9, we would claim that we do not have enough evidence to suggest that the unknown population parameter β_1 is significantly different from 0. Since the test statistic does not fall into the shaded region, our sample statistic $\hat{\beta}_1$ is not different enough from one that would be drawn from a population whose true value is β_1.

Thus, by using the estimated regression equation, we can infer that the years of education does not have a significant effect in predicting yearly executive salaries (holding years of experience fixed).

Similarly, to test whether years of experience is significant in predicting executive salaries (holding education fixed), we compare the following test statistic to the t-distribution with 12 degrees of freedom, as illustrated in Figure 8.9:

$$T = \frac{\hat{\beta}_1}{\text{SE}\left(\hat{\beta}_1\right)} = \frac{4.8824}{0.9215} \approx 5.30$$

Since the value of $T = 5.30$ falls in the shaded region as defined by $t = \pm 2.179$, we can claim that the number of years of experience is significant in predicting yearly executive salaries. To interpret this within the context of our problem, we can infer that for every additional year of experience, the yearly salary would increase by an average of approximately $4,882 dollars per year, holding the variable of education fixed.

In addition to testing the effect of each of the individual predictor variables, we can calculate a confidence interval for any of the individual population regression parameters by using the estimated values as follows:

$$\left(\hat{\beta}_i - t_{\alpha/2} \cdot \text{SE}\left(\hat{\beta}_i\right), \hat{\beta}_i + t_{\alpha/2} \cdot \text{SE}\left(\hat{\beta}_i\right)\right)$$

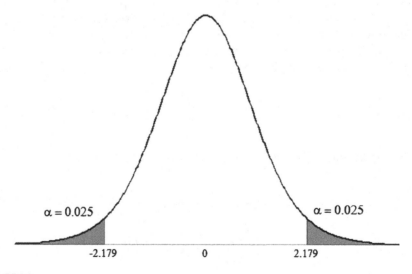

FIGURE 8.9
The t-distribution for a significance level of 0.05 with 12 degrees of freedom.

where β_i is the estimated coefficient for the ith predictor variable, $\text{SE}\left(\hat{\beta}_i\right)$ is the estimated standard error for the ith predictor variable, and $t_{\alpha/2}$ is the t-value as found by the desired level of significance with $n - p$ degrees of freedom, where p is the total number of beta parameters being estimated in the model and n is the number of observations.

Example 8.5

Suppose that we want to calculate a 95% confidence interval for the unknown population parameter β_2, which represents the effect of the years of experience on executive salaries. To do this, we add and subtract the standard error of the estimate for the appropriate variable from the value of the estimated parameter, as follows:

$$\left(\hat{\beta}_2 - t_{\alpha/2} \cdot \text{SE}\left(\hat{\beta}_2\right), \hat{\beta}_2 + t_{\alpha/2} \cdot \text{SE}\left(\hat{\beta}_2\right)\right)$$

$$\approx \left(4.882 - (2.179)(0.922), 4.882 + (2.179)(0.922)\right) \approx \left(2.873, 6.891\right)$$

Interpreting this confidence interval would suggest that we are 95% confident that for every additional year of experience, this results in a salary increase between approximately \$2,874 and \$6,890 per year, holding the variable of education fixed.

If we calculated a 95% confidence interval for β_1, the effect of the years of education on executive salaries, we would find that our confidence interval covers both positive and negative values, as follows:

$$\left(-4.18 - (2.179 \cdot 4.49), -4.18 + (2.179 \cdot 4.49)\right) \approx \left(-13.96, 5.60\right)$$

And just as you might expect, because this confidence interval contains both positive and negative values, and hence the value of 0, there is no significant effect of the years of education in predicting the yearly salary of executives.

8.8 Using Minitab to Test Individual Regression Parameters

Whenever a regression analysis is run using Minitab, it automatically tests whether each of the individual regression parameters is significantly different from 0. This is given in the Coefficients section of the Minitab printout given in Figure 8.10. Notice that the values of the test statistics for each of the regression parameters are the same as those calculated manually,

Coefficients

Term	Coef	SE Coef	T-Value	P-Value	VIF
Constant	108.2	54.0	2.00	0.068	
Education	-4.18	4.49	-0.93	0.370	1.39
Experience	4.882	0.922	5.30	0.000	1.39

FIGURE 8.10
Portion of the Minitab printout highlighting the *t*-test for each of the individual beta parameters for the executive salary data in Table 8.1.

and the p-values correspond to the same conclusions that the number of years of experience has a significant effect on yearly salaries, but the number of years of education does not.

8.9 Multicollinearity

The reason for including numerous predictor variables in a multiple regression model is because these variables are believed to be related to the response (or dependent) variable. And in order to develop a good model, we need to be sure that we have included all possible predictor variables that could have a significant impact on the response variable. However, we need to be careful that the individual predictor variables themselves are *not highly correlated with each other. Multicollinearity* occurs when two or more predictor variables are highly correlated with each other.

Some very serious consequences may occur if a regression model consists of predictor variables that are highly correlated with each other. In particular, the estimates of the population regression parameters (the betas) become unstable. This means that if highly correlated variables are included as predictor variables, the regression parameter estimates may fluctuate dramatically when such terms are added to or dropped from the model. This tends to happen when the standard errors become very large.

Using the executive salary data, we are now going to look at the impact that including highly correlated predictor variables can have on a regression analysis.

Example 8.6

Consider the data set provided in Table 8.3. This data set consists of some of the same data given in Table 8.1, but with the addition of the variable of age in years (x_3).

TABLE 8.3

Executive Salary Data Consisting of the Variables of Years of Education (x_1), Years of Experience (x_2), Age in Years (x_3), and Yearly Salary in Thousands of Dollars (y)

Observation Number	Education in Years (x_1)	Experience in Years (x_2)	Age in Years (x_3)	Yearly Salary in Thousands of Dollars (y)
1	12	12	42	95
2	13	19	48	145
3	16	20	51	164
4	14	24	53	186
5	15	30	61	197
6	12	16	46	139
7	13	19	50	163
8	15	15	45	125
9	16	25	56	122
10	13	26	56	173
11	14	25	55	152
12	12	5	35	75
13	13	19	48	165
14	13	22	52	167
15	14	28	57	187

Notice what happens to the estimate of the coefficients and the estimated standard errors when Minitab is used to run two different regression analyses—one (Figure 8.11) is the Coefficients portion of the Minitab printout that includes the predictor variables for education and experience but does not include the variable of age and the other (Figure 8.12) is the Coefficients portion of the Minitab printout that includes the predictor variables for education and experience along with the variable of age.

Table 8.4 summarizes the results of the regression analyses from both Figures 8.11 and 8.12, which compare the parameter estimates, standard errors, and *p*-values for these two separate models.

As can be seen in Table 8.4, the effect of years of experience on yearly salary was significant in the model that did not include the variable of age (Figure 8.11), but this same predictor was not significant in the model that included the variable of age (Figure 8.12). The reason for this inconsistency is because by adding the variable of age, which is highly correlated with years of experience, the standard errors became much larger and the effects of the independent variables

TABLE 8.4

Parameter Estimates, Standard Errors, and *p*-values for the Regression Model that Does Not Include the Variable Age (Figure 8.11) versus the Regression Model that Does Include the Variable Age (Figure 8.12)

	Parameter Estimate		Standard Error		*p*-value	
	Without Age Figure 8.11	With Age Figure 8.12	Without Age Figure 8.11	With Age Figure 8.12	Without Age Figure 8.11	With Age Figure 8.12
Constant	108.2	240	54.0	214	0.068	0.285
Education	−4.18	−2.65	4.49	5.19	0.370	0.620
Experience	4.882	9.76	0.922	7.69	0.000	0.230
Age	—	−5.01	—	7.84	—	0.535

Coefficients

Term	Coef	SE Coef	T-Value	P-Value	VIF
Constant	108.2	54.0	2.00	0.068	
Education	-4.18	4.49	-0.93	0.370	1.39
Experience	4.882	0.922	5.30	0.000	1.39

FIGURE 8.11
Portion of the regression analysis of years of education and years of experience on yearly salaries.

Coefficients

Term	Coef	SE Coef	T-Value	P-Value	VIF
Constant	240	214	1.12	0.285	
Education	-2.65	5.19	-0.51	0.620	1.77
Experience	9.76	7.69	1.27	0.230	92.24
Age	-5.01	7.84	-0.64	0.535	98.47

FIGURE 8.12
Portion of the regression analysis of years of education, years of experience, and age in years on a yearly salary.

can become insignificant. Therefore, by adding variables that are highly correlated with each other, we can generate different interpretations of an effect. The standard error columns in Table 8.4 illustrate just how large the standard errors became when the predictor age was added to the regression model and it is interesting to note how the parameter estimates themselves have also changed.

8.10 Variance Inflation Factors

There is a measure called the *variance inflation factor* (often abbreviated as VIF) that can be used to determine if there are strong correlations between the predictor variables and whether these correlations are strong enough that they may be affecting the estimates of the standard errors. A high VIF suggests that the parameter estimates may become unstable because the predictor variables that are highly correlated with each other can generate very large standard errors for the individual regression parameters. A low VIF would suggest that the correlation between the predictor variables is not strong enough to significantly impact the estimates of the standard error for the estimated parameters.

Variance inflation factors for each individual predictor variable (excluding β_0) are calculated as follows:

$$\text{VIF}_j = \frac{1}{1 - R_j^2}$$

where j is the individual predictor variable of interest β_j, and R^2 is the coefficient of determination for the model in which the variable x_j is used as the *response variable* and all the other variables in the model are included as predictor variables.

To calculate the VIFs, we first need to calculate the value of R^2 for each of the regression analyses *where each predictor variable is used as the response variable, with all the other variables as predictor variables.*

For example, to find the variance inflation for each of the predictor variables of years of education (x_1), years of experience (x_2), and age (x_3) using the executive salary data given in Table 8.3, we would have to run three separate regression analyses to find the value of the R^2 statistic for each of the three models that are described in Table 8.5.

Example 8.7

To find the VIF for the variable of years of education (x_1), we will use Minitab to run a regression analysis in which education is treated as the response variable and experience (x_2) and age (x_3) are included as predictor variables. The results from this regression analysis are presented in Figure 8.13.

TABLE 8.5

Regression Models to Calculate the VIFs

Response Variable	Predictor Variables	
x_1	x_2	x_3
x_2	x_1	x_3
x_3	x_1	x_2

Regression Analysis: Education versus Experience, Age

Analysis of Variance

Source	DF	Adj SS	Adj MS	F-Value	P-Value
Regression	2	11.0525	5.52623	4.64	0.032
Experience	1	2.7326	2.73263	2.30	0.156
Age	1	3.8815	3.88155	3.26	0.096
Error	12	14.2809	1.19007		
Lack-of-Fit	11	14.2809	1.29826	*	*
Pure Error	1	0.0000	0.00000		
Total	14	25.3333			

Model Summary

S	R-sq	R-sq(adj)	R-sq(pred)
1.09090	43.63%	34.23%	20.23%

Coefficients

Term	Coef	SE Coef	T-Value	P-Value	VIF
Constant	-9.4	11.6	-0.81	0.433	
Experience	-0.593	0.391	-1.52	0.156	77.42
Age	0.698	0.386	1.81	0.096	77.42

Regression Equation

Education = -9.4 - 0.593 Experience + 0.698 Age

FIGURE 8.13

Minitab printout of the regression analysis where education is treated as the response variable and experience (x_2) and age (x_3) are included as predictor variables.

Then, using the value of $R^2 = 43.63\% = 0.4363$ from Figure 8.13, we can calculate the variance inflation for the variable of education as follows:

$$\text{VIF}_{\text{Education}} = \frac{1}{1 - R^2_{\text{Education}}} = \frac{1}{1 - 0.4363} \approx 1.774$$

A similar analysis can be used to find the VIFs for the variables that represent years of experience and age.

8.11 Using Minitab to Calculate Variance Inflation Factors

Minitab automatically generates VIFs every time you run a regression analysis. The Coefficients portion of the Minitab printout for the original regression model that uses education (x_1), experience (x_2), and age (x_3) to predict yearly salary is provided in Figure 8.14, where the variance inflation factors are given for each of the three predictor variables.

Recall that the R^2 statistic is the ratio of the sum of squares of the regression analysis to the total sum of squares, and this provides a measure of the proportion of the variability in y that is explained by the given regression model. If R^2 is close to 0, then the given regression model does not contribute to predicting y, and thus the VIF will be a small number.

Coefficients

Term	Coef	SE Coef	T-Value	P-Value	VIF
Constant	240	214	1.12	0.285	
Education	-2.65	5.19	-0.51	0.620	1.77
Experience	9.76	7.69	1.27	0.230	92.24
Age	-5.01	7.84	-0.64	0.535	98.47

FIGURE 8.14
Portion of the Minitab printout for the original regression model that uses education (x_1), experience (x_2), and age (x_3) to predict yearly salary showing the VIFs for each of the three predictor variables.

However, if R^2 is close to 1 (or 100%), then the given regression model does contribute to predicting y, and thus the value of the VIF can be very large. Notice in Figure 8.14 that the VIFs for experience and age are quite large. Large values of VIFs indicate that these variables are highly correlated with each other and such a strong correlation could be having an impact on the estimated standard errors and the parameter estimates. Thus, including both highly correlated predictor variables of experience and age in a regression model can generate very large standard errors, as can be seen by comparing the parameter estimates, standard errors, and p-values displayed in Figures 8.11 and 8.12. Also, note that the manual calculations for the variance inflation factors may not always exactly match the value of the variance inflation factors generated by Minitab because of round-off error.

But how do you know when a VIF is too high? As a general rule of thumb, a VIF that is less than 5 suggests that the predictor variables may not be highly correlated with each other. A VIF between 5 and 10 indicates that there may be some strong correlations between the predictor variables and that such correlations could be affecting the estimated standard errors and the parameter estimates. Even modest correlations amongst the predictor variables can attribute to the estimated standard errors becoming very large and the parameter estimates becoming unstable. For this reason, VIFs that are above 10 usually require further investigation.

8.12 Multiple Regression Model Assumptions

Just like with simple linear regression analysis, before we can use an estimated model to make inferences regarding an unknown population model, we need to make sure that the following model assumptions are reasonably met. These assumptions are based on the distribution of the random error component and are identical to those model assumptions for a simple regression analysis:

1. The observations are independent of each other.
2. The error component is normally distributed.
3. The error component has constant variance.
4. The functional relationship between the set of predictor variables (the x-variables) and the response variable (y) can be established.

Just as with simple linear regression analysis, we can use Minitab to create residual plots and perform some formal tests of these assumptions.

8.13 Using Minitab to Check Multiple Regression Model Assumptions

In a similar fashion, as with simple regression analysis, the **Graphs** tab on the **Regression** dialog box provides a four-in-one plot that can be used to assess the assumptions of normality and constant variance of the error component. The four-in-one plot for the model that predicts executive salaries based on the two predictor variables of education and experience is given in Figure 8.15.

We can also use the Ryan–Joiner test as a formal check on the assumption of normality of the error component. To do this, we first need to select the **Storage** tab from the **Regression** dialog box and then check the **Residuals** box in order to store the residuals in the current worksheet.

Then, running the Ryan–Joiner test on the residuals from the **Normality Test** selection under **Stat** and **Basic Statistics** gives the normal probability plot, the Ryan–Joiner test statistic, and associated p-value, as illustrated in Figure 8.16.

Recall that the null and alternative hypotheses for the Ryan–Joiner test are as follows:

H_0 : The population error component follows a normal distribution.

H_1 : The population error component does not follow a normal distribution.

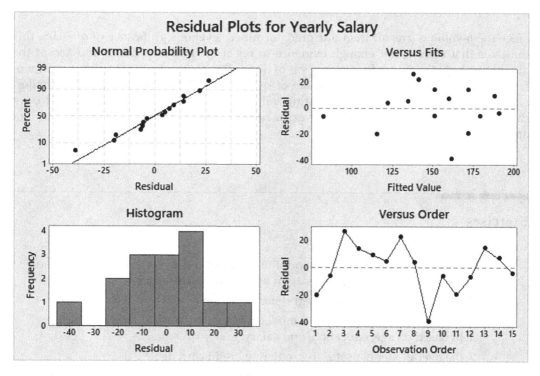

FIGURE 8.15
Four-in-one plot for the regression analysis that uses education and experience to predict yearly salaries.

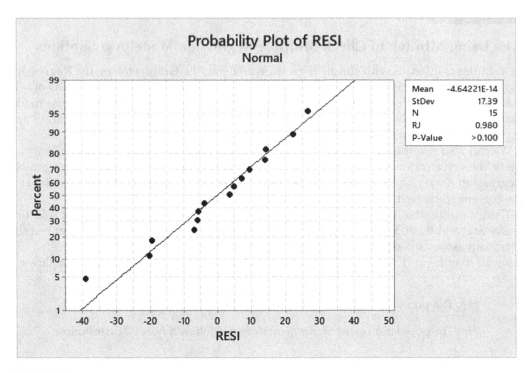

FIGURE 8.16
Normal probability plot, Ryan–Joiner test statistic, and *p*-value.

Since the *p*-value is greater than our predetermined level of significance of $\alpha = 0.05$, this indicates that there is not enough evidence to reject the null hypothesis and accept the alternative hypothesis, so the assumption of normality does not appear to have been violated. In other words, because the value of the test statistic is greater than the smallest correlation defined by the corresponding value *rj* that defines the rejection region, we can infer that the assumption that the error component is normally distributed does not appear to have been violated.

Exercises

Unless otherwise specified, let $\alpha = 0.05$.

1. The data in Table 8.6 gives the amount of credit card debt for a sample of 24 individuals along with their age, education level, and yearly salary.

   a. Write the population linear regression equation.

   b. Using this data set, develop a model that can be used to predict credit card debt based on age, education, and salary.

   c. Is the overall model useful in predicting credit card debt?

   d. Draw a matrix plot that shows the relationship between the response and each of the predictor variables.

TABLE 8.6

Credit Card Debt, Age, Education, and Yearly Salary

Observation	Credit Card Debt (in Dollars)	Age (in Years)	Education Level (in Years)	Yearly Salary (in Thousands of Dollars)
1	5,020	45	16	55
2	8,550	50	19	75
3	6,750	48	18	68
4	1,255	29	18	42
5	0	19	12	25
6	0	20	13	38
7	4,685	35	16	50
8	5,790	33	14	57
9	2,150	30	16	42
10	550	19	12	35
11	6,850	22	16	42
12	1,250	45	16	58
13	460	50	16	62
14	1,280	35	12	39
15	500	28	14	52
16	430	32	14	58
17	250	44	12	35
18	650	49	12	41
19	1,285	32	14	57
20	690	50	12	35
21	500	18	13	41
22	6,000	30	17	90
23	7,500	39	16	54
24	12,850	32	20	85

e. Use exploratory techniques and any formal tests to check relevant model assumptions and check for multicollinearity.

f. What factors are significant in predicting credit card debt? Interpret your findings in the context of the problem, and comment on whether you think your conclusions make sense.

g. Using your estimated regression equation, estimate what the average credit card debt would be for an individual who is 25 years old, with 20 years of education, and whose salary is $50,000 per year.

h. Can you think of any other factors that could influence the amount of credit card debt?

2. Does your age and weight affect your blood pressure? To test this, a medical researcher collected measures of systolic blood pressure (in mmHg, millimeters of mercury), age (in years), and weight (in pounds) for a random sample of 17 individuals. These measures are presented in Table 8.7.

a. Write the population linear regression equation.

b. Using this data set, develop a model that can be used to predict systolic blood pressure based on age and weight.

TABLE 8.7

Systolic Blood Pressure, Age, and Weight for a Random Sample of 17
Individuals

Systolic Blood Pressure (mmHg)	Age (in Years)	Weight (in Pounds)
128	58	190
136	63	171
141	74	139
134	68	186
123	47	187
128	50	182
141	58	176
124	53	137
140	37	154
121	29	185
124	34	158
132	49	180
134	47	145
145	58	184
122	28	145
119	23	158
119	20	159

   c. Draw a matrix plot that shows the relationship between the response and each of the predictor variables.

   d. Is the overall model useful in predicting systolic blood pressure?

   e. Use exploratory techniques and any formal tests to check relevant model assumptions and check for multicollinearity.

   f. What factors are significant in predicting systolic blood pressure? Interpret your findings in the context of the problem and comment on whether you think your conclusions make sense.

   g. Using your estimated regression equation, estimate what the average systolic blood pressure would be for an individual who is 35 years old and weighs 154 pounds.

   h. Can you think of any other factors that could impact systolic blood pressure?

3. The asking price of a home is influenced by many different factors such as the number of bedrooms, number of bathrooms, square footage, and the lot size. A random sample of 13 recent home sales collected including the aforementioned variables is given in Table 8.8.

   a. Write the population linear regression equation.

   b. Using this data set, develop a model that can be used to predict the asking price of a home based on the number of bedrooms, number of bathrooms, square footage, and the lot size.

   c. Draw a matrix plot that shows the relationship between the response and each of the predictor variables.

TABLE 8.8

Asking Price (in Thousands of Dollars), Number of Bedrooms, Number of Bathrooms, Square Footage (in Feet), and Lot Size (in Acres) for a Random Sample of 13 Recent Home Sales

Observation	Asking Price (in Dollars)	Number of Bedrooms	Number of Bathrooms	Square Footage (in Feet)	Lot Size (in Acres)
1	209	5	13	77	57
2	226	9	15	142	52
3	224	3	12	51	27
4	275	7	16	108	41
5	334	8	19	127	51
6	155	6	8	100	48
7	184	5	11	81	38
8	266	12	16	189	44
9	303	16	19	251	17
10	228	3	12	49	60
11	275	4	16	68	53
12	291	12	19	191	23
13	332	7	20	112	64

d. Is the overall model useful in predicting the asking price?

e. Use exploratory techniques and any formal tests to check relevant model assumptions and check for multicollinearity.

f. What factors are significant in predicting the asking price? Interpret your findings in the context of the problem and comment on whether you think your conclusions make sense.

g. Using your estimated regression equation, estimate what the average asking price would be for a home with four bedrooms, three bathrooms, which is 3,500 square feet and is on 1.25 acres.

h. Can you think of any other factors that could impact the asking price of a home?

4. What factors influence how much people are willing to spend on a mortgage or rent payment? The data set in Table 8.9 consists of a random sample of 20 residences and it contains measures of the monthly rent/mortgage payment, total household income, monthly utilities, the number in the household, total commute time, the number of vehicles, and the number of workers.

a. Draw a matrix plot that shows the relationship between the response and each of the predictor variables.

b. Use this data set to develop a linear model that can be used to predict which factors, if any, impact the amount spent on mortgage or rent payments ($\alpha = 0.10$).

c. Using exploratory plots and formal tests, check the regression model assumptions and comment on whether you believe these model assumptions appear to hold true.

TABLE 8.9

Monthly Rent/Mortgage Payment, Total Household Income, Monthly Utilities, Number in the Household, Total Commute Time, Number of Vehicles, and Number of Workers

Rent/Mortgage (in Dollars)	Monthly Income (in Dollars)	Monthly Utilities (in Dollars)	Number in Household	Total Commute Time (in Minutes)	Number of Vehicles	Number of Workers
1,845	5,593	290	4	82	2	2
1,870	2,959	139	1			1
1,160	2,859	187		31	1	1
950	894	175	2	25	1	1
2,000	1,444	119	4	30	2	1
0	800	125	2		0	0
2,640	3,470	277	3	15	3	2
1,250	959	135	2		1	
1,650	2,584	199	4	50		1
3,450	3,865	459	2	35	2	1
5,980	13,057	561	3		1	3
2,765	3,358	338		25	1	1
3,430	5,337	401	1		0	
2,080	2,902	288	2	35	2	2
1,360	2,778	220	1	30	0	
1,860	2,896	172	3	45	2	2
1,160	4,113	345	1	45	1	
1,070	4,821	376		30	1	
3,980	6,851	222	4	35	3	2
1,280	786	110	2		1	

  d. Is the overall model useful in predicting the amount spent on mortgage or rent payments?

  e. What variables impact the amount spent on mortgage or rent payments?

5. You may have noticed in the data set in Table 8.9 that many of the variables have missing values. In fact, only eight of the rows have measures on all six predictor variables. Data sets with missing values are not uncommon and tend to be due to individuals not providing responses to all the questions in a survey. One strategy that can be used to deal with missing values is *imputation*. Imputation entails imputing a value in place of each missing value. One typical imputation strategy would be to replace all of the missing values of a variable with the mean of the variable or the median of the variable.

  a. For the data in Table 8.9, use the strategy of imputing the mean for the missing values for each of the variables that have missing data, and then run a regression analysis.

  b. Comment on what you think has changed by imputing the mean for the missing values of each variable.

c. For the data in Table 8.9, use the strategy of imputing the median for the missing values for each of the variables that have missing data, and then run a regression analysis ($\alpha = 0.10$).

d. Comment on what you think has changed by imputing the median for the missing values of each variable.

e. Do you think imputing the mean or the median is more appropriate for this problem?

9

More on Multiple Regression

9.1 Introduction

As we have seen from the different multiple regression analyses performed thus far, in some ways it may seem like we are on a fishing expedition trying to discover a useful relationship between the response and the set of predictor variables. Not only do we need to find the appropriate model that describes the relationship between the set of predictor variables and the response variable, but we also need to be sure that the regression assumptions are reasonably met and that we have not included predictor variables in the regression model that are highly correlated with each other. You may have also noticed that all of the predictor and response variables that we have considered thus far have been continuous.

In this chapter, we will describe ways to include categorical predictor variables in a regression model and describe a more systematic approach to finding a regression model with the best fit. This chapter will also describe how confidence and prediction intervals can be found with a multiple regression analysis and how to access outliers.

9.2 Using Categorical Predictor Variables

There may be situations when you want to include a predictor variable that describes some characteristic of interest, but such a characteristic can only be represented as a *categorical variable*. Recall from Chapter 1 that a categorical variable can be used to represent different categories. For instance, the variable "gender" is a categorical variable because it represents two different categories—male and female. We can code this variable in such a way that the number 1 indicates the presence of the characteristic of interest (being female) and the number 0 indicates the absence of the characteristic of interest (being male). Even though gender can be coded numerically, you may recall from Chapter 1 that the numbers used to code a categorical variable are only used for reference and they cannot be manipulated by using basic arithmetic. In other words, a coding scheme for a categorical variable only represents a way to label different categories with numeric codes, but such codes have no mathematical properties.

Categorical predictor variables can be included in a multiple regression analysis provided that they are described in a precise manner, as will be illustrated in the next two examples.

Example 9.1

Consider the data set presented in Table 9.1. This data set is the same one we used in the last chapter (Table 8.1) but with the addition of the categorical

TABLE 9.1

Executive Salary Data Including a Variable to Represent Profit Status Where
Nonprofit = 1, if the Subject Works for a Nonprofit Company; and Nonprofit = 0, if the
Subject Works for a For-Profit Company

Observation Number	Education in Years (x_1)	Experience in Years (x_2)	Nonprofit (x_3)	Yearly Salary in Thousands of Dollars (y)
1	12	12	1	95
2	13	19	0	145
3	16	20	0	164
4	14	24	0	186
5	15	30	0	197
6	12	16	1	139
7	13	19	0	163
8	15	15	1	125
9	16	25	1	122
10	13	26	0	173
11	14	25	1	152
12	12	5	1	75
13	13	19	0	165
14	13	22	0	167
15	14	28	1	187

variable "Nonprofit." The variable "Nonprofit" in Table 9.1 is used to represent whether the company is a nonprofit or a for-profit company. This variable is coded in such a way that if Nonprofit = 1, then the subject works for a nonprofit company, and if Nonprofit = 0, then the subject works for a for-profit company.

Note that for any observation in the data set, the assignment of the value Nonprofit = 1 represents the presence of the characteristic of interest (working for a nonprofit company), and the assignment of the value Nonprofit = 0 represents the absence of the characteristic of interest (working for a for-profit company). Using the numbers 1 and 0 to represent the presence or absence of some characteristic of interest will allow us to include such variables in a multiple regression model. Such variables that only consist of the numbers 1 or 0 to represent the presence or absence of some characteristic are called *binary or indicator variables*. A binary or indicator variable is a variable that can take on only one of two possible values, either a 1 or a 0.

The population linear model that includes the variables of years of education (x_1), years of experience (x_2), along with the profit status of the company the executive works for (x_3), can be written as follows:

$$y = \beta_0 + \beta_1 x_1 + \beta_2 x_2 + \beta_3 x_3 + \varepsilon$$

where the variable x_3 now represents the categorical variable "Nonprofit," and this is a binary variable that indicates whether or not the subject works for a nonprofit company.

We will use Minitab to estimate and interpret the effect of a categorical predictor variable.

9.3 Using Minitab for Categorical Predictor Variables

Figure 9.1 shows the data from Table 9.1 entered into Minitab. Note that the variable "Nonprofit" is represented as a numeric variable because we are using the numeric values of 1 to represent if the company is nonprofit and 0 to represent if the company is for-profit; there is no text or date associated with this variable.

To include a categorical predictor variable in a multiple regression model, all we need to do is run a multiple regression analysis in the exact same way as for only continuous predictor variables. That is, the predictor variable "Nonprofit" needs to be included along with all of the other predictor variables, as illustrated in Figure 9.2. Running the regression analysis yields the Minitab printout presented in Figure 9.3.

Notice from Figure 9.3 that the estimated effect of profit status on executive salaries is given by the coefficient −24.95. And since the p-value = 0.011 is less than $\alpha = 0.05$, this suggests that the profit status is a significant predictor of executive salaries. Interpreting the parameter estimates for a categorical predictor variable is very similar to what we have done in interpreting the parameter estimates for numeric predictor variables. But when the parameter estimate is for a categorical predictor that consists of only two possible values, the category that has the value of 1 is interpreted with respect to the *baseline category* that has the value of 0. For this example, working at a nonprofit company (Nonprofit = 1) would suggest that the yearly salary is approximately $24,950 *less than* those executives who work at for-profit companies (holding all other variables fixed). Because profit status is represented by a binary or indicator variable, we can only interpret the finding for the characteristic that has been assigned the value of 1, which in this case represents working at a nonprofit company. The interpretation of a parameter estimate for a categorical predictor variable is always given with respect to the baseline category. The baseline category is the category that has been assigned the value of 0.

You can also see the effect of profit status by looking at the estimated regression equation:

	C1	C2	C3	C4	C5
	Observation	Education	Experience	Non-Profit	Yearly Salary
1	1	12	12	1	95
2	2	13	19	0	145
3	3	16	20	0	164
4	4	14	24	0	186
5	5	15	30	0	197
6	6	12	16	1	139
7	7	13	19	0	163
8	8	15	15	1	125
9	9	16	25	1	122
10	10	13	26	0	173
11	11	14	25	1	152

Worksheet 1 ***

FIGURE 9.1
Executive salary data including qualitative variable "Nonprofit" in column C4 of a Minitab worksheet.

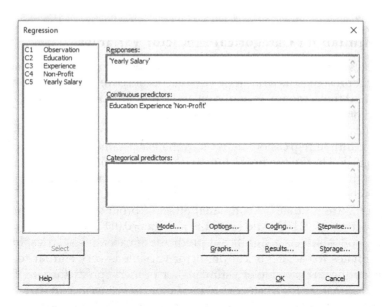

FIGURE 9.2
Minitab dialog box including the qualitative predictor for profit status.

$$\hat{y} = 116.1 - 2.65x_1 + 4.037x_2 - 24.95x_3$$

If you set $x_3 = 1$ (i.e., Nonprofit = 1), then you can see that the estimated effect is $24,950 *less* than when $x_3 = 0$ (i.e., Nonprofit = 0), holding all other variables fixed.

Although it may seem that you can only compare two categories because the category that is described by the value 1 is compared to the category that is described by the value 0, this does not necessarily have to be the case. There may be instances when you want to include a categorical predictor variable that has *more than two distinct categories*. We can extend using indicator variables to include categorical predictors that have more than two distinct categories. However, when a variable has more than two categories, it is important to note that only the values of 0 and 1 can be used to describe the characteristics of a categorical predictor variable. The following example illustrates how to include a categorical predictor variable in a multiple regression analysis when there are three distinct categories, along with a description of how to interpret the parameter estimate for such a categorical predictor variable.

Example 9.2

Suppose that in addition to education, experience, and profit status, we want to include a variable that describes an executive's field of training, as described in Table 9.2. For this example, the categorical predictor variable has three distinct categories: "N" represents that the executive completed no training at all, "M" represents that the executive completed management training, and "F" represents that the executive completed a training program in finance.

Since there are three categories of the training field instead of two, when the data are entered into Minitab, the variable that describes an executive's field of training is represented as a text column. This can be seen in column C5-T of Figure 9.4. A text column is created because we are using the letters F, M, and N to represent the three different categories instead of numbers. Be careful not

Regression Analysis: Yearly Salary versus Education, ... nce, Non-Profit

Analysis of Variance

Source	DF	Adj SS	Adj MS	F-Value	P-Value
Regression	3	14101.6	4700.5	21.65	0.000
Education	1	124.6	124.6	0.57	0.465
Experience	1	6090.6	6090.6	28.06	0.000
Non-Profit	1	2005.7	2005.7	9.24	0.011
Error	11	2387.7	217.1		
Lack-of-Fit	9	2145.0	238.3	1.96	0.383
Pure Error	2	242.7	121.3		
Total	14	16489.3			

Model Summary

S	R-sq	R-sq(adj)	R-sq(pred)
14.7331	85.52%	81.57%	67.84%

Coefficients

Term	Coef	SE Coef	T-Value	P-Value	VIF
Constant	116.1	41.7	2.79	0.018	
Education	-2.65	3.49	-0.76	0.465	1.42
Experience	4.037	0.762	5.30	0.000	1.61
Non-Profit	-24.95	8.21	-3.04	0.011	1.16

Regression Equation

Yearly Salary = 116.1 - 2.65 Education + 4.037 Experience - 24.95 Non-Profit

Fits and Diagnostics for Unusual Observations

Obs	Yearly Salary	Fit	Resid	Std Resid	
9	122.00	149.69	-27.69	-2.39	R

R Large residual

FIGURE 9.3
Minitab output from a regression analysis modeling executive salaries as a function of education, experience, and the categorical predictor for profit status.

to just change the letters to numbers and run a regression analysis since the numbers themselves will not be meaningful within the context that they are being used.

There are two ways to include categorical predictor variables with more than two distinct categories in a regression model. The first way is to create indicator variables for each of the different categories and the second way is to have Minitab create them for you.

In order to include a categorical predictor variable that has more than two distinct categories in a multiple regression model, a set of indicator variables can be created for each of the different categories. In order to do this, we will use the **Make Indicator Variables** command from the **Calc** menu in Minitab. This gives us the indicator dialog box that is illustrated in Figure 9.5.

TABLE 9.2

Executive Salary Data That Includes Years of Education (x_1), Years of Experience (x_2), Profit Status (x_3), and Executive Training (x_4)

Observation Number	Education in Years (x_1)	Experience in Years (x_2)	Nonprofit (x_3)	Training (x_4)	Yearly Salary in Thousands of Dollars (y)
1	12	12	1	N	95
2	13	19	0	M	145
3	16	20	0	M	164
4	14	24	0	F	186
5	15	30	0	F	197
6	12	16	1	M	139
7	13	19	0	M	163
8	15	15	1	N	125
9	16	25	1	N	122
10	13	26	0	F	173
11	14	25	1	M	152
12	12	5	1	N	75
13	13	19	0	F	165
14	13	22	0	F	167
15	14	28	1	F	187

Note that we have specified the variable that is to be converted into a set of indicator variables, which is the variable that represents an executive's training (C5 Training).

Upon specifying this variable in the top portion of the dialog box, Minitab will then create the appropriate number of indicator variables depending on the number of categories of this variable, as seen in Figure 9.6.

Note that a set of three different indicator variables for each of the three categories was generated by Minitab, as can be seen in Figure 9.6. In column C7 (Training_F), a "1" is assigned to an executive who has completed a training program in finance and "0" otherwise. Similarly, column C8 (Training_M) is

Worksheet 1 ***						
↓	C1	C2	C3	C4	C5-T	C6
	Observation	Education	Experience	Non-Profit	Training	Yearly Salary
1	1	12	12	1	N	95
2	2	13	19	0	M	145
3	3	16	20	0	M	164
4	4	14	24	0	F	186
5	5	15	30	0	F	197
6	6	12	16	1	M	139
7	7	13	19	0	M	163
8	8	15	15	1	N	125

FIGURE 9.4
Executive salary data including the categorical variable that describes an executive's field of training as a text variable in column 5 (C5-T).

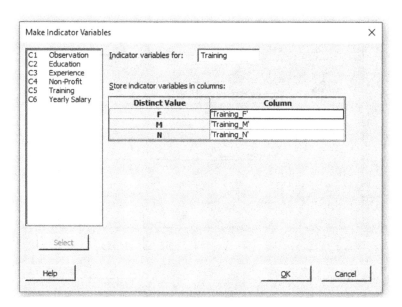

FIGURE 9.5
Minitab dialog box to create indicator variables for the variable specified as "Training".

↓	C1	C2	C3	C4	C5-T	C6	C7	C8	C9
	Observation	Education	Experience	Non-Profit	Training	Yearly Salary	Training_F	Training_M	Training_N
1	1	12	12	1	N	95	0	0	1
2	2	13	19	0	M	145	0	1	0
3	3	16	20	0	M	164	0	1	0
4	4	14	24	0	F	186	1	0	0
5	5	15	30	0	F	197	1	0	0
6	6	12	16	1	M	139	0	1	0
7	7	13	19	0	M	163	0	1	0
8	8	15	15	1	N	125	0	0	1
9	9	16	25	1	N	122	0	0	1

FIGURE 9.6
Minitab worksheet that includes indicator variables for each of the three distinct categories of executive training.

where a "1" represents that the executive has received management training and a "0" otherwise, and for column C9 (Training_N), the value of "1" is assigned to an executive who has had no training and "0" otherwise.

To include these categorical variables in our regression analysis, we must *only include two* of these three columns of indicator variables as can be seen in Figure 9.7.

The results of this regression analysis are presented in Figure 9.8.

It is worth noting that if you were to include all three columns of these indicator variables, then Minitab will automatically discard one of the columns and give you a warning message. Minitab will generate a warning message because if you include all of the indicator variables, each one of these variables

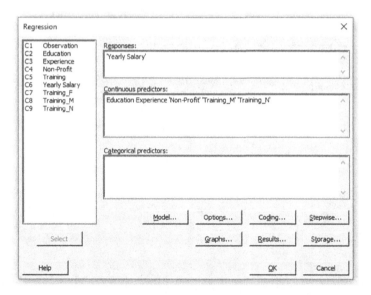

FIGURE 9.7
Minitab regression dialog box including only the two categorical predictor variables for management training and no training.

is perfectly negatively correlated with the sum of the other two indicator variables. For our example, notice that financial training will always be perfectly negatively correlated with the sum of the other two indicator variables. To see what this means, notice in Figure 9.6 that the variable representing the category for financial training has a "0" where either no training or management training has a "1", and the category financial training has a "1" where either no training or management training has a "0".

The category that is not included in the regression model becomes the baseline or reference category that can be used for making comparisons. What this means is that the effect of the other indicator variables is compared to the omitted category. For our example, we would interpret the findings for each of the two categories entered in the model with respect to the reference or omitted category, which in this case is financial training. The effect of having no training on executive salaries, as seen in the regression printout in Figure 9.8, is estimated to be −60.5, which is significant at $\alpha = 0.05$, as compared to executives who received financial training. This suggests that executives with no training tend to make approximately $60,500 *less than* executives who have completed financial training (holding all other factors fixed). Similarly, executives who have completed management training tend to make $19,620 *less than* executives who have completed financial training.

When there are more than two categories for a given categorical variable, you can make any category the baseline or reference category. This entails creating the appropriate number of indicator variables to match the number of categories and excluding the baseline category from the regression analysis, and all the other categories will then be compared to the omitted category. For instance, if we wanted management training to be the reference category, we would include the indicator variables for both no training and financial training and

Regression Analysis: Yearly Salary versus Education, ... _M, Training_N

Analysis of Variance

Source	DF	Adj SS	Adj MS	F-Value	P-Value
Regression	5	15784.7	3156.94	40.32	0.000
Education	1	346.9	346.87	4.43	0.065
Experience	1	168.4	168.40	2.15	0.177
Non-Profit	1	17.2	17.21	0.22	0.650
Training_M	1	595.9	595.91	7.61	0.022
Training_N	1	1632.2	1632.22	20.85	0.001
Error	9	704.6	78.29		
Lack-of-Fit	8	542.6	67.83	0.42	0.839
Pure Error	1	162.0	162.00		
Total	14	16489.3			

Model Summary

S	R-sq	R-sq(adj)	R-sq(pred)
8.84827	95.73%	93.35%	86.19%

Coefficients

Term	Coef	SE Coef	T-Value	P-Value	VIF
Constant	69.9	26.9	2.59	0.029	
Education	5.94	2.82	2.10	0.065	2.57
Experience	1.154	0.787	1.47	0.177	4.76
Non-Profit	-3.19	6.81	-0.47	0.650	2.21
Training_M	-19.62	7.11	-2.76	0.022	2.15
Training_N	-60.5	13.3	-4.57	0.001	6.59

Regression Equation

Yearly Salary = 69.9 + 5.94 Education + 1.154 Experience - 3.19 Non-Profit - 19.62 Training_M
- 60.5 Training_N

FIGURE 9.8
Minitab printout including indicator variables for management training and no training.

exclude the indicator variable for management training. Then any inferences made regarding the other categories would be with respect to the omitted category, which would be those executives who received management training.

We can also have Minitab do all the work for us by including the original categorical predictor variable (C5 Training) in the categorical predictor box of regression model as is shown in Figure 9.9.

The corresponding regression printout is given in Figure 9.10.

9.4 Adjusted R^2

One key point to remember when creating a regression model is to not include variables in the regression model that are highly correlated with each other and to not include

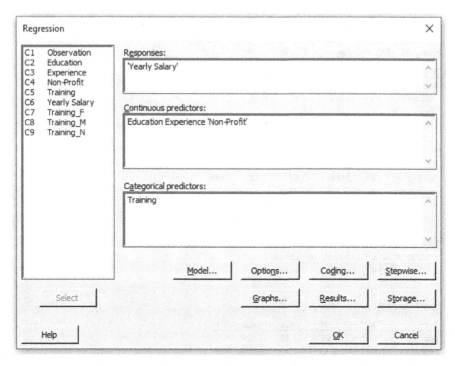

FIGURE 9.9
Minitab dialog box to include the type of training as a categorical predictor (without transforming the categorical variable into a collection of binary or indicator variables).

too many variables in the regression model such that there are more variables than there are observations in the data set. Adding more variables to a regression model will always cause the R^2 statistic to increase or stay the same, regardless of whether or not the additional variables are significant in influencing the response variable y. Furthermore, when there are more variables than observations, the R^2 statistic could be forced to equal 100%. A value of $R_2 = 1.00$ or 100%, would suggest that the underlying regression model perfectly predicts the response variable and this is not likely a situation that would ever be found in practice.

If there are numerous variables included in a regression model, then using another statistic, called the *adjusted* R^2, denoted as R_{adj}^2, can be a more appropriate measure for assessing which variables to include in a regression model.

The R_{adj}^2 statistic adjusts the value of the R^2 statistic to account for both the total number of parameters being estimated in the model and the sample size. This is done by dividing SST and SSE by their respective degrees of freedom, as seen in the following equation:

$$R_{adj}^2 = \frac{\dfrac{\text{SST}}{(n-1)} - \dfrac{\text{SSE}}{(n-p)}}{\dfrac{\text{SST}}{(n-1)}}$$

where SST is the total sum of squares, SSE is the error sum of squares, n is the number of observations, and p is the total number of beta parameters being estimated in the model.

Regression Analysis: Yearly Salary versus Education, ... -Profit, Training

Method

Categorical predictor coding (1, 0)

Analysis of Variance

Source	DF	Adj SS	Adj MS	F-Value	P-Value
Regression	5	15784.7	3156.94	40.32	0.000
Education	1	346.9	346.87	4.43	0.065
Experience	1	168.4	168.40	2.15	0.177
Non-Profit	1	17.2	17.21	0.22	0.650
Training	2	1683.1	841.54	10.75	0.004
Error	9	704.6	78.29		
Lack-of-Fit	8	542.6	67.83	0.42	0.839
Pure Error	1	162.0	162.00		
Total	14	16489.3			

Model Summary

S	R-sq	R-sq(adj)	R-sq(pred)
8.84827	95.73%	93.35%	86.19%

Coefficients

Term	Coef	SE Coef	T-Value	P-Value	VIF
Constant	69.9	26.9	2.59	0.029	
Education	5.94	2.82	2.10	0.065	2.57
Experience	1.154	0.787	1.47	0.177	4.76
Non-Profit	-3.19	6.81	-0.47	0.650	2.21
Training					
M	-19.62	7.11	-2.76	0.022	2.15
N	-60.5	13.3	-4.57	0.001	6.59

Regression Equation

Training	
F	Yearly Salary = 69.9 + 5.94 Education + 1.154 Experience - 3.19 Non-Profit
M	Yearly Salary = 50.3 + 5.94 Education + 1.154 Experience - 3.19 Non-Profit
N	Yearly Salary = 9.4 + 5.94 Education + 1.154 Experience - 3.19 Non-Profit

FIGURE 9.10
Minitab printout including the original categorical variable for management training.

Example 9.3

For the regression analysis in Figure 9.8 that includes the variables of education, experience, profit status, management training, and no training, the approximate values of R^2 and R^2_{adj} can be calculated as follows:

$$R^2 = \frac{\text{SST} - \text{SSE}}{\text{SST}} = \frac{16489.3 - 704.6}{16489.3} \approx 0.9573$$

$$R^2_{adj} = \frac{\dfrac{\text{SST}}{(n-1)} - \dfrac{\text{SSE}}{(n-p)}}{\dfrac{\text{SST}}{(n-1)}} = \frac{\dfrac{16489.3}{15-1} - \dfrac{704.6}{15-6}}{\dfrac{16489.3}{15-1}} \approx 0.9335 = 93.35\%$$

For this example, $n = 15$ and $p = 6$, because there are a total of six beta parameters being estimated in the model (including the constant term β_0).

The R^2 statistic will increase or remain relatively the same when additional variables are added to the model, and this will hold true even if adding other variables only contributes a minimal amount in predicting y. However, the R_{adj}^2 statistic will only increase if the addition of the new variable improves the fit of the model by more than just a small amount. In fact, the adjusted R-squared statistic can actually decrease if a variable is added to a model that does not improve the fit by more than a small amount. Thus, it is possible that negative values of R_{adj}^2 could be obtained, and if this were the case (although this rarely happens), then Minitab would report the value of $R_{adj}^2 = 0.0$.

9.5 Best Subsets Regression

There are many different techniques that can be used to take a more systematic approach to selecting which variables to include in a multiple regression model other than by simply using trial and error. One such technique that we will be considering is called *best subsets regression*. Best subsets regression is a collection of regression analyses that is based on different subsets of the predictor variables. The best-fitting regression model can then be selected based on three selection criteria.

The three criteria for model selection that can be used to distinguish between competing models in a best subsets regression analysis are as follows:

1. R^2
2. R_{adj}^2
3. C_p statistic (also called Mallow's C_p statistic)

The values of R^2 and R_{adj}^2 are the familiar R-squared and adjusted R-squared statistics that have been previously described. Recall that the R^2 statistic tells us the proportion of variability in the response that is attributed to the given model and the higher values of R^2 imply that more variability in y is attributed to the model and less variability is due to error. The R_{adj}^2 statistic can be used to see if including additional predictor variables influences the response variable by more than just a small amount.

The C_p statistic is used as a measure to help decide which predictor variables to include in a regression model. The C_p statistic compares the full model, which includes all of the predictor variables, to those models that include various subsets of the predictor variables. The C_p statistic provides a way of balancing too many predictor variables in the model (over-fitting the model) versus including too few predictor variables in the model (under-fitting the model). *Over-fitting* a model can cause the parameter estimates to be unstable by introducing large standard errors, and *under-fitting* the model can introduce bias in the parameter estimates. Note that the full model is defined to be the model that includes the entire set of predictor variables of interest.

In a best subsets regression, the C_p statistic can be used such that the model that minimizes the value of C_p that is also close to the total number of beta parameters estimated in the model is the "best" with respect to balancing over-fitting and under-fitting of the model. In other words, the smallest value of C_p that is also close to the number of beta parameters being estimated in the model is the model that minimizes the standard errors that can be introduced by including too many predictor variables (over-fitting the model) and minimizes the bias that can be introduced if there are too few predictor variables (under-fitting the model).

The C_p statistic is calculated as follows:

$$C_p = \frac{\text{SSE}_p}{\text{MSE}_{\text{FULL}}} - n + 2p$$

where p is the total number of beta parameters being estimated in the *subset model*, n is the total number of observations, SSE_p is the sum of squares of the error for the regression that has p beta variables in the *subset model*, and MSE_{FULL} is the mean square error for the *full model*, which includes all of the predictor variables that we are interested in.

We will use best subsets regression as a statistical modeling technique to provide a systematic way to select a model with a given number of predictor variables that maximizes R^2 and minimizes C_p, where C_p is also close to the total number of beta parameters being estimated in the model. We can use the R^2_{adj} statistic to see if including additional predictor variables improves the fit of the model.

Example 9.4

Using the data from Table 9.1, suppose we are interested in finding the best-fitting model for one, two, and three predictors. Recall that x_1 represents years of education, x_2 represents years of experience, and x_3 represents profit status. To find the best-fitting model consisting of only a *single predictor*, we could run three separate regression analyses and estimate each of the following (subset) population regression equations:

1. $y = \beta_0 + \beta_1 \text{Education} + \varepsilon$

2. $y = \beta_0 + \beta_1 \text{Experience} + \varepsilon$

3. $y = \beta_0 + \beta_1 \text{Non-Profit} + \varepsilon$

Figures 9.11–9.13 give the Minitab printouts for each of these regression analyses using the executive salary data from Table 9.1.

From each of these three estimated subset models, we could collect the values for R^2 and R^2_{adj}. We can also calculate the value of C_p for each subset model by running the full model to find $\text{MSE}_{\text{FULL}} = 217.10$ (see Figure 9.3 for the Minitab printout of the full model) and then calculating C_p for each subset model as follows:

For Model 1:

$$C_p = \frac{\text{SSE}_p}{\text{MSE}_{\text{FULL}}} - n + 2p = \frac{14670.0}{217.10} - 15 + 2(2) \approx 56.57$$

For Model 2:

$$C_p = \frac{\text{SSE}_p}{\text{MSE}_{\text{FULL}}} - n + 2p = \frac{4711.0}{217.10} - 15 + 2(2) \approx 10.70$$

Regression Analysis: Yearly Salary versus Education

Analysis of Variance

Source	DF	Adj SS	Adj MS	F-Value	P-Value
Regression	1	1819	1819.0	1.61	0.226
Education	1	1819	1819.0	1.61	0.226
Error	13	14670	1128.5		
Lack-of-Fit	3	7815	2605.0	3.80	0.047
Pure Error	10	6855	685.5		
Total	14	16489			

Model Summary

S	R-sq	R-sq(adj)	R-sq(pred)
33.5929	11.03%	4.19%	0.00%

Coefficients

Term	Coef	SE Coef	T-Value	P-Value	VIF
Constant	34.5	91.6	0.38	0.712	
Education	8.47	6.67	1.27	0.226	1.00

Regression Equation

Yearly Salary = 34.5 + 8.47 Education

Fits and Diagnostics for Unusual Observations

Obs	Yearly Salary	Fit	Resid	Std Resid	
12	75.0	136.2	-61.2	-2.01	R

R Large residual

FIGURE 9.11
Regression analysis that uses only the variable education.

For Model 3:

$$C_p = \frac{SSE_p}{MSE_{FULL}} - n + 2p = \frac{9859}{217.10} - 15 + 2(2) \approx 34.41$$

The values of R^2, R_{adj}^2, and C_p for the three models with one predictor are summarized in Table 9.3.

Note that for these last three calculations $p = 2$, because there are two variables being estimated in each of the subset models, namely, one predictor variable (β_i) and the constant term (β_0).

Using the values of R^2 and C_p provided in Table 9.3, the best one-variable model would be Model 2 because this model is the one that has the maximum value of R^2, and the value of C_p for this model is the smallest, and it is also the model such that the value of the C_p statistic is closest to the number of parameters that are being estimated in the subset models (namely, $p = 2$).

Regression Analysis: Yearly Salary versus Experience

Analysis of Variance

Source	DF	Adj SS	Adj MS	F-Value	P-Value
Regression	1	11778.3	11778.3	32.50	0.000
Experience	1	11778.3	11778.3	32.50	0.000
Error	13	4711.0	362.4		
Lack-of-Fit	10	4018.3	401.8	1.74	0.355
Pure Error	3	692.7	230.9		
Total	14	16489.3			

Model Summary

S	R-sq	R-sq(adj)	R-sq(pred)
19.0364	71.43%	69.23%	63.87%

Coefficients

Term	Coef	SE Coef	T-Value	P-Value	VIF
Constant	60.3	16.5	3.65	0.003	
Experience	4.426	0.776	5.70	0.000	1.00

Regression Equation

Yearly Salary = 60.3 + 4.426 Experience

Fits and Diagnostics for Unusual Observations

Obs	Yearly Salary	Fit	Resid	Std Resid	
9	122.00	170.99	-48.99	-2.72	R
12	75.00	82.47	-7.47	-0.53	X

R Large residual
X Unusual X

FIGURE 9.12
Regression analysis that uses only the variable experience.

We can repeat this process for the following collection of all possible *two-variable* subset models:

$$4. \quad y = \beta_0 + \beta_1 \text{Education} + \beta_2 \text{Experience} + \varepsilon$$

$$5. \quad y = \beta_0 + \beta_1 \text{Education} + \beta_2 \text{Non-Profit} + \varepsilon$$

$$6. \quad y = \beta_0 + \beta_1 \text{Experience} + \beta_2 \text{Non-Profit} + \varepsilon$$

For this scenario, $p = 3$ because there are three predictors in each subset model (two variables and the constant term β_0).

The values of R^2, R^2_{adj}, and C_p for Models 4, 5, and 6 are summarized in Table 9.4.

Using R^2 and C_p from Table 9.4, the best two-variable subset model would be Model 6, which includes the variables of experience (x_2) and profit status (x_3). This is because Model 6 has the maximum R^2, and the value of the C_p statistic is the minimum. Furthermore, this model fits better than the best one-variable

Regression Analysis: Yearly Salary versus Non-Profit

Analysis of Variance

Source	DF	Adj SS	Adj MS	F-Value	P-Value
Regression	1	6630	6630.5	8.74	0.011
Non-Profit	1	6630	6630.5	8.74	0.011
Error	13	9859	758.4		
Total	14	16489			

Model Summary

S	R-sq	R-sq(adj)	R-sq(pred)
27.5386	40.21%	35.61%	19.20%

Coefficients

Term	Coef	SE Coef	T-Value	P-Value	VIF
Constant	170.00	9.74	17.46	0.000	
Non-Profit	-42.1	14.3	-2.96	0.011	1.00

Regression Equation

Yearly Salary = 170.00 - 42.1 Non-Profit

Fits and Diagnostics for Unusual Observations

Obs	Yearly Salary	Fit	Resid	Std Resid	
12	75.0	127.9	-52.9	-2.07	R
15	187.0	127.9	59.1	2.32	R

R Large residual

FIGURE 9.13
Regression analysis that uses only the variable for-profit status.

model because the value of C_p is much closer to the number of beta parameters being estimated (which in this case is $p = 3$).

Since there is only one three-variable model as (represented by Model 7), the values of R^2, R^2_{adj}, and C_p are in Table 9.5.

$$7.\ y = \beta_0 + \beta_1 x_1 + \beta_2 x_2 + \beta_3 x_3 + \varepsilon$$

By combining all three of these subset regressions, which look at all the different one-, two-, and three-variable models, we can now select which of these different subset models best fits the data with respect to the R^2, R^2_{adj}, and C_p statistics.

On the basis of all the different subset models for one, two, and three variables, we conclude that there are two models that seem to have the best fit—Model 6, which has two variables, and Model 7, the three-variable model. However, distinguishing between Model 6 and Model 7 requires more than just a simple matter of finding the model that best meets the three-model criteria; it is also a matter of professional judgment and known theory. For instance, the best two-variable model includes the predictors of experience and profit status, whereas

TABLE 9.3

R^2, R^2_{adj}, and C_p Statistics for the One-Variable Models of Education, Experience, and Profit Status for the Executive Salary Data

Model Equation	Variable	R^2	R^2_{adj}	C_p
(1)	Education (x_1)	11.03	4.19	56.57
(2)	Experience (x_2)	71.43	69.23	10.70
(3)	Nonprofit (x_3)	40.21	35.61	34.41

TABLE 9.4

R^2, R^2_{adj}, and C_p Statistics for All Two-Variable Models for the Executive Salary Data

Model Equation	Variables	R^2	R^2_{adj}	C_p
(4)	Education, experience	73.36	68.92	11.24
(5)	Education, nonprofit	48.58	40.01	30.05
(6)	Experience, nonprofit	84.76	82.22	2.57

TABLE 9.5

R^2, R^2_{adj}, and C_p Statistics for the Three-Variable Model for the Executive Salary Data

Model Equation	Variables	R^2	R^2_{adj}	C_p
(7)	Education, experience, non-profit	85.52	81.57	4.00

the best three-variable model consists of education, experience, and profit status. Here is where the R^2_{adj} statistic comes into play. Notice that the R^2_{adj} statistic *decreased* from 82.22 in Model 6 to 81.57 in Model 7 when we added the predictor for education. Including this predictor did not improve the model by more than a small amount. Thus, we can use the R^2_{adj} statistic to see if including additional predictors improves the model fit by more than just a small amount.

9.6 Using Minitab for Best Subsets Regression

Minitab can be used to perform all of the aforementioned calculations for the one-, two-, and three-variable models. Under the **Stat** menu, select **Regression** and then **Best Subsets** to get the dialog box in Figure 9.14, where the response and predictor variables need to be specified.

The **Free predictors** box is used to specify those predictors that are free to be added or dropped when running a best subsets regression analyses with different numbers of predictors. The **Predictors in all models** box is used to specify those predictors that will be included in every subset model.

Notice in Figure 9.14 that the best subsets regression can be run by including the three variables of education, experience, and female in the **Free predictors** box to give the Minitab printout in Figure 9.15.

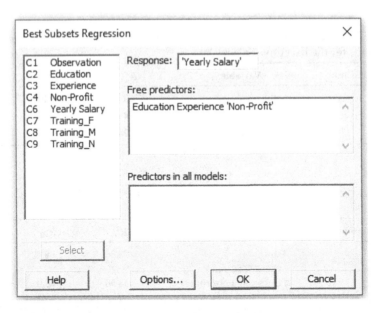

FIGURE 9.14
Minitab best subsets regression dialog box.

The rows of the Minitab printout in Figure 9.15 give the model selection criteria for the different subset models. Also notice that the left-most column gives the number of variables (Vars) that are included in each subset model (the Minitab default is to display the two best models with respect to the different number of variables in the model being considered based on the maximum value of the R^2 statistic). The Minitab printout in Figure 9.15 also gives the values of R^2, R_{adj}^2, and C_p along with the root mean square error, S, for each of the subset models. The last three columns represent the different variables that are included in each of the subset models. For instance, the first two-variable model includes the variables of experience and profit status, and the second two-variable model includes the variables of education and experience. Lining up the "X" values under the given variable with the number of variables in the model tells you which variables are included in each of the different subset models.

Best Subsets Regression: Yearly Salary versus Education, ... Non-Profit

Response is Yearly Salary

Vars	R-Sq	R-Sq (adj)	R-Sq (pred)	Mallows Cp	S	Education	Experience	Non-Profit
1	71.4	69.2	63.9	10.7	19.036		X	
1	40.2	35.6	19.2	34.4	27.539			X
2	84.8	82.2	74.0	2.6	14.469		X	X
2	73.4	68.9	52.4	11.2	19.134	X	X	
3	85.5	81.6	67.8	4.0	14.733	X	X	X

FIGURE 9.15
Minitab output of best subsets regression analysis for the executive salary data in Table 9.1.

Also notice in Figure 9.15 that the subset model that includes the predictors of experience and profit status found that $R_{adj}^2 = 82.2$. But in the subset model that includes all three predictors, $R_{adj}^2 = 81.6$. So by including the variable of education in the regression model, this did not improve the fit of the model since R_{adj}^2 decreased.

Choosing between whether to use the two- or three-variable model depends on whether education is a variable that needs to be included in the model. The choice of whether to include the variable for education would be guided by what previous studies have done and what theory suggests, and also by analyzing what the impact would be from including this extra predictor variable in the model. Thus, good model fitting requires a lot more than simply running a best subsets regression analysis and using the three-model criteria to find the best-fitting model; it relies on the researcher investigating the significance and practical impact of including all the relevant variables in a model.

Although it is very simple to run a best subsets regression analysis using a statistics package to find the best-fitting model, a mechanical process of model fitting such as best-subsets regression, does not account for any of the other aspects of regression analysis, such as checking the model assumptions and verifying that there is not a problem with multicollinearity. These other aspects of model fitting will always need to be considered.

9.7 Confidence and Prediction Intervals for Multiple Regression

Recall when we were studying simple linear regression analysis that we described two types of intervals: *confidence intervals* that can be used for estimating the *mean* population response for *all* observations with a specific value of the predictor variable, and *prediction intervals* that can be used for *predicting* the response for a *single observation* with a specific value.

We can also calculate confidence and prediction intervals in a multiple regression analysis. In calculating confidence and prediction intervals for multiple regression analysis, the specific value has to consist of specific values for *all* of the predictor variables that are included in the model. In other words, the "specific value" is really a vector of values in a multiple regression analysis that represents a specific value for each of the individual predictor variables that are included in the model.

For instance, suppose we want to find confidence and prediction intervals using the executive salary data given in Table 9.1. We are modeling the effects that education, experience, and profit status have on yearly salaries. If we are interested in calculating confidence and prediction intervals for executives with 16 years of education (this would be equivalent to having a bachelor's degree), 10 years of experience, and those who work in a for-profit company, our specific value would correspond to the vector $(x_1, x_2, x_3) = (16, 10, 0)$.

9.8 Using Minitab to Calculate Confidence and Prediction Intervals for a Multiple Regression Analysis

After running a regression analysis, from the **Stat > Regression > Regression > Predict** dialog box, select the **Options** tab. To have Minitab calculate confidence and prediction

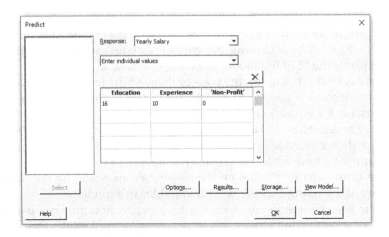

FIGURE 9.16
Minitab options dialog box for specifying the specific values for the set of predictor variables.

Prediction for Yearly Salary

Regression Equation

Yearly Salary   =   116.1 - 2.65 Education + 4.037 Experience - 24.95 Non-Profit

Settings

Variable	Setting
Education	16
Experience	10
Non-Profit	0

Prediction

Fit	SE Fit	95% CI	95% PI	
114.085	16.0740	(78.7063, 149.464)	(66.0935, 162.076)	X

X denotes an unusual point relative to predictor levels used to fit the model.

FIGURE 9.17
Minitab regression printout with 95% prediction and confidence intervals for executives with 16 years of education, 10 years of experience, and who work in for-profit companies.

intervals, the specific values for the set of predictor variables need to be entered in the appropriate column. So for our example, to find confidence and prediction intervals for executives with 16 years of education, 10 years of experience, and who work at a for-profit company, we would enter the values as $(x_1, x_2, x_3) = (16, 10, 0)$, which can be seen in Figure 9.16.

Figure 9.17 gives the Minitab printout with the confidence and prediction intervals.

The confidence interval in Figure 9.17 suggests that we are 95% confident that for *all* executives with 16 years of education and 10 years of experience who work at for-profit companies, the *mean* population yearly salary will be between $78,706 and $149,464. For the prediction interval, we can predict with 95% confidence that *one* executive who has 16 years of education and 10 years of experience who works at for-profit company will earn a yearly salary between $66,094 and $162,076. Similar to simple linear regression analysis, prediction intervals are much wider than confidence intervals for the same specific values of the predictor variables. This is because there is more uncertainty in estimating a single response from a single observation versus estimating the mean (or average) response by using all the observations with a specific set of values.

Also notice in Figure 9.17 that there is a symbol "X" on the outside of the prediction interval. This symbol means that there is a point in the data set that is an extreme outlier with respect to the predictors. When using confidence and prediction intervals, the "X" does not mean that the leverage value is high, as is indicated by the "X" symbol in the unusual observations portion of the Minitab printout.

Similar to simple linear regression analysis, when making inferences with confidence and prediction intervals, it should be noted that it is not usually a good idea to make inferences where the specific values of the predictor variables are outside of the range of the data that was used when fitting the regression model. Using a set of predictor values where even one of the predictor values is out of range can cause the confidence and prediction intervals to become very wide and unstable.

9.9 Assessing Outliers

We can use Minitab in the exact same manner as with simple linear regression to calculate outlier statistics such as leverage values, standardized residuals, and Cook's distances (see Sections 7.11–7.18). For instance, we can create a Cook distance plot by plotting the observation number against the value of the Cook distance statistic for the regression model that predicts executives' yearly salary based on years of education, years of experience, and profit status, as can be seen in Figure 9.18.

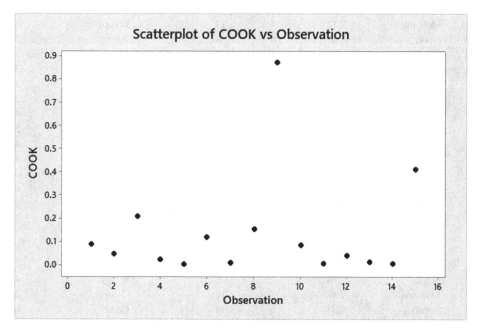

FIGURE 9.18
Cook's distance plot for the regression analysis that models executive yearly salaries based on years of education, years of experience, and profit status.

As expected, observation 9 is an outlier and may impact the estimates of the regression parameters. Furthermore, the value of the standardized residual for observation 9 is less than −2 as can be seen in the regression printout that is given in Figure 9.3.

Exercises

Unless otherwise specified, use $\alpha = 0.05$.

1. An island ferry company wants to determine what factors impact the revenue it makes from running its ferry boats to a local island. The passengers are charged to ride the ferry and also for delivering large items to the island, such as automobiles and heavy construction equipment. The data set presented in Table 9.6 consists of the revenue (in dollars), time of day the ferry left the dock, number of passengers on the ferry, number of large objects on the ferry, weather conditions, and number of crew for a random sample of 27 ferry runs over a 1-month period.

TABLE 9.6

Revenue (in Dollars), Time of Day, Number of Passengers, Number of Large Objects, Weather Conditions, and Number of Crew for a Random Sample of 27 Ferry Runs

Revenue	Time of Day	Number of Passengers	Number of Large Objects	Weather Conditions	Number of Crew
7,812	Morning	380	6	Calm	4
6,856	Noon	284	4	Calm	4
9,568	Evening	348	10	Calm	3
10,856	Morning	257	16	Rough	5
8,565	Morning	212	12	Rough	4
8,734	Noon	387	6	Calm	3
9,106	Evening	269	13	Calm	5
8,269	Evening	407	8	Rough	3
6,373	Noon	385	7	Calm	4
5,126	Noon	347	4	Calm	3
6,967	Noon	319	9	Rough	4
8,518	Morning	297	7	Rough	3
7,229	Evening	345	6	Rough	4
6,564	Morning	287	8	Rough	4
8,168	Morning	348	7	Calm	4
9,879	Evening	189	20	Calm	6
10,288	Evening	215	22	Calm	5
11,509	Morning	247	21	Rough	5
5,254	Noon	345	9	Rough	3
9,895	Morning	348	4	Calm	3
8,434	Evening	451	6	Calm	3
7,528	Noon	378	8	Rough	4
7,667	Morning	346	10	Calm	3
9,591	Evening	287	18	Calm	6
8,543	Evening	245	12	Rough	5
9,862	Morning	274	10	Calm	5
10,579	Morning	245	20	Rough	7

a. Using Minitab, run a multiple regression analysis on the total revenue based on time of day, number of passengers, number of large objects, weather conditions, and the number of crew members.

b. Is the overall model useful in predicting revenue?

c. Check any relevant model assumptions and comment on whether you believe such assumptions have been violated.

d. Identify any outliers.

e. Using the results from the regression analysis, determine which predictor(s) have an effect on the amount of revenue generated.

f. Do you think that multicollinearity is a problem?

g. Using a best subsets regression analysis, determine two models that best fits the data according to the three criteria for model selection. Justify why you chose these particular models.

h. Would the ferry operators make more money if they only ran their ferry boats in either the morning or the evening as compared to the afternoon?

2. The data in Figure 9.19 are from running a multiple regression analysis to develop a model that predicts the selling price of homes based on the number of bedrooms, the number of bathrooms, square footage, the age of the home, lot size, the number of garages, and the number of stories.

a. Is the overall model useful in predicting the sale price of homes?

b. Write the population regression equation.

c. Write the estimated regression equation.

d. Find and interpret the coefficient of determination.

e. Using the results from the regression analysis, determine if newer homes have higher selling prices and if larger homes have higher selling prices.

f. Do you think that multicollinearity is a concern?

g. A printout of a best subsets regression is given in Figure 9.20. Find a best-fitting model using the three-model selection criteria of R^2, R^2_{adj}, and C_p and justify why you selected this model.

3. A financial advisor wants to determine what factors influence a mutual funds rate of return. A random sample of 1-year return rates (expressed as a percentage) for 78 mutual funds was collected along with measures of the age of the fund (rounded to the nearest year), the risk profile of the fund (high or low risk), the amount invested in the fund (in millions of dollars), and the expense to manage the fund (expressed as a percentage). A portion of the data is given in Table 9.7.

Use the information from the printout in Figure 9.21 and answer each of the following questions ($\alpha = 0.10$):

a. Which variables are continuous and which variables are discrete (nominal/ordinal)?

b. Write the population regression equation.

c. Write the estimated regression equation.

d. Is the overall model useful in predicting the rate of return?

Regression Analysis: Selling Price versus Number of ... mber of Stories

Analysis of Variance

Source	DF	Adj SS	Adj MS	F-Value	P-Value
Regression	7	2863.65	409.093	28.94	0.000
Number of Bedrooms	1	109.47	109.472	7.75	0.006
Number of Bathrooms	1	27.02	27.023	1.91	0.170
Square Footage	1	633.89	633.892	44.85	0.000
Age of Home	1	2.51	2.512	0.18	0.674
Lot Size	1	27.24	27.240	1.93	0.168
Number of Garages	1	29.29	29.286	2.07	0.153
Number of Stories	1	60.45	60.448	4.28	0.041
Error	107	1512.32	14.134		
Total	114	4375.97			

Model Summary

S	R-sq	R-sq(adj)	R-sq(pred)
3.75950	65.44%	63.18%	44.45%

Coefficients

Term	Coef	SE Coef	T-Value	P-Value	VIF
Constant	-3.70	1.69	-2.19	0.031	
Number of Bedrooms	1.414	0.508	2.78	0.006	5.57
Number of Bathrooms	-0.663	0.480	-1.38	0.170	8.31
Square Footage	0.000945	0.000141	6.70	0.000	2.88
Age of Home	0.0047	0.0112	0.42	0.674	1.25
Lot Size	0.167	0.120	1.39	0.168	1.23
Number of Garages	0.688	0.478	1.44	0.153	2.26
Number of Stories	-1.249	0.604	-2.07	0.041	1.31

Regression Equation

Selling Price = -3.70 + 1.414 Number of Bedrooms - 0.663 Number of Bathrooms
+ 0.000945 Square Footage + 0.0047 Age of Home + 0.167 Lot Size
+ 0.688 Number of Garages - 1.249 Number of Stories

Fits and Diagnostics for Unusual Observations

Obs	Selling Price	Fit	Resid	Std Resid		
23	1.93	6.37	-4.44	-2.49	R	X
49	2.17	5.31	-3.14	-1.03		X
59	3.06	7.05	-3.99	-1.28		X
94	7.79	16.27	-8.47	-2.40	R	
101	11.28	4.19	7.09	2.03	R	
110	14.38	6.17	8.21	2.27	R	
113	19.48	26.41	-6.93	-2.35	R	X
114	39.77	17.52	22.26	6.31	R	
115	39.50	22.80	16.70	5.04	R	X

R Large residual
X Unusual X

FIGURE 9.19
Regression analysis to predict the selling price (in thousands of dollars), number of bedrooms, number of bathrooms, square footage, age of home, lot size, and number of garages for a random sample of 115 recent home sales.

e. What are the effects of each of the different predictor (independent) variables? Interpret any and all significant effects.

f. Find and interpret the coefficient of determination.

g. Determine you believe that whether any of your predictor variables are highly correlated with each other.

Best Subsets Regression: Selling Price versus Number of ... r of Stories

Response is Selling Price

Vars	R-Sq	R-Sq (adj)	R-Sq (pred)	Mallows Cp	S	Number of Bedrooms	Number of Bathrooms	Square Footage	Age	Lot Size	Number of Garages	Number of Stories
1	58.5	58.2	54.5	17.4	4.0068			X				
1	42.4	41.9	38.2	67.3	4.7224	X						
2	61.8	61.1	55.2	9.2	3.8626	X		X				
2	60.0	59.2	41.0	15.0	3.9551			X	X			
3	63.6	62.6	54.7	5.8	3.7896	X		X				X
3	62.7	61.7	44.0	8.4	3.8335	X		X	X			
4	64.4	63.1	45.7	5.3	3.7651	X		X	X			X
4	64.0	62.7	54.9	6.4	3.7837	X		X			X	X
5	64.8	63.1	53.8	6.1	3.7617	X	X	X			X	X
5	64.7	63.1	44.1	6.2	3.7624	X	X	X		X		X
6	65.4	63.5	45.3	6.2	3.7452	X	X	X		X	X	X
6	64.8	62.9	45.4	7.9	3.7753	X		X	X	X	X	X
7	65.4	63.2	44.4	8.0	3.7595	X	X	X	X	X	X	X

FIGURE 9.20
Best subsets regression for the regression analysis in Figure 9.20.

TABLE 9.7

Portion of Data Collected for Return Rate, Age, Risk Profile, Amount Invested, and Expense for a Random Sample of 78 Mutual Funds

Return Rate	Age	Risk Profile	Amount Invested	Expense
13.04	25	High	67	0.78
14.04	11	High	25	0.14
7.97	24	Low	40	1.65
4.78	10	Low	38	0.12
15.51	18	High	59	2.00
12.06	31	High	72	1.68
2.26	25	Low	25	1.96
12.62	33	Low	96	0.94

h. Do you think the given model is a good model for predicting the rate of return? Justify your answer.

4. Can you predict final examination scores based on the number of hours studying, major, gender, and current GPA? For a random sample of 17 students, their exam grade (on a scale of 0–100), the number of hours they spent studying for the final

Regression Analysis: Return Rate versus Age, Amount ... e, Risk Profile

Method

Categorical predictor coding   (1, 0)

Analysis of Variance

Source	DF	Adj SS	Adj MS	F-Value	P-Value
Regression	4	1011.46	252.866	24.10	0.000
Age	1	0.40	0.396	0.04	0.847
Amount Invested	1	2.79	2.788	0.27	0.608
Expense	1	9.10	9.104	0.87	0.355
Risk Profile	1	982.14	982.140	93.60	0.000
Error	73	765.99	10.493		
Total	77	1777.45			

Model Summary

S	R-sq	R-sq(adj)	R-sq(pred)
3.23929	56.91%	54.54%	51.14%

Coefficients

Term	Coef	SE Coef	T-Value	P-Value	VIF
Constant	11.83	1.49	7.95	0.000	
Age	0.0066	0.0340	0.19	0.847	1.02
Amount Invested	-0.0082	0.0160	-0.52	0.608	1.02
Expense	0.568	0.609	0.93	0.355	1.04
Risk Profile					
Low	-7.334	0.758	-9.67	0.000	1.06

Regression Equation

Risk Profile			
High	Return Rate	=	11.83 + 0.0066 Age - 0.0082 Amount Invested + 0.568 Expense
Low	Return Rate	=	4.49 + 0.0066 Age - 0.0082 Amount Invested + 0.568 Expense

Fits and Diagnostics for Unusual Observations

Obs	Return Rate	Fit	Resid	Std Resid	
8	12.620	4.670	7.950	2.51	R
9	14.720	5.316	9.404	2.97	R

R Large residual

FIGURE 9.21
Minitab printout to predict 1-year mutual fund return rates based on the age of the fund, the amount invested, the expense to manage the fund, and the risk profile.

exam (rounded to the nearest hour), their gender, and current GPA (on a scale of 0–4.00) are presented in Table 9.8.

a. Which variables are continuous and which variables are discrete (nominal/ ordinal)?

b. Write the population regression equation.

c. Write the estimated regression equation.

TABLE 9.8

Final Examination Grade, Number of Hours they Spent Studying, Major, Gender, and
Current GPA for a Random Sample of 17 Students

Exam Grade	Hours Studying	Major	Gender	Current GPA
83	10	Business	Male	3.41
70	8	Engineering and science	Male	2.98
49	3	Liberal arts	Female	2.64
64	7	Liberal arts	Male	3.12
87	12	Liberal arts	Female	3.68
74	6	Engineering and science	Female	3.45
90	14	Business	Male	3.8
48	2	Business	Male	1.87
73	8	Engineering and science	Male	2.74
97	18	Liberal arts	Female	3.28
51	2	Business	Female	1.48
82	6	Business	Male	3.1
85	8	Liberal arts	Female	3.3
86	10	Liberal arts	Male	3.56
57	2	Engineering and science	Male	1.98
63	5	Engineering and science	Female	2.4
66	4	Liberal arts	Male	2.95

d. Is the overall model useful in predicting the final exam grades?

e. What are the effects of each of the different predictor (independent) variables? Interpret any and all significant effects.

f. Find and interpret the coefficient of determination.

g. Determine whether any of your predictor variables are highly correlated with each other.

h. Do you think the given model is a good model for predicting final exam grades? Justify your answer.

10

Analysis of Variance (ANOVA)

10.1 Introduction

Till now, we have been concerned with making inferences about population parameters using statistics obtained from sample data by calculating confidence intervals and conducting hypothesis tests. We have also used regression analysis to develop models to predict one variable (y) based on another (x), or a set of variables $(x_1, x_2, \ldots, x_k)$. In this chapter, we will be discussing ways to use sample data to make inferences for *more than two* population means. But before we begin a discussion about how to compare more than two population means, we first need to describe some basic concepts of designing an experiment.

10.2 Basic Experimental Design

Often, one needs to design an experiment to address some question of interest. For example, suppose that we are interested in comparing the effect of several different brands of lawn fertilizer on the growth rate of grass. To make such a comparison, we could conduct an experiment by partitioning a plot of grass into several different sections and then applying different brands of the fertilizer to each section. After a period of time has elapsed, we could measure the growth rate of the grass for each of the different brands of the fertilizer. In such an experiment, there are no other factors that could have contributed to the growth rate of the grass other than the different types of fertilizer. Thus, the goal in designing an experiment is to be able to isolate the effect of a single factor or set of factors of interest.

There are different types of experimental designs that deal with isolating the factor or factors of interest. We will be describing two basic types of experimental designs: *randomized designs* and *randomized block designs*. A *randomized design* is when all factors except for a *single factor* of interest are held constant. For instance, a randomized design can be used to compare the effect of different brands of fertilizer on the growth rate of grass by taking a large plot of grass, dividing it into sections that represent the number of fertilizer brands that are being compared, and then randomly applying the brands of fertilizer to each of the different sections. Then, if all other factors, such as sunlight, water, pests, and so on, were the same for each of these sections, this would constitute a randomized design.

Figure 10.1 illustrates what a randomized design would look like if we were to partition a plot of land into six equal sections and then randomly assign six different brands of lawn fertilizer to each of the six sections.

Brand 3	Brand 5	Brand 2	Brand 6	Brand 1	Brand 4

FIGURE 10.1
Random assignment of six different brands of fertilizer to six different sections of land.

However, one problem with using a randomized design occurs when there is more than a single factor that may be different among the various sections. When this is the case, we may need to consider using a *randomized block design*.

A *randomized block design* can be used when there may be more than one factor of interest that may not be constant. If this were the case, then we would need to block (or control) these other factors so that they do not affect the outcome of interest.

For example, suppose that our plot of grass was half in the sun and half in the shade. If we were to use the randomized design illustrated in Figure 10.1, then we could not be sure that we are teasing out the effect of the fertilizer because we cannot be certain whether it was the sun or the lack of sun that had an impact on the growth rate of the grass.

In order to control for the difference in sunlight among the different sections of grass, we could use a randomized block design that entails partitioning the section of grass into both sunny and shady portions, and then randomly assigning all of the different brands of fertilizer to both of these sections, as illustrated in Figure 10.2.

Once we have designed an experiment where we can control for any extraneous factors that could affect our outcome of interest (which, for our example, is the growth rate of grass), we can then begin to compare whether there is a difference in the unknown population means based on the given factor. So for our example, we can test to see if there is a difference in the mean growth rate of grass between the different brands of fertilizer. The statistical method that we will use to compare whether *three or more population means* are significantly different from each other with respect to one factor is called a *one-way analysis of variance* (ANOVA).

10.3 One-Way ANOVA

You may recall that in a regression analysis, we used an ANOVA table to describe how the variability was partitioned into the sum of squares based on the regression and the

Sun	Shade
Brand 3	Brand 2
Brand 4	Brand 6
Brand 1	Brand 3
Brand 2	Brand 1
Brand 5	Brand 4
Brand 6	Brand 5

FIGURE 10.2
Randomized block design blocking out the effect of sun and shade.

sum of squares based on the error. Similarly, we can also perform an ANOVA to determine whether there is a difference between the unknown population means from three or more populations by partitioning the variability based on some factor of interest and the error.

A one-way ANOVA is a special case of a regression analysis where the only independent variables in the model are indicator (or binary) variables that represent the treatment. A *one-way ANOVA* compares the means of a variable that are classified by only one other variable, which is called the *factor*. The possible values of the factor are called the *levels of the factor*. We will be dealing with a *fixed-effects one-way ANOVA*, which means that we will be considering the effect of only a single factor and assume that the factor is fixed.

For instance, we could use a one-way ANOVA to compare the effect of several brands of lawn fertilizer on the mean growth rate of grass. We could consider the variable "fertilizer" as the factor or treatment and the different brands of the fertilizer could correspond to the different levels of the factor.

In general, when comparing the population means based on k different levels of some factor or treatment, we could draw a random sample from each of these k different levels of the population, as illustrated in Table 10.1.

In Table 10.1, $y_{i,j}$ represents the ith observation from the jth level. Notice that we do not necessarily need to take equal sample sizes from each level. An *unbalanced one-way ANOVA* is when at least two of the different levels of the factor do not have equal sample sizes. A *balanced one-way ANOVA* is when the levels of the factor all have the exact same sample size.

We want to know whether there is a difference in the population means among the k factor levels. To do this, we can describe a special case of a regression analysis, as seen in the following model:

$$y_{i,j} = \mu + \tau_j + \varepsilon_{i,j}$$

where j ranges from 1, ..., k, where k is the number of factor levels; $y_{i,j}$ is the ith response observation from level j; μ is the mean common to all levels of the factor; τ_j is the effect of the jth factor level; and $\varepsilon_{i,j}$ is the error component for the ith observation from level j. Thus, the treatment means for each of the k different levels can be described as follows:

TABLE 10.1

Samples Drawn from k Different Populations

Sample 1	Sample 2	Sample 3	...	Sample k
$y_{1,1}$	$y_{1,2}$	$y_{1,3}$	...	$y_{1,k}$
$y_{2,1}$	$y_{2,2}$	$y_{2,3}$	...	$y_{2,k}$
$y_{3,1}$	...	$y_{3,3}$	...	...
...	...	$y_{4,3}$	...	...
...	$y_{n_2,2}$	...	...	...
...		...	...	$y_{n_k,k}$
$y_{n_1,1}$		$y_{n_3,3}$	...	

$$\mu_1 = \mu + \tau_1$$
$$\mu_2 = \mu + \tau_2$$
$$\vdots$$
$$\mu_k = \mu + \tau_k$$

For instance, in considering the effect of k different brands of lawn fertilizer on the growth rate of grass, μ would be the average growth rate common to all the brands of fertilizer, and τ_j would be the effect of the jth brand of fertilizer on the growth rate. For instance, suppose that the mean growth rate common to all brands of fertilizer is 25 mm/week. If brand j contributes an additional 5 mm/week to the growth rate, then $\mu_j = \mu + \tau_j = 25 + 5 = 30$ mm would be the mean growth rate for the grass treated with the fertilizer brand j.

The main idea behind a one-way ANOVA is to determine whether there is a difference in the unknown population treatment effects for the k different levels of the treatment as described by the following null and alternative hypotheses:

$$H_0 : \tau_1 = \tau_2 = \tau_3 = \ldots = \tau_k$$

H_A : At least one pair of population effects are different.

The null hypothesis states that all of the unknown *population effects* are equal to each other, whereas the alternative hypothesis states that at least two population effects are different.

Similarly, the difference in treatment effects can also be stated in terms of the *population means* for the k different levels of the treatment. In other words, we can test whether at least one pair of population means are significantly different. The null and alternative hypotheses would then be stated as follows:

$$H_0 : \mu_1 = \mu_2 = \mu_3 = \ldots = \mu_k$$

H_A : At least one pair of populations means are different.

In performing a one-way ANOVA, we will be using the F-distribution. Recall that a random variable X follows the F-distribution if it has the shape of a skewed curve, as presented in Figure 10.3.

Recall that F-distributions are identified by two parameters to represent the degrees of freedom. One is called the *degrees of freedom for the numerator* and the other is called the *degrees of freedom for the denominator*.

For a one-way ANOVA, we use the following notation to describe the rejection region for the F-distribution:

$$f_\alpha\left(k-1, n-k\right)$$

where $k - 1$ is the number of degrees of freedom for the numerator, $n - k$ is the number of degrees of freedom for the denominator, and α is the level of significance. The variable k is

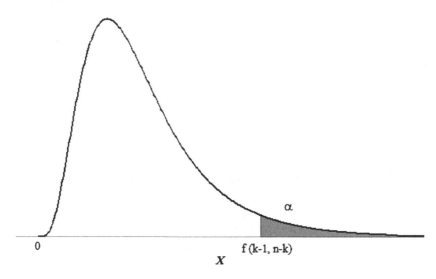

FIGURE 10.3
F-distribution.

the number of populations we are sampling from, which is the different levels of the factor, and n is the total sample size, which represents the total number of observations collected across all the different levels of the factor.

The following test statistic, which follows the F-distribution, is used for a one-way ANOVA:

$$F = \frac{\text{MSTR}}{\text{MSE}}$$

The test statistic, F, describes the ratio of two mean square terms, where the term MSTR corresponds to the mean square due to the treatment or factor and MSE corresponds to the mean square due to the error. If some of the means (or treatment effects) from the populations we are sampling from are significantly different from each other, then the MSTR would not likely be due to sampling error and would tend to be larger than the MSE. This would lead to F being large, thus allowing us to reject the null hypothesis and accept the alternative hypothesis. That is why the rejection region is in the right-tail for an ANOVA.

The test statistic F is then compared to the value of f with $(k - 1, n - k)$ degrees of freedom, where k is the number of populations being sampled from and n is the total number of observations collected.

Example 10.1

Suppose that we are interested in whether there is a difference in the mean driving distance between three different brands of golf balls. The data set presented in Table 10.2 consists of the driving distances (in yards) for three different brands of golf balls, which were collected under identical conditions using an automatic professional driving machine.

TABLE 10.2

Random Sample of the Driving Distances for Three Different Brands of Golf Balls

Brand 1	Brand 2	Brand 3
271	272	269
285	279	273
288	283	291
290	292	295

Because our response variable is the driving distance, we can find the mean driving distance for each brand as follows:

$$\bar{y}_1 = \frac{271 + 285 + 288 + 290}{4} = 283.50$$

$$\bar{y}_2 = \frac{272 + 279 + 283 + 292}{4} = 281.50$$

$$\bar{y}_3 = \frac{269 + 273 + 291 + 295}{4} = 282.00$$

The sample size for each of the three different brands is:

$$n_1 = 4$$

$$n_2 = 4$$

$$n_3 = 4$$

To use a one-way ANOVA to see if there is a significant difference in the driving distances between the three different brands of golf balls, the null and alternative hypotheses would be stated as follows:

$$H_0 : \mu_1 = \mu_2 = \mu_3$$

H_1 : At least one pair of populations means are different.

The alternative hypothesis can also be expressed as showing that at least one pair of population means is different, as follows:

$$H_1 : \mu_1 \neq \mu_2 \text{ or } \mu_1 \neq \mu_3 \text{ or } \mu_2 \neq \mu_3$$

To calculate the test statistic, we need to find the value of MSTR (the mean square for the treatment) by first calculating the *overall or grand mean* for all of the observations, denoted by $\bar{y}_G$, as follows:

$$\bar{y}_G = \frac{\text{sum of all observations}}{n_1 + n_2 + \ldots + n_k}$$

where n_j is the sample size from level j.

To find the MSTR, we first need to find the sum of squares for the treatment (SSTR). The sum of squares for the treatment can be found by taking the sum of the squared differences *between* the mean from each factor level and the grand mean multiplied by the sample size for each of the given factor levels, as follows:

$$\text{SSTR} = \sum_{i=1}^{k} n_i \left(\bar{y}_i - \bar{y}_G \right)^2 = n_1 \left(\bar{y}_1 - \bar{y}_G \right)^2 + n_2 \left(\bar{y}_2 - \bar{y}_G \right)^2 + \cdots + n_k \left(\bar{y}_k - \bar{y}_G \right)^2$$

where n_i is the sample size from factor level i, $\bar{y}_i$ is the mean of the observations from factor level i, and $\bar{y}_G$ is the grand mean of all the observations.

Because SSTR represents the sum of squares for the treatment that measures the difference between the mean of the jth group and the grand mean, SSTR can be used as an estimate of the variability between the different populations.

The MSTR is then found by dividing the SSTR by $k-1$ degrees of freedom, as follows:

$$\text{MSTR} = \frac{\text{SSTR}}{k-1}$$

where k is the number of levels.

For our example, there are three populations we are sampling from (the three different brands of golf balls that correspond to three different factor levels), $k = 3$, and a total of 12 observations collected, $n = 12$. Thus, the grand mean for all the observations can be calculated as follows:

$$\bar{y}_G = \frac{271 + 285 + 288 + 290 + 272 + 279 + 283 + 292 + 269 + 273 + 291 + 295}{4 + 4 + 4} \approx 282.33$$

Since $n_1 = n_2 = n_3 = 4$, we have that:

$$\text{SSTR} = 4(283.50 - 282.33)^2 + 4(281.50 - 282.33)^2 + 4(282.00 - 282.33)^2$$

$$= 5.48 + 2.76 + 0.44 \approx 8.68$$

$$\text{So, MSTR} = \frac{\text{SSTR}}{k-1} = \frac{8.68}{3-1} \approx 4.34$$

The mean square error (MSE) is defined as

$$\text{MSE} = \frac{\text{SSE}}{n-k}$$

$$\text{SSE} = \sum_{j=1}^{k} \left(n_j - 1 \right) s_j^2 = (n_1 - 1) s_1^2 + (n_2 - 1) s_2^2 + \cdots + (n_k - 1) s_k^2$$

where s_j^2 is the variance for the jth group.

For our example, the variances $\left(s^2 = \dfrac{\sum (x_i - \bar{x})^2}{n-1} \right)$ from each of the three levels is as follows:

$$s_1^2 \approx 73.67$$

$$s_2^2 \approx 69.67$$

$$s_3^2 \approx 166.67$$

So, $\text{SSE} = (4-1) \cdot 73.67 + (4-1) \cdot 69.69 + (4-1) \cdot 166.67 \approx 930.03$.

SSE can be used to estimate the variability *within* each of the k populations we are sampling from.

Then:

$$\text{MSE} = \frac{\text{SSE}}{n-k} = \frac{930.03}{12-3} \approx 103.34$$

And finally,

$$F = \frac{\text{MSTR}}{\text{MSE}} = \frac{4.34}{103.34} \approx 0.042$$

We can now compare the value of this test statistic with the value obtained from the F-distribution ($\alpha = 0.05$). The degrees of freedom for the numerator is $k - 1 = 3 - 1 = 2$, and the degrees of freedom for the denominator is $n - k = 12 - 3 = 9$. Using the F-table (Table A.3 or Minitab to generate the distribution) gives $f = 4.256$, as can be seen in Figure 10.4. And since $F = 0.042$ does

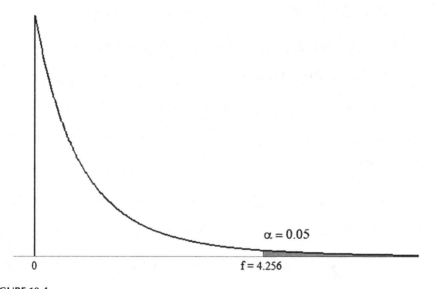

FIGURE 10.4
Rejection region for the F-distribution with 2 degrees of freedom for the numerator and 9 degrees of freedom for the denominator ($\alpha = 0.05$).

TABLE 10.3

General Form for a One-Way ANOVA Table

Source	Degrees of Freedom	Sum of Squares	Mean Square	F	p-value
Treatment variation (between groups)	$k-1$	SSTR	MSTR	$\dfrac{\text{MSTR}}{\text{MSE}}$	p
Error variation (within groups)	$n-k$	SSE	MSE	—	—
Total	$n-1$	SSTR + SSE	—	—	—

TABLE 10.4

A One-Way ANOVA Table for the Golf Ball Example

Source	Degrees of Freedom	Sum of Squares	Mean Square	F	p-value
Treatment variation	2	8.68	4.34	0.04	$p > 0.05$
Error variation	9	930.03	103.34	—	—
Total	11	**938.71**	—	—	—

not fall in the shaded region as established by $f = 4.26$, we do not have enough evidence to reject the null hypothesis and accept the alternative hypothesis. Therefore, there is no significant difference in the population mean driving distance between any pair of the different brands of golf balls.

One way to summarize all of the different calculations that are needed to conduct a one-way ANOVA is to put the calculations in the form of a table. The general form of a one-way ANOVA table illustrates how the variability is partitioned based on the factor (or treatment) and the error, along with the appropriate degrees of freedom, as illustrated in Table 10.3.

For the last example, the one-way ANOVA table is given in Table 10.4.

10.4 One-Way ANOVA Model Assumptions

As with most analyses that we have considered thus far, there are some model assumptions for a one-way ANOVA that need to be considered before any relevant inferences can be made. Because a one-way ANOVA is just a special case of a regression analysis, the model assumptions for an ANOVA are equivalent to those model assumptions that we considered regarding the distribution of the error component in a regression analysis. This is because the one-way ANOVA model, $y_{i,j} = \mu + \tau_j + \varepsilon_{i,j}$, is represented by a fixed component $(\mu + \tau_j)$ and a random component $(\varepsilon_{i,j})$.

The assumptions for a one-way ANOVA are as follows:

1. Any two observed values are independent of each other.
2. The error component is normally distributed.
3. The error component has constant variance.

The first assumption (independence) is not always easy to check. Typically, if a random sample is collected, then one can be relatively sure that the independence assumption has not been violated. However, if there is any reason to believe that the independence assumption could have been violated, then there are some exploratory techniques that can be done to see if this is an issue. For instance, if the data are collected over time and each observation depends on another, this would clearly be a violation of the independence assumption.

One of the more common ways to check the normality and constant variance assumptions for a one-way ANOVA model is to use residual plots and formal tests just like we did with regression analysis. Residual plots can be useful in determining whether some of the one-way ANOVA model assumptions may have been violated. The next section will provide some details on how to conduct various formal analyses for some of the ANOVA assumptions.

10.5 Assumption of Constant Variance

There are two formal tests that can be used to determine if the population variances for each of the k levels are equal; these are Bartlett's test (1937) and Levene's test (1960). For both of these tests, the null and alternative hypotheses for comparing the population variance from k different factor levels are as follows:

$$H_0 : \sigma_1^2 = \sigma_2^2 = \sigma_3^2 = \cdots = \sigma_k^2$$

$$H_1 : \text{At least two population variances are unequal.}$$

where σ_i^2 is the population variance from level i. For samples taken from k populations that are normally distributed, the test statistic for Bartlett's test is essentially the weighted arithmetic average and the weighted geometric average of each sample variance. The greater the difference is for these two averages, the more likely the variances of at least two of the populations are unequal (i.e., rejecting the null hypothesis and accepting the alternative hypothesis).

When equal sample sizes are taken from each population (i.e., the data is balanced), the test statistic for Bartlett's test is as follows:

$$B = \frac{3k(n_s - 1)^2 \left[k \cdot \ln(\bar{s}^2) - \sum_{i=1}^{k} \ln(s_i^2) \right]}{3k(n_s) - 2k + 1}$$

where k is the number of different populations being sampled from, n_s is the sample size taken from *each* population (since we have equal sample sizes from each of the k populations, this means that $n_s = n_1 = n_2 = \ldots = n_k$), s_i^2 is the sample variance for sample i,

$\bar{s}^2 = \dfrac{\sum_{i=1}^{k} s_i^2}{k}$ is the average of all k sample variances, and ln is the natural logarithm.

For unequal sample sizes (i.e., unbalanced data), the test statistic for Bartlett's test becomes a bit more complicated:

$$B = \frac{(n-k)\ln\left(\bar{s}^2\right) - \sum_{i=1}^{k}(n_i - 1)\ln\left(s_i^2\right)}{1 + \dfrac{1}{3(k-1)}\left[\displaystyle\sum_{i=1}^{k}\dfrac{1}{(n_i - 1)} - \dfrac{1}{n-k}\right]}$$

where k is the number of populations being sampled from, n_i is the sample size for sample i, n is the total number of observations collected across all samples ($n = n_1 + n_2 + n_3 + \ldots + n_k$),

s_i^2 is the sample variance for sample i, and $\bar{s}^2 = \dfrac{\displaystyle\sum_{i=1}^{k}(n_i - 1)s_i^2}{n-k}$ is a weighted average of the sample variances for all k populations.

Notice that in rejecting the null hypothesis and accepting the alternative hypothesis for Bartlett's test, this means that there are at least two variances that are unequal. Because we want to check for constant variance, we hope to find that there is *not* enough evidence to reject the null hypothesis and accept the alternative hypothesis. For the assumption of constant variance to hold true, *we do not want the test statistic to fall in the rejection region.* If we cannot claim that there are at least two variances that are unequal (in other words, if we cannot accept the alternative hypothesis), then we have reason to believe that the assumption of constant variance has not been violated.

For both equal and unequal sample sizes, we compare the appropriate test statistic with the critical value of the χ^2 distribution with $k - 1$ degrees of freedom. (See Table A.4). The shape of the χ^2 distribution is determined by $k - 1$ degrees of freedom, where k is the number of populations we are sampling from, as illustrated in Figure 10.5.

When we have equal sample sizes, n_s is the sample size taken from *each* population, and when we are considering unequal sample sizes, n is the *total number* of observations collected across all samples.

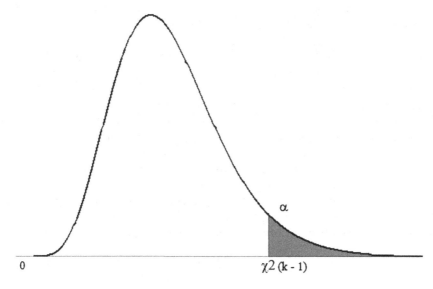

FIGURE 10.5
χ^2 distribution with $k - 1$ degrees of freedom.

Example 10.2

We will use Bartlett's test for the golf ball data in Table 10.2 as a formal test for constant variance. Because we have equal sample sizes from each of the populations we are sampling from, we can use Bartlett's test statistic for equal sample sizes. For this set of data, we have the following descriptive statistics:

$n_s = 4$ (there are four observations from each population)
$k = 3$ (we are sampling from three different populations)

$$\left. \begin{array}{l} s_1^2 \approx 73.67 \\[2ex] s_2^2 \approx 69.67 \\[2ex] s_3^2 \approx 166.67 \end{array} \right\} \qquad \text{(sample variances for the samples from each population)}$$

$$\bar{s}^2 = \frac{\sum\limits_{i=1}^{k} s_i^2}{k} = \frac{73.67 + 69.67 + 166.67}{3} \approx 103.34$$

And $\sum\limits_{i=1}^{k} \ln\left(s_i^2\right) = \ln\left(s_1^2\right) + \ln\left(s_2^2\right) + \ln\left(s_3^2\right) = \ln\left(73.67\right) + \ln\left(69.67\right) + \ln\left(166.67\right) \approx 13.66$

Thus, we can now calculate Bartlett's test statistic as follows:

$$B = \frac{3k(n_s - 1)^2 \left[k \cdot \ln\left(\bar{s}^2\right) - \sum\limits_{i=1}^{k} \ln\left(s_i^2\right) \right]}{3k(n_s) - 2k + 1} = \frac{3(3)(4-1)^2 \left[3 \cdot \ln\left(103.34\right) - 13.66 \right]}{3(3(4)) - 2(3) + 1} \approx 0.66$$

In comparing this to the value of the χ^2 distribution ($\alpha = 0.05$) for $k - 1 = 3 - 1 = 2$ degrees of freedom ($\chi^2 = 5.991$), we do not have enough evidence to suggest that at least two variances are different because the test statistic $B = 0.66$ does not fall in the rejection region as can be seen in Figure 10.6. Therefore, we have reason to believe that the assumption of constant variance has not been violated.

Another test of constant variance is Levene's test as modified by Brown and Forsythe (1974). The basic idea behind Levene's modified test is to use the absolute value of the difference between the observations in each treatment group and the median of the treatment. It then tests whether or not the mean of these differences is equal for all of the k treatment levels.

The test statistic for Levene's test is the same as the F-statistic for a one-way ANOVA, but the analysis is applied to the absolute value of the difference between the observation and the treatment median for each of the k treatments. If the average differences are equal for all of the treatments, this suggests that the variances of the observations in all of the treatment groups will be the same.

$$F = \frac{\text{MSTR}}{\text{MSE}}$$

which is compared to the F-distribution with $(k - 1, n - k)$ degrees of freedom.

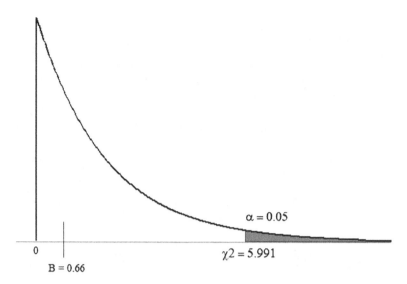

FIGURE 10.6
χ^2 distribution with $k - 1 = 3 - 1 = 2$ degrees of freedom ($\alpha = 0.05$).

Example 10.3

Let's use Levene's test on the golf ball data from Table 10.2 to test for constant variance. In order to do this, we first need to transform the response variable by taking the absolute value of the difference between each observation and the median for all the sample data from the three different levels of treatment. In other words:

$$\left| y_{i,j} - \hat{\eta}_j \right|$$

where $y_{i,j}$ is the ith observation from sample j and $\hat{\eta}_j$ is the median for sample j.

The medians of the response variables for each of the three different brands of golf balls are as follows:

$$\hat{\eta}_1 = 286.50$$

$$\hat{\eta}_2 = 281.00$$

$$\hat{\eta}_3 = 282.00$$

The transformed response data can be found by taking the absolute value of the difference between each individual observation and the median for each treatment level, as in Table 10.5.

For instance, the first observation for Brand 1 is 285 yards, and the median of Brand 1 is 286.50 yards. Therefore, the entry in the first row for column $k = 1$ is as follows:

$$\left| 285 - 286.50 \right| = 1.5$$

Then, the means of the transformed variables for each treatment level are 5.5, 6, and 11, respectively (in other words, these are the averages for the columns $k = 1$, $k = 2$, and $k = 3$ in Table 10.5).

TABLE 10.5

Difference Between Median and Response Value for Each Observation in Each Treatment

	$k = 1$	$k = 2$	$k = 3$
	$\lvert 285 - 286.50 \rvert = 1.5$	$\lvert 292 - 286.50 \rvert = 11$	$\lvert 291 - 286.50 \rvert = 9$
	$\lvert 290 - 286.50 \rvert = 3.5$	$\lvert 272 - 286.50 \rvert = 9$	$\lvert 269 - 286.50 \rvert = 13$
	$\lvert 271 - 286.50 \rvert = 15.5$	$\lvert 279 - 286.50 \rvert = 2$	$\lvert 295 - 286.50 \rvert = 13$
	$\lvert 288 - 286.50 \rvert = 1.5$	$\lvert 283 - 286.50 \rvert = 2$	$\lvert 273 - 286.50 \rvert = 9$
Mean of each level	$\dfrac{1.5+3.5+15.5+1.5}{4}=5.5$	$\dfrac{11+9+2+2}{4}=6$	$\dfrac{9+13+13+9}{4}=11$

To find the value of MSTR for the transformed data, we need to first find the grand mean for the data in Table 10.5:

$$\bar{y}_G = \frac{1.5+3.5+15.5+1.5+11+9+2+2+9+13+13+9}{4+4+4} = 7.5$$

So,

$$\text{SSTR} = \sum_{i=1}^{k} n_i\left(\hat{\eta}_i - \bar{y}\right)^2 = 4(5.5-7.5)^2 + 4(6-7.5)^2 + 4(11-7.5)^2 = 16+9+49 = 74$$

$$\text{MSTR} = \frac{\text{SSTR}}{k-1} = \frac{74}{3-1} = 37$$

The variances for each of the treatments of the transformed variables in Table 10.5 are as follows:

$$s_1^2 \approx 45.33$$

$$s_2^2 = 22.00$$

$$s_1^2 \approx 5.33$$

Then $\text{SSE} = \sum_{i=1}^{k}(n_i - 1)s_i^2 = (4-1)45.33 + (4-1)22.00 + (4-1)5.33 \approx 217.98$, and to find the MSE we divide by $n - k$:

$$\text{MSE} = \frac{\text{SSE}}{n-k} = \frac{217.98}{12-3} \approx 24.22$$

We can now calculate the test statistic as follows:

$$F = \frac{\text{MSTR}}{\text{MSE}} = \frac{37}{24.22} \approx 1.53$$

We now compare the value of this test statistic to $f_{0.05} = 4.26(\alpha = 0.05)$, as can be seen in Figure 10.7.

Since the test statistic does not fall in the rejection region as specified by $f = 4.256$, this suggests that the assumption of constant variance does not appear to have been violated.

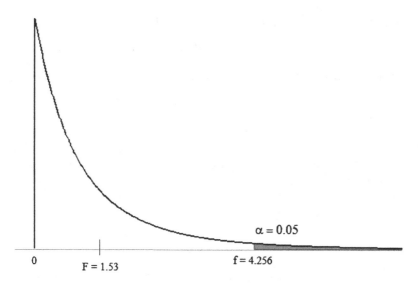

FIGURE 10.7
F-distribution with the test statistic $F = 1.53$ and the value $f(2,9) = 4.256$ that defines the rejection region.

Typically, Bartlett's test is used for data that comes from populations having normal distributions. Levene's test tends to be the preferred test for constant variance because it does not rely on the populations having normal distributions. Because Levene's test considers the distances between the observations and their sample medians rather than their sample means, it makes the test more appropriate for smaller sample sizes or when outliers are present.

10.6 Normality Assumption

Recall in Chapter 7 that we used the Ryan-Joiner test (Ryan and Joiner, 1976) to compare the unknown distribution of the population error component with a normal distribution to see if they are different. We compared the residuals with data from a normal distribution and used this to create a test statistic that serves to measure the correlation between the residuals and data that comes from a normal distribution. If this correlation measure is close to 1.0, then we are likely to believe that the assumption of normality for the error component does not appear to have been violated. Recall that the Ryan-Joiner test statistic can be calculated as follows:

$$RJ = \frac{\sum_{i=1}^{n} \hat{\varepsilon}_i \cdot z_i}{\sqrt{s^2 (n-1) \sum_{i=1}^{n} z_i^2}}$$

where $\hat{\varepsilon}_i$ is the ith residual, z_i is the normal score for the ith residual (see Section 7.8 for a review of how to calculate normal scores), n is the sample size, and s^2 is the variance of the residuals.

The null and alternative hypotheses for the Ryan-Joiner test are given as follows:

H_0 : The population error component follows a normal distribution.

H_1 : The population error component does not follow a normal distribution.

Recall that the Ryan-Joiner test statistic represents a measure of correlation between the residuals and data obtained from a normal distribution, and this statistic is compared to the rejection region for a given level of significance that defines the smallest correlation between the residuals and data obtained from a normal distribution (see Table A.5). The Ryan–Joiner test is a formal test of the normality assumption for the population error component and this test can be very useful when the exploratory plots of the residual are difficult to assess. The residual plots may not be very revealing when considering small sample sizes or when outliers may be present.

Example 10.4

For the data regarding the driving distance by the brand of golf ball that is given in Table 10.2, Table 10.6 gives the driving distance, brand, and the residuals along with their respective normal scores. The data in Table 10.6 is provided in stacked form because the driving distance (the response variable) is provided in one single column, and the brand (the factor) is provided in another single column.

Using the data provided in Table 10.6, we can see that:

$$\sum \hat{\varepsilon}_i \cdot z_i \approx 93.2712, \; s^2 \approx 84.5455, \; \text{and} \; \sum z_i^2 \approx 9.8514,$$

so the Ryan–Joiner test statistic, RJ, can be calculated as follows:

$$RJ = \frac{\sum \hat{\varepsilon}_i \cdot z_i}{\sqrt{s^2 (n-1) \sum z_i^2}} = \frac{93.2712}{\sqrt{84.5455(12-1)(9.8514)}} \approx 0.9744$$

TABLE 10.6

Residuals and Normal Scores for the Golf Ball Data Given in Table 10.2

Driving Distance	Brand	Residual $\hat{\varepsilon}_i$	Approximate Normal Score z_i	z_i^2	$\hat{\varepsilon}_i \cdot z_i$
285	1	1.5	0.00000	0.00000	0.00000
290	1	6.5	0.53618	0.28749	3.48517
271	1	−12.5	−1.11394	1.24086	13.92425
288	1	4.5	0.31192	0.09729	1.40364
292	2	10.5	1.11394	1.24086	11.69637
272	2	−9.5	−0.79164	0.62669	7.52058
279	2	−2.5	−0.31192	0.09729	0.77980
283	2	1.5	0.00000	0.00000	0.00000
291	3	9.0	0.79164	0.62669	7.12476
269	3	−13.0	−1.63504	2.67336	21.25552
295	3	13.0	1.63504	2.67336	21.25552
273	3	−9.0	−0.53618	0.28749	4.82562
			Total	9.8514	93.2712

Table A.5 gives the values for rejecting the null hypothesis. Comparing this test statistic to the value that falls between 0.9180 and 0.9383 for $n = 10$ and $n = 15$ with $\alpha = 0.05$ suggests that there is not enough evidence to imply that the error component does not come from a normal distribution (there is not enough evidence to reject the null hypothesis and accept the alternative hypothesis). This is because our test statistic is greater than the smallest correlation defined by the corresponding values of rj. Thus, we can infer that the assumption regarding the error component being normally distributed does not appear to have been violated.

Recall that the assumption of normality would be rejected if the value of the Ryan-Joiner test statistic was *less than* the value that defines the rejection region for a given level of significance and sample size, because this would imply that the correlation between the residuals and the values obtained from a normal distribution is less than the minimum amount that is specified by the value rj.

10.7 Using Minitab for One-Way ANOVAs

As you can see, it can be quite cumbersome to perform all of the various calculations for a one-way ANOVA manually, so we will now use Minitab to perform a one-way ANOVA. Using statistical software such as Minitab to run an ANOVA allows for a graphical assessment of the model assumptions for a one-way ANOVA in addition to performing the formal tests of some of the assumptions that were just described.

Using Minitab for a one-way ANOVA is fairly straightforward and requires putting all the collected data into either a *stacked* or an *unstacked* form. In using *unstacked form*, we can enter the sample data collected from each population into separate columns, where each column consists of the data from each factor (similar to how the golf ball data are presented in Table 10.2). In using *stacked form*, the data are entered using only two columns, where one column consists of the response variable and the other column describes the levels of the factor. For our example, the response variable would be the driving distance in yards, and the levels of the factor will be Brand 1, Brand 2, and Brand 3 to denote the three different brands of golf balls. In stacked form, the first two columns in our Minitab worksheet are presented in Figure 10.8.

To run a one-way ANOVA using Minitab, select **Stat, ANOVA, One-way**, to get the ANOVA dialog box that is illustrated in Figure 10.9.

The driving distance is the response variable and the brand is the factor, as seen in Figure 10.9. This gives the Minitab printout in Figure 10.10.

Note that we arrived at approximately the same values for the SSTR, SSE, MSTR, and MSE as when we performed the calculations manually (the slight differences are due to round-off error). On the basis of the p-value (0.959), as is illustrated in Figure 10.10, we can infer that we do not have enough evidence to suggest that any of the population mean driving distances for the three different brands of golf balls are significantly different from each other (since $p > 0.05$). Furthermore, on the bottom portion of the Minitab printout, there is a collection of confidence intervals of the population means for each of the different brands of golf balls. On the basis of the width and position of these confidence intervals, you can see that they are not much different from each other since all of these intervals are overlapping.

We can also use Minitab to determine if the assumptions for the ANOVA have been reasonably met. Similar to multiple regression analysis, checking the distribution of the error

Worksheet 1 ***		
↓	C1	C2
	Distance	Brand
1	285	1
2	290	1
3	271	1
4	288	1
5	292	2
6	272	2
7	279	2
8	283	2
9	291	3
10	269	3
11	295	3
12	273	3

FIGURE 10.8
Minitab worksheet with the golf ball data entered in stacked form.

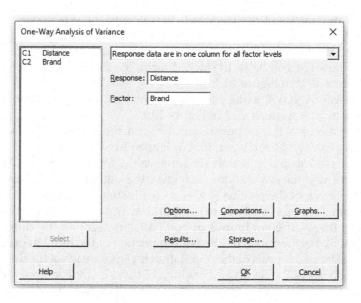

FIGURE 10.9
Minitab dialog box for a stacked one-way ANOVA.

One-way ANOVA: Distance versus Brand

Method

Null hypothesis	All means are equal
Alternative hypothesis	Not all means are equal
Significance level	$\alpha = 0.05$

Equal variances were assumed for the analysis.

Factor Information

Factor	Levels	Values
Brand	3	1, 2, 3

Analysis of Variance

Source	DF	Adj SS	Adj MS	F-Value	P-Value
Brand	2	8.667	4.333	0.04	0.959
Error	9	930.000	103.333		
Total	11	938.667			

Model Summary

S	R-sq	R-sq(adj)	R-sq(pred)
10.1653	0.92%	0.00%	0.00%

Means

Brand	N	Mean	StDev	95% CI
1	4	283.50	8.58	(272.00, 295.00)
2	4	281.50	8.35	(270.00, 293.00)
3	4	282.00	12.91	(270.50, 293.50)

Pooled StDev = 10.1653

FIGURE 10.10
Minitab printout for one-way ANOVA of driving distance versus brand of golf ball.

component for a one-way ANOVA can be done using Minitab by opening the **Graphs** dialog box within the one-way ANOVA box, as in Figure 10.11. As an additional check on the assumption of constant variance, we can also check the interval plots in the same dialog box.

The interval plot in Figure 10.12 illustrates the variability for each of the different brands of golf balls.

By examining the interval plot, it appears that the variance is approximately the same for all the different brands of golf balls.

The four-in-one plot of the residuals in Figure 10.13 can be used as an empirical check of the ANOVA model assumptions.

The normal probability plot, as presented in the top-left panel of Figure 10.13, shows the data points appearing along the diagonal if the error component is normally distributed. The bottom-left panel in Figure 10.13 should resemble a bell-shaped curve if the error component is normally distributed.

We can also graphically check the assumption of constant variance by plotting the residual values versus the fitted values, as in the top-right panel of Figure 10.13. The residuals should appear to be scattered randomly about the 0 line and there should be no obvious patterns or clusters. However, with small sample sizes, this plot is usually not too

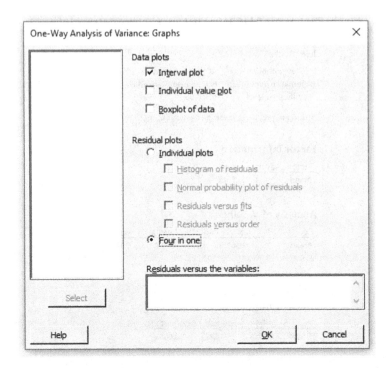

FIGURE 10.11
Minitab dialog box to graph an interval plot and check model assumptions with a four-in-one plot.

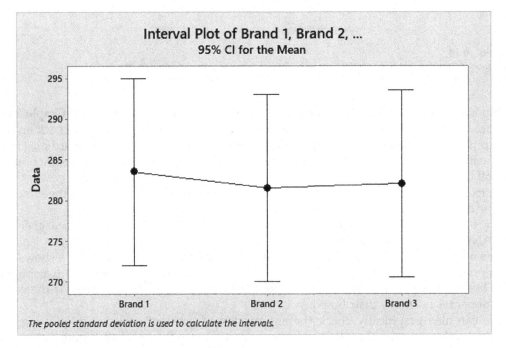

FIGURE 10.12
Interval plot of the driving distance for each brand of golf ball.

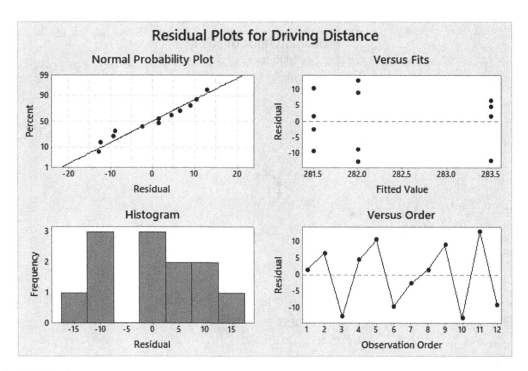

FIGURE 10.13
Four-in-one plot of the residuals for the golf ball data.

revealing because the observations will cluster based on the number of levels of the factor. The key is to look for approximately the same number of observations that lie above as below the 0 line.

Finally, the residuals versus the order of the observations plot in the bottom-right panel of Figure 10.13 can be used to assess whether the values of the residuals for any of the observations differ based on if the data were collected sequentially. If the pattern of this line diagram does not appear to be random, we could have reason to believe that the samples may not be independent of each other.

Because we have such a small sample size, it can be very challenging to use residual plots to assess whether the assumptions for a one-way ANOVA have been violated. However, we can use Minitab to conduct the Ryan-Joiner test of normality, along with Bartlett's and Levene's tests of constant variance.

Instead of generating a normal probability plot directly from the one-way ANOVA dialog box to check whether we are sampling from a normal distribution, we can use Minitab to conduct the Ryan-Joiner test of normality. To do this, we first need to save the residuals obtained from running the one-way ANOVA. To store the residuals using Minitab, simply check the **Storage** box on the one-way ANOVA dialog box to store the residuals. We can then perform the Ryan-Joiner test under **Basic Statistics > Normality Test**.

Selecting Ryan-Joiner in the dialog box gives the graph that is presented in Figure 10.14. Figure 10.14 is the normal probability plot along with the value of the Ryan-Joiner test statistic (RJ) in the box on the right of the graph. Since the p-value for the Ryan-Joiner test is greater than 0.05 ($p > 0.100$), this suggests that there is not enough evidence at the $\alpha = 0.05$ level of significance to imply that the assumption of the normality of the error component has been violated.

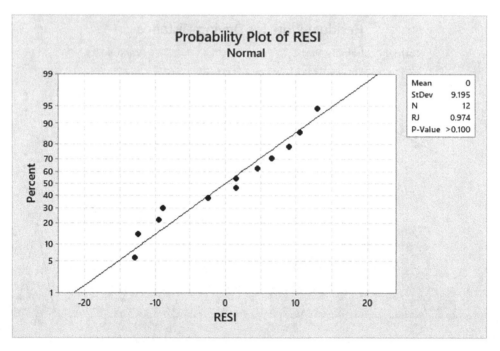

FIGURE 10.14
Minitab normality plot including the Ryan-Joiner test.

Minitab can also perform Bartlett's and Levene's tests for equal variances. Using the **Test for Equal Variance** tab from the **ANOVA** menu, as in Figure 10.15, gives the Minitab dialog box in Figure 10.16.

Checking the **Options** tab in Figure 10.16 gives the options dialog box that is presented in Figure 10.17. In this dialog box, you can select the confidence level and if you want to use

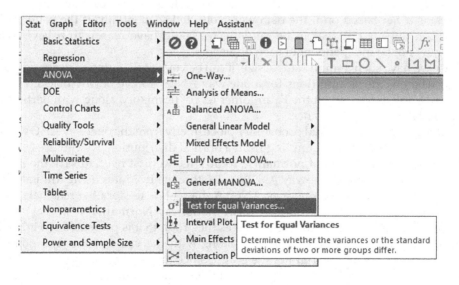

FIGURE 10.15
Minitab commands to test for equal variances.

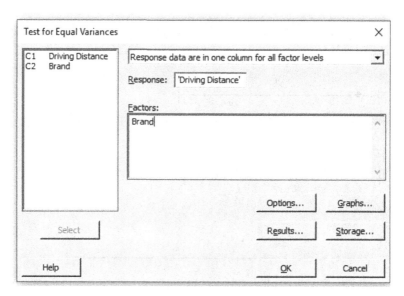

FIGURE 10.16
Minitab dialog box to test for equal variances.

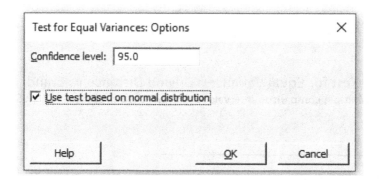

FIGURE 10.17
Options box to specify the level of confidence and whether to use Bartlett's test (which is based on the normal distribution).

Bartlett's test, you need to check the box that uses the test based on the normal distribution since Bartlett's test for equal variances is based on the normal distribution. If you want to use Levene's test for equal variances, do not check the box for the test to be based on the normal distribution.

Figure 10.18 gives the confidence intervals for the standard deviations for each level of the factor and the p-value for Bartlett's test. Figure 10.19 gives the confidence intervals of the standard deviations for each level of the factor and the p-value for Levene's test.

Minitab will also provide the test statistics for Bartlett's and Levene's test in the session output window. The session output for Bartlett's test can be seen in Figure 10.20, and the session output for Levene's test can be seen in Figure 10.21.

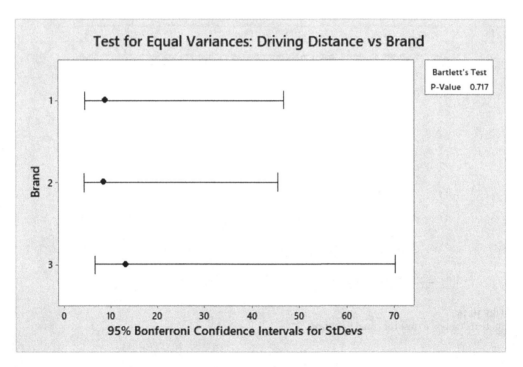

FIGURE 10.18
Minitab graph of confidence intervals and *p*-value for Bartlett's test.

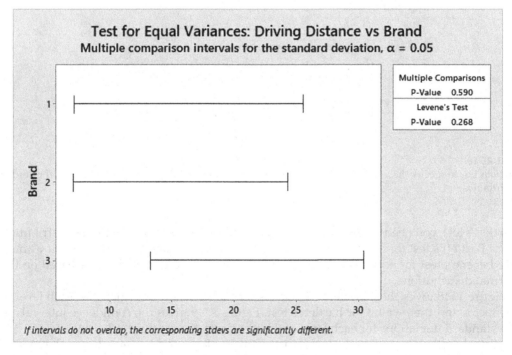

FIGURE 10.19
Minitab graph of confidence intervals and *p*-value for Levene's test.

Test for Equal Variances: Driving Distance versus Brand

Method

Null hypothesis	All variances are equal
Alternative hypothesis	At least one variance is different
Significance level	$\alpha = 0.05$

Bartlett's method is used. This method is accurate for normal data only.

95% Bonferroni Confidence Intervals for Standard Deviations

Brand	N	StDev	CI
1	4	8.5829	(4.33893, 46.6807)
2	4	8.3467	(4.21949, 45.3957)
3	4	12.9099	(6.52637, 70.2145)

Individual confidence level = 98.3333%

Tests

Method	Test Statistic	P-Value
Bartlett	0.67	0.717

FIGURE 10.20
Minitab output for Bartlett's test for equal variances.

Test for Equal Variances: Driving Distance versus Brand

Method

Null hypothesis	All variances are equal
Alternative hypothesis	At least one variance is different
Significance level	$\alpha = 0.05$

95% Bonferroni Confidence Intervals for Standard Deviations

Brand	N	StDev	CI
1	4	8.5829	(1.23991, 147.975)
2	4	8.3467	(1.48010, 117.231)
3	4	12.9099	(4.07639, 101.832)

Individual confidence level = 98.3333%

Tests

Method	Test Statistic	P-Value
Multiple comparisons	—	0.590
Levene	1.53	0.268

FIGURE 10.21
Minitab output for Levene's test for equal variances.

The *p*-values for both Barlett's and Levene's tests are greater than 0.05, suggesting that we do not have enough evidence to believe that the assumption of constant variance has been violated.

Example 10.5

The data set provided in Table 10.7 consists of the weekly growth rate of grass for six different brands of lawn fertilizer. Suppose that we want to test whether there is a difference in the mean, weekly growth rates of grass between the six different brands of fertilizer. The six columns in Table 10.7 represent the weekly growth rates of the grass in millimeters for each of the different brands of fertilizer.

Notice that for this data set, the sample sizes drawn from each of the populations are not all equal to each other, so this would be considered an *unbalanced* ANOVA. This is also an example of a data set that is in *unstacked* form. This is because each column represents a sample taken from a different population. By running a one-way ANOVA in the unstacked form using Minitab, the results are presented in Figure 10.22.

The interval plot and the graphs to assess the model assumptions for a one-way ANOVA are in Figures 10.23–10.27.

When you use the unstacked option for an ANOVA, there will not be a residual versus order plot and you only get a three-in-one plot. If you have reason to believe that the order in which the data were collected may affect the independence assumption, you would want to put the data into Minitab in stacked form so you could obtain the plot of the residual values versus the order of the observations. In most practical situations, data for an ANOVA is usually put in the stacked form in Minitab.

Note in Figure 10.22 that the *p*-value is 0.000. This suggests that we can reject the null hypothesis and accept the alternative hypothesis. In other words, at least two brands of lawn fertilizer have significantly different mean growth rates.

Also, the confidence intervals for each brand of fertilizer given in the bottom portion of the Minitab printout in Figure 10.22 and displayed on the interval plot in Figure 10.23, the growth rate for Brand 2 appears to be significantly higher than the growth rate for all of the other brands of fertilizer. This is because the confidence interval for Brand 2 does not overlap the confidence intervals for any of the other brands and also suggests a higher mean growth rate.

TABLE 10.7

Weekly Growth Rate of Grass (in Millimeters) for Six Different Brands of Fertilizer

Brand 1	Brand 2	Brand 3	Brand 4	Brand 5	Brand 6
26	35	26	25	26	27
31	36	27	34	30	27
34	42	28	37	32	28
35	47	29	38	34	29
	49	30	39	35	30
	50	34	42	42	
	52		45		
	54				

One-way ANOVA: Brand 1, Brand 2, Brand 3, Brand 4, Brand 5, Brand 6

Method

Null hypothesis	All means are equal
Alternative hypothesis	Not all means are equal
Significance level	$\alpha = 0.05$
Rows unused	12

Equal variances were assumed for the analysis.

Factor Information

Factor	Levels	Values
Factor	6	Brand 1, Brand 2, Brand 3, Brand 4, Brand 5, Brand 6

Analysis of Variance

Source	DF	Adj SS	Adj MS	F-Value	P-Value
Factor	5	1450.9	290.19	10.25	0.000
Error	30	849.4	28.31		
Total	35	2300.3			

Model Summary

S	R-sq	R-sq(adj)	R-sq(pred)
5.32092	63.08%	56.92%	49.03%

Means

Factor	N	Mean	StDev	95% CI
Brand 1	4	31.50	4.04	(26.07, 36.93)
Brand 2	8	45.63	7.19	(41.78, 49.47)
Brand 3	6	29.00	2.83	(24.56, 33.44)
Brand 4	7	37.14	6.41	(33.04, 41.25)
Brand 5	6	33.17	5.38	(28.73, 37.60)
Brand 6	5	28.200	1.304	(23.340, 33.060)

Pooled StDev = 5.32092

FIGURE 10.22
Minitab printout for a one-way ANOVA testing the mean difference in the growth rate of six different brands of fertilizer.

Although we have a larger sample size than we did for the golf ball example, it can still be difficult to assess whether the assumptions for an ANOVA may have been violated by relying solely on the residual plots. Thus, we may want to conduct some formal tests of the model assumptions.

To estimate if the normality assumption has been violated, we can save the residuals after running an ANOVA and conduct the Ryan–Joiner test. The graph of the normality plot from the Ryan–Joiner test is given in Figure 10.25.

The p-value for the Ryan-Joiner test is greater than our predetermined level of significance of $\alpha = 0.05$, and this suggests that there is not enough evidence to imply that the assumption of the normality of the error component has been violated.

Now using Bartlett's and Levene's tests for constant variance, the Minitab output is presented in Figures 10.26 and 10.27.

The box on the top right-hand side of Figures 10.26 and 10.27 gives the results for Bartlett's and Levenes's tests, respectively. For Barlett's test, the p-value of 0.040 suggests that the variance is not the same for at least two brands of lawn fertilizer (since the p-value is less than 0.05, we can reject the null hypothesis and accept the alternative hypothesis that at least two population variances are unequal). However, Levene's test reports a p-value greater than 0.05, which suggests that we do not have enough evidence to imply that the assumption

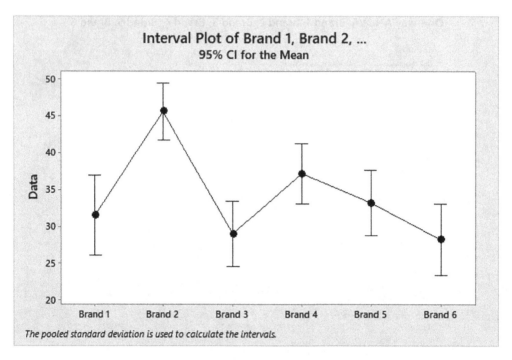

FIGURE 10.23
Interval plot of growth rate for the six different brands of fertilizer.

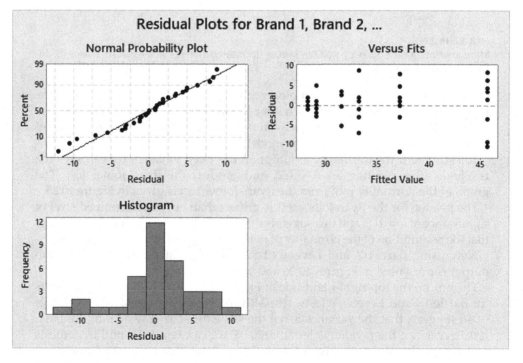

FIGURE 10.24
Three-in-one plot of the residuals.

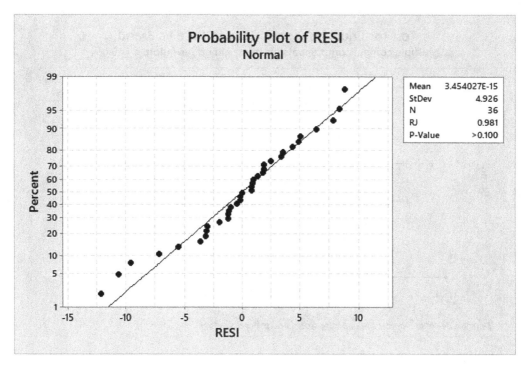

FIGURE 10.25
Ryan-Joiner test of the normality assumption.

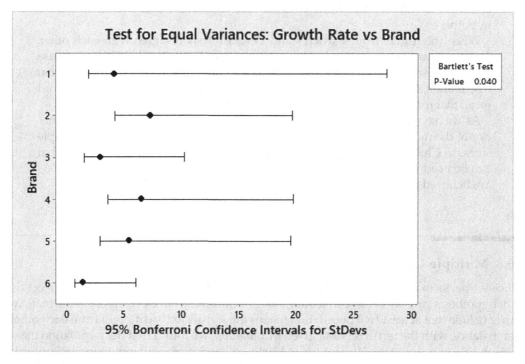

FIGURE 10.26
Minitab output for Bartlett's test of constant variance.

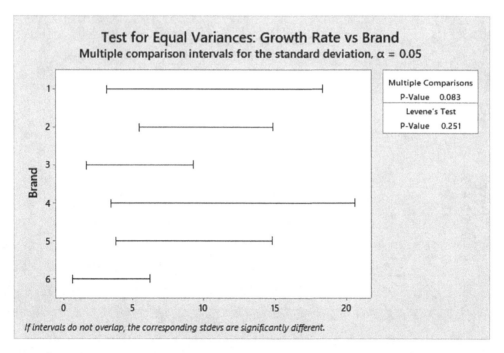

FIGURE 10.27
Minitab output for Levene's test of constant variance.

of constant variance has been violated (i.e., we cannot accept the alternative hypothesis).

When the results from Bartlett's and Levene's tests disagree with each other, you may want to consider using Levene's test over Bartlett's test. This is because Bartlett's test is much more sensitive to outliers because it relies on the normal distribution as opposed to Levene's test, which is based on median distances. And recall that for small samples, outliers can often have a more pronounced effect.

As we have seen on numerous occasions, it can be difficult to assess whether or not the model assumptions have been violated when we have small sample sizes. In Chapter 11, we will be describing alternative statistical methods that can be used when considering small sample sizes or if the model assumptions are believed to have been violated.

10.8 Multiple Comparison Techniques

Once we perform a one-way ANOVA and accept the alternative hypothesis and reject the null hypothesis (and we believe that our model assumptions have been reasonably met), we can conclude that at least two population means are significantly different from each other. For instance, with the fertilizer data given in Table 10.7, we found that the population mean growth rate of grass was different for at least two brands of fertilizer. Once a significant difference is found, we may want to identify which specific pair (or pairs) of population means are different from each other, and we also may want to estimate the magnitude and direction of such a difference or differences. We will use *multiple comparison techniques* to make such assessments.

In order to generate and compare a pairwise collection of means, we will use *Fisher's Least Significant Difference* (LSD). This method is based on the t-distribution with $n - k$ degrees of freedom, where n is the total sample size collected across all factor levels, and k is the number of populations being sampled from (i.e., the number of levels of the factor).

We can create confidence intervals to estimate the difference between any pair of means, $\mu_l - \mu_m$, where l and m are the indices for the two individual populations we are interested in comparing. The formula for the confidence interval to estimate the difference between two means for an ANOVA is as follows:

$$\left((\bar{x}_l - \bar{x}_m) - t_{\alpha/2}\sqrt{\text{MSE}\left(\frac{1}{n_l} + \frac{1}{n_m}\right)}, (\bar{x}_l - \bar{x}_m) + t_{\alpha/2}\sqrt{\text{MSE}\left(\frac{1}{n_l} + \frac{1}{n_m}\right)} \right)$$

where $\bar{x}_l$ is the mean of the sample from population l, $\bar{x}_m$ is the mean of the sample from population m, n_l is the size of the sample from population l, n_m is the size of the sample from population m, $t_{\alpha/2}$ is the t-value for an area of $\alpha/2$ in the right tail for a $(100 - \alpha)\%$ confidence interval, and MSE is the mean square error from the ANOVA.

This confidence interval represents the range of values of the difference in the means for populations l and m. If this confidence interval does not contain the value 0, then we can say that the two means from populations l and m are significantly different from each other. Otherwise, if the interval contains the value of 0, we cannot claim that the two means are different.

Example 10.6

Suppose we are interested in comparing the mean growth rate for the six different brands of fertilizer that is presented in Table 10.7 ($\alpha = 0.05$). Given that we already ran a one-way ANOVA, we found that there is at least one pair of means that are significantly different from each other. We can use Fisher's LSD to estimate the magnitude and direction of the difference in the growth rates between the different brands. Let's start by looking at the difference between Brand 1 and Brand 2.

On the basis of having six levels of our factor and a total sample size of 36, the degrees of freedom for the t-distribution would be $n - k = 36 - 6 = 30$, thus $t_{\alpha/2} = t_{0.05/2} = t_{0.025} = 2.042$, as can be seen in Figure 10.28.

The confidence interval that compares the mean growth rate for Brand 2 as compared to the mean growth rate for Brand 1 can be calculated as follows:

$$\left((\bar{x}_1 - \bar{x}_2) - t_{\alpha/2}\sqrt{\text{MSE}\left(\frac{1}{n_1} + \frac{1}{n_2}\right)}, (\bar{x}_1 - \bar{x}_2) + t_{\alpha/2}\sqrt{\text{MSE}\left(\frac{1}{n_1} + \frac{1}{n_2}\right)} \right)$$

$$= \left((31.5 - 45.63) - 2.042\sqrt{28.31\left(\frac{1}{4} + \frac{1}{8}\right)}, (31.5 - 45.63) + 2.042\sqrt{28.31\left(\frac{1}{4} + \frac{1}{8}\right)} \right)$$

$$\approx (-14.13 - 6.65, -14.13 + 6.65) \approx (-20.78, -7.48)$$

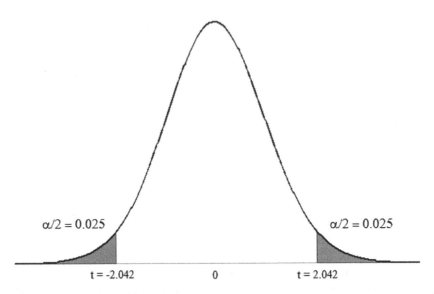

FIGURE 10.28
The t-distribution for $n - k = 36 - 6 = 30$ degrees of freedom for a 95% confidence interval.

Therefore, we are 95% confident that the growth rate for Brand 1 is 7.48–20.78 mm *less* than Brand 2.

Interpreting this confidence interval says that in subtracting the mean of Brand 2 from the mean of Brand 1, we are 95% confident that the difference in the population mean weekly growth rate will fall in this interval. In other words, we can claim that there is a significant difference in the mean weekly growth rates between Brand 2 and Brand 1. Specifically, Brand 1 has a significantly lower mean growth rate as compared to Brand 2. This is because the mean difference, Brand 1 – Brand 2, is less than 0. Since the confidence interval for the mean of Brand 2 subtracted from the mean of Brand 1 does not contain the value of 0, we can conclude that the mean weekly growth rates for Brands 2 and 1 are significantly different from each other. Furthermore, we can estimate that the magnitude of the difference between these two means to be between 7.48 and 20.78 mm. In other words:

$$-20.78 < \mu_1 - \mu_2 < -7.48$$

Similar calculations for the difference between the mean growth rate of Brand 4 and the mean growth rate of Brand 2 is as follows:

$$\left((\bar{x}_4 - \bar{x}_2) - t_{\alpha/2} \sqrt{\text{MSE}\left(\frac{1}{n_1} + \frac{1}{n_2}\right)}, (\bar{x}_4 - \bar{x}_2) + t_{\alpha/2} \sqrt{\text{MSE}\left(\frac{1}{n_1} + \frac{1}{n_2}\right)} \right)$$

$$= \left((37.14 - 45.63) - 2.042 \sqrt{28.31\left(\frac{1}{7} + \frac{1}{8}\right)}, (37.14 - 45.63) + 2.042 \sqrt{28.31\left(\frac{1}{7} + \frac{1}{8}\right)} \right)$$

$$\approx \left(-8.49 - 5.62, -8.49 + 5.62 \right) \approx \left(-14.11, -2.87 \right)$$

We would claim to be 95% confident that the mean growth rate for Brand 4 is 2.87–14.11 mm *less* than the mean growth rate of Brand 2.

Now let's consider the difference between Brand 4 and Brand 5. The calculations for a 95% confidence interval for the difference between these two means would be as follows:

$$\left((\bar{x}_4 - \bar{x}_5) - t_{\alpha/2} \sqrt{MSE\left(\frac{1}{n_1} + \frac{1}{n_2}\right)}, (\bar{x}_4 - \bar{x}_5) + t_{\alpha/2} \sqrt{MSE\left(\frac{1}{n_1} + \frac{1}{n_2}\right)} \right)$$

$$= \left((37.14 - 33.17) - 2.042\sqrt{28.31\left(\frac{1}{7} + \frac{1}{6}\right)}, (37.14 - 33.17) + 2.042\sqrt{28.31\left(\frac{1}{7} + \frac{1}{6}\right)} \right)$$

$$\approx (3.97 - 6.04, 3.97 + 6.04) \approx (-2.07, 10.01)$$

This confidence interval suggests that the population mean difference between the growth rates for Brand 5 subtracted from Brand 4 are between -2.07 and 10.01 mm. Since this confidence interval contains the value of 0, we cannot claim that there is a significant difference between the mean growth rates for these two brands of fertilizer.

Generally, multiple comparison techniques are only performed following a one-way ANOVA when there is enough evidence to accept the alternative hypothesis and reject the null hypothesis, and we can conclude that at least two means are significantly different from each other.

10.9 Using Minitab for Multiple Comparisons

We can use Minitab to perform Fisher's LSD multiple comparison techniques for a one-way ANOVA. This is done by selecting the **Comparison** tab on the one-way ANOVA dialog box (see Figure 10.9) to get the multiple comparison dialog box that is presented in Figure 10.29.

Then checking the Fisher box and entering the desired level of significance (by default, a value of 5%, which assumes a 0.05 level of significance), we get the printout in Figure 10.30 in addition to the one-way ANOVA printout. We can also get an interval plot for differences of means as can be seen in Figure 10.31 by checking the **Interval plot for differences of means** in Figure 10.29. We can also get grouping information and the results of hypothesis tests by checking the **Grouping information** and **Tests** boxes in Figure 10.29.

The top portion of the printout in Figure 10.30 gives the grouping information where means that do not share a grouping letter are significantly different. For instance, Brand 2 (which has the letter A) does not share a letter with any other brand (all the other brands have letters B and/or C), and therefore, Brand 2 is significantly different from every other brand.

Confidence intervals for each pair of differences between the mean growth rates for all of the different brands of fertilizer and an adjusted *p*-value are given in the bottom portion of Figure 10.30. Thus, by using Fisher's LSD multiple comparison method, we can identify and compare the individual pairs of means that are significantly different from each other, and we can also distinguish which specific brands of fertilizer are generating a higher

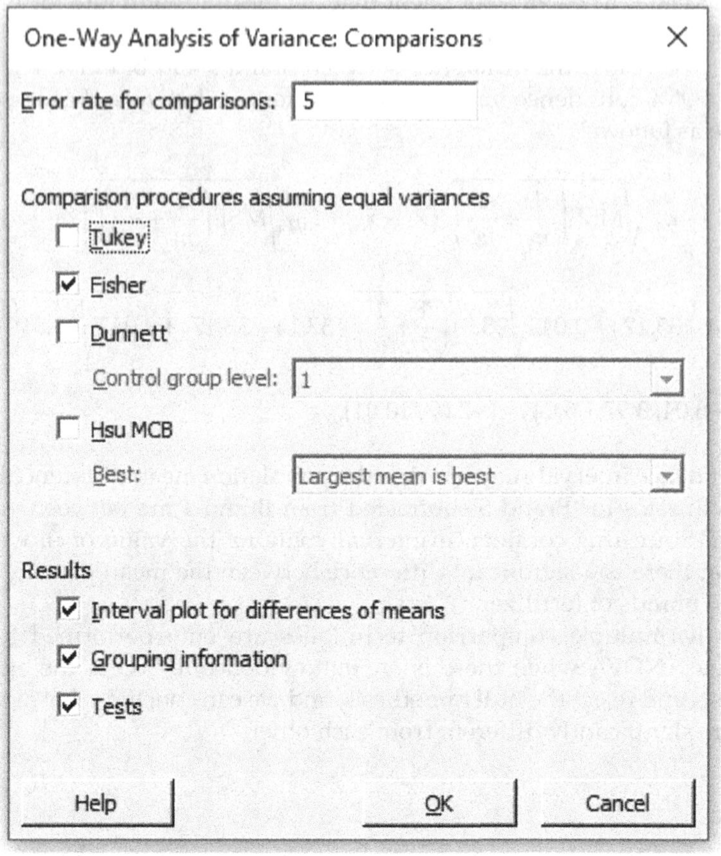

FIGURE 10.29
Minitab dialog box for running multiple comparison techniques.

(or lower) mean growth rate, along with a range of values that represent the magnitude of the difference. The adjusted p-value tells us which confidence intervals are significant.

Figure 10.31 is the graphical display of the Minitab printout that illustrates all the possible pairwise comparisons. If an interval does not cross over the 0 line, then the means are significantly different.

10.10 Power Analysis and One-Way ANOVA

Recall that a power analysis allows you to determine how large of a sample is needed to make reasonable inferences and how likely it is that the statistical test you are considering will be able to detect a minimum difference of a given size. Also recall that the power of a test is the probability of correctly rejecting the null hypothesis when it is, in fact, false. For example, a power value of 0.80 would suggest that there is an 80% chance of correctly rejecting the null hypothesis and that there is a 20% chance of failing to detect a difference if such a difference actually exists.

Fisher Pairwise Comparisons

Grouping Information Using the Fisher LSD Method and 95% Confidence

Brand	N	Mean	Grouping		
2	8	45.63	A		
4	7	37.14		B	
5	6	33.17		B	C
1	4	31.50		B	C
3	6	29.00			C
6	5	28.200			C

Means that do not share a letter are significantly different.

Fisher Individual Tests for Differences of Means

Difference of Levels	Difference of Means	SE of Difference	95% CI	T-Value	Adjusted P-Value
2 - 1	14.13	3.26	(7.47, 20.78)	4.33	0.000
3 - 1	-2.50	3.43	(-9.51, 4.51)	-0.73	0.472
4 - 1	5.64	3.34	(-1.17, 12.45)	1.69	0.101
5 - 1	1.67	3.43	(-5.35, 8.68)	0.49	0.631
6 - 1	-3.30	3.57	(-10.59, 3.99)	-0.92	0.363
3 - 2	-16.63	2.87	(-22.49, -10.76)	-5.79	0.000
4 - 2	-8.48	2.75	(-14.11, -2.86)	-3.08	0.004
5 - 2	-12.46	2.87	(-18.33, -6.59)	-4.34	0.000
6 - 2	-17.43	3.03	(-23.62, -11.23)	-5.74	0.000
4 - 3	8.14	2.96	(2.10, 14.19)	2.75	0.010
5 - 3	4.17	3.07	(-2.11, 10.44)	1.36	0.185
6 - 3	-0.80	3.22	(-7.38, 5.78)	-0.25	0.806
5 - 4	-3.98	2.96	(-10.02, 2.07)	-1.34	0.189
6 - 4	-8.94	3.12	(-15.31, -2.58)	-2.87	0.007
6 - 5	-4.97	3.22	(-11.55, 1.61)	-1.54	0.134

Simultaneous confidence level = 65.64%

FIGURE 10.30
Fisher pairwise comparisons for the Fertilizer data.

Sample size estimation and power analysis are very important aspects of any empirical study because if the sample size is too small, the experiment may lack precision, and thus not allow the investigator to make a reasonable inference. Also, if it is costly to obtain a large sample size, it can be a waste of resources if a smaller sample can adequately detect a given difference.

In Chapter 4 we used Minitab to determine the sample size needed to achieve an adequate level of power or to determine the power level for a given sample size for some basic statistical inferences. In order to conduct a power analysis, we had to specify two of three values, namely, the sample size, difference, or power. Similarly, for a one-way ANOVA, we also need to specify two of the following: the power, the sample size (the number of observations from each level of the factor), or the difference between the smallest and largest factor means. However, with a one-way ANOVA, we have different levels of a factor, so we also need to specify the number of levels of the factor and provide an estimate of the standard deviation.

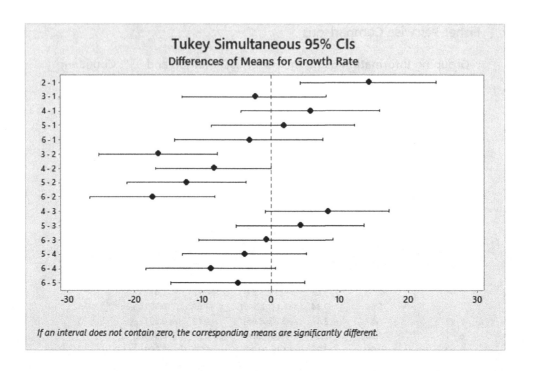

FIGURE 10.31
Fisher confidence interval plots for the Fertilizer data.

Example 10.7

Suppose we want to determine if the sample we collected for our golf ball experiment has at least 80% power and we want to detect a difference of at least 10 yards between the means among the different brands. From our one-way ANOVA, we found that the MSE was approximately 103, so the estimate of our standard deviation would be approximately 10 (the root mean square error is the square root of the MSE). The sample size from each factor level is 4, so the power analysis would be entered into Minitab, as illustrated in Figure 10.32.

The results of this power analysis are presented in Figure 10.33.

Thus, in order to detect at least a 10 yards difference between the smallest and the largest factor level means, a sample size of 4 from each level only achieves about 17% power. Hence, our one-way ANOVA is *underpowered*, which means that we are not likely to detect a difference of 10 yards if such a difference actually does exist with a sample of size 4 from each group.

To determine how large of a sample we would need to achieve 80% power for a difference of at least 10 yards, we would need to sample at least 21 observations from each factor level, as described in Figure 10.34.

One thing to keep in mind is that Minitab describes the maximum difference as the minimum detectable difference between the *smallest and largest factor means*. It can also be described as the smallest difference between the two most extreme means.

Similarly, if we wanted to find a larger difference, say 50 yards, then we would need at least three observations from each level, as presented in Figure 10.35.

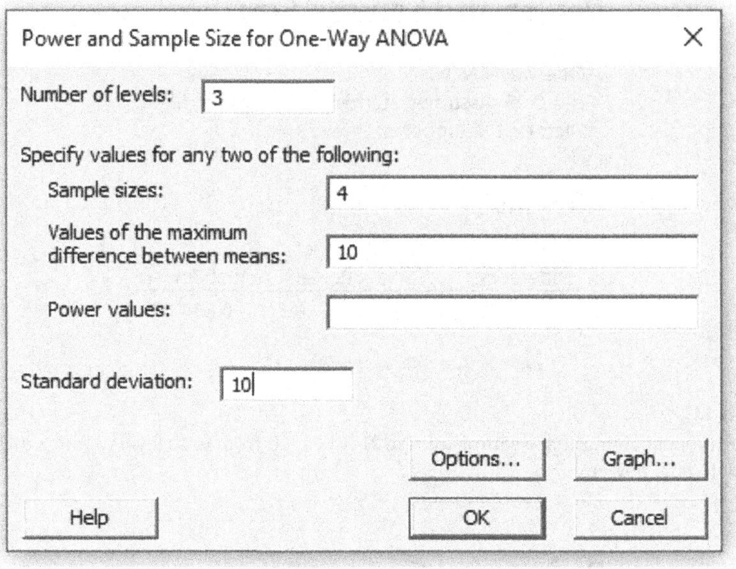

FIGURE 10.32
Minitab dialog box for a power analysis for a one-way ANOVA.

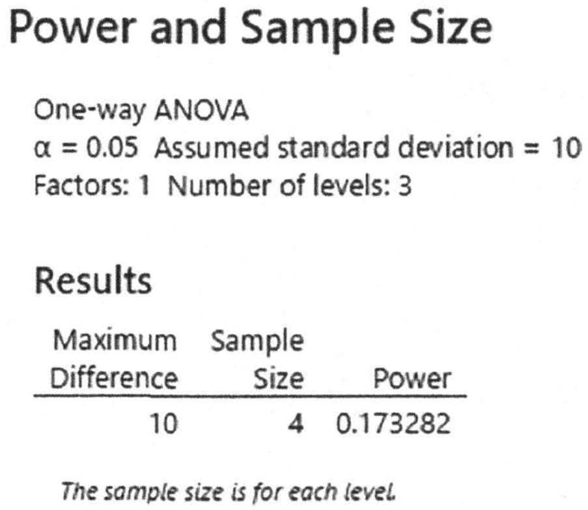

Power and Sample Size

One-way ANOVA
$\alpha = 0.05$ Assumed standard deviation = 10
Factors: 1 Number of levels: 3

Results

Maximum Difference	Sample Size	Power
10	4	0.173282

The sample size is for each level.

FIGURE 10.33
Minitab printout for the golf ball data in Table 10.2 showing that a sample size of 4 from each level with three different levels and a difference of at least 10 yards will achieve approximately 17% power.

As Figure 10.35 illustrates, larger differences require smaller sample sizes to achieve adequate power, whereas smaller differences require larger sample sizes to achieve adequate power. This is because larger differences are easier to detect than are smaller differences.

In conducting a power analysis for a one-way ANOVA, a conservative approach is to collect the *smallest sample size* from any of the levels. And just as was described in earlier chapters regarding the power of a test, power analyses are typically done before data collection even begins (in other words, by conducting a prospective power analysis). By estimating how large a sample

Power and Sample Size

One-way ANOVA
$\alpha = 0.05$ Assumed standard deviation = 10
Factors: 1 Number of levels: 3

Results

Maximum Difference	Sample Size	Target Power	Actual Power
10	21	0.8	0.814770

The sample size is for each level.

FIGURE 10.34
Minitab printout showing that a sample of size 21 is needed from each level to find a difference of 10 yards with 80% power.

Power and Sample Size

One-way ANOVA
$\alpha = 0.05$ Assumed standard deviation = 10
Factors: 1 Number of levels: 3

Results

Maximum Difference	Sample Size	Target Power	Actual Power
50	3	0.8	0.988251

The sample size is for each level.

FIGURE 10.35
Power analysis to find a difference of 50 yards with 80% power at three levels.

is needed to find a certain difference allows you to be assured that the sample sizes you plan to collect will achieve adequate power for your study, so you do not have to be concerned that the reason you did not find any effect is because your study was underpowered.

Exercises

Unless otherwise specified, use a significance level of $\alpha = 0.05$ and a power level of 80%.

1. Can your diet make you happier? A researcher wants to know if the food you eat has an effect on your happiness. A random sample of 41 participants was asked to complete a survey on happiness and a questionnaire on their diet. The happiness survey is scored on a scale of 50–100, where higher scores are associated with

someone who has a higher degree of happiness and satisfaction with their life. The data from the survey and the type of diet are provided in Table 10.8.

Is there a difference in the responses to the happiness survey based on the type of diet? In order to answer this question:

a. State the appropriate null and alternative hypotheses.

b. How would you describe the factor? And how many factor levels are there?

c. Find the SSE and the MSE.

d. Find the SSTR and the MSTR.

e. Find the number of degrees of freedom for the numerator and the number of degrees of freedom for the denominator.

f. Find the p-value and interpret it within the context of the problem.

g. Check any relevant model assumptions.

2. In problem #1, we used the given sample data to estimate whether there is a difference in happiness scores based on the type of diet. Using Fisher's LSD, describe and interpret any and all differences between the type of diet and happiness score.

3. A researcher wants to investigate whether the amount of alcohol consumed by first-year college students has an impact on their grades. To study this, the researcher first categorized alcohol consumption as one of the following four categories: abstain (no alcohol consumption), light consumption (less than 2 drinks/week), medium consumption (3–6 drinks/week), and heavy consumption (more than 6 drinks/week). A survey was then given out to 34 random freshmen asking them their current grade point average (GPA) as well as an estimate of the amount of alcohol they consumed each week. The results are provided in Table 10.9.

Is there a difference in the amount of alcohol consumed and the GPA for first-year college students? In order to answer this question,

a. State the appropriate null and alternative hypotheses.

b. How would you describe the factor? And how many factor levels are there?

c. Find the SSE and the MSE.

d. Find the SSTR and the MSTR.

TABLE 10.8

Responses to Happiness Survey by Type of Diet

No Specific Diet	Vegetarian	Vegan	Low Carb	Low Fat
70	92	78	74	58
55	77	74	73	94
56	84	94	72	97
50	92	62	69	73
53	74	61	69	55
75	76	86	—	86
71	81	90	—	67
50	86	66	—	—
53	72	73	—	—
66	—	58	—	—

TABLE 10.9

First-Year GPA By Amount of Alcohol Consumption

Abstain	Light	Medium	Heavy
2.35	3.45	2.48	2.04
3.41	2.68	2.65	2.65
3.26	2.48	2.47	2.31
2.58	3.54	1.54	1.65
3.98	2.98	1.69	0.89
3.45	2.38	2.45	3.12
3.26	3.25	3.01	3.00
3.58	2.89	2.45	—
—	2.36	3.39	—
—	—	2.65	—

e. Find the number of degrees of freedom for the numerator and the number of degrees of freedom for the denominator.

f. Find the *p*-value and interpret it within the context of the problem.

g. Check any relevant model assumptions.

h. How big of a sample would be needed to find a minimum difference of 0.50 GPA points?

4. In problem #3, we used the given sample data to estimate whether there is a difference in the first-year GPAs based on the amount of alcohol consumed per week. Using Fisher's LSD, describe and interpret any and all differences between GPA and alcohol consumption.

5. A researcher wants to know if there is a difference in the average number of hours a person sleeps based on four different types of natural sleep supplements (melatonin, herbal tea, tart cherry juice, and lemon balm). A random sample of 104 participants was assigned to use one of these supplements for a 2-week period and asked to estimate the average number of hours they slept each night during this period. Table 10.10 gives the number of hours of sleep (rounded to the nearest hour) based on the type of natural sleep supplement used.

   Is there a difference in the mean number of hours slept based on the different types of natural supplements? In order to answer this question,

a. State the appropriate null and alternative hypotheses.

b. How would you describe the factor? And how many factor levels are there?

c. Find the SSE and the MSE.

d. Find the SSTR and the MSTR.

e. Find the number of degrees of freedom for the numerator and the number of degrees of freedom for the denominator.

f. Find the *p*-value and interpret it within the context of the problem.

g. Check any relevant model assumptions.

h. What is the minimum difference in the number of hours slept between natural sleep aids that the given sample can find, provided such a difference does actually exist?

TABLE 10.10

Number of Hours Slept (Rounded to the Nearest Hour) by the Type of Natural Sleep Supplement

Melatonin (1)	Herbal Tea (2)	Tart Cherry Juice (3)	Lemon Balm (4)
8	5	7	8
7	5	9	5
8	6	6	6
9	6	6	9
8	7	8	6
8	5	8	8
9	7	8	5
7	7	8	9
7	9	9	9
8	8	9	6
9	9	7	6
9	6	9	6
8	9	9	9
8	5	5	8
8	7	5	8
8	7	7	5
8	8	4	9
6	7	4	7
8	7	5	6
9	5	7	5
6	9	6	7
9	5	5	7
9	8	7	9
8	5	7	8
7	5	9	5
8	6	6	6

6. In problem #5, we used the given sample data to estimate whether there is a difference in the mean number of hours slept based on the type of natural supplement used. Using Fisher's LSD, describe and interpret any and all differences between the type of natural supplement used and the mean number of hours slept.

7. A study was conducted where 36 patients were randomly assigned to use one of the three over-the-counter pain medications and were asked how long the medicine provided pain relief (rounded to the nearest hour). The data is provided in Table 10.11.

    Is there a significant difference in the mean relief time between the typical over-the-counter pain medications? Find and interpret any and all differences.

8. Is there a difference in the final examination grades for courses that are offered on-ground, online, or as a hybrid class (a hybrid class is half on-ground and half online)? The data provided in Table 10.12 gives the final examination grades (out of a maximum possible score of 100) received for 40 students randomly assigned an introductory statistics course in one of three different formats—on-ground, online, or as a hybrid.

    a. State the appropriate null and alternative hypotheses.

TABLE 10.11

Number of Hours of Pain Relief by Type of Over-the-Counter Medication

Aspirin	Acetaminophen	Ibuprofen
4	6	6
5	5	8
3	7	9
5	8	7
4	4	8
6	5	9
4	7	6
5	8	8
4	4	7
5	7	7
6	4	8
8	4	9

TABLE 10.12

Final Examination Grades for 40 Students Randomly Assigned to Participate in an On-Ground, Online, or Hybrid Introductory Statistics Course

On-Ground	Online	Hybrid
83	54	65
61	59	51
67	55	67
51	72	79
73	58	53
88	77	85
94	55	87
96	73	68
80	84	72
61	89	75
76	62	82
65	93	69
93	99	—
—	72	—
—	90	—

b. How would you describe the factor? And how many factor levels are there?

c. Find the SSE and the MSE.

d. Find the SSTR and the MSTR.

e. Find the number of degrees of freedom for the numerator and the number of degrees of freedom for the denominator.

f. Find the p-value and interpret it within the context of the problem.

g. Find and interpret any and all significant differences.

h. Check any relevant model assumptions.

i. Is this sample large enough to be able to detect at least a five-point difference in final examination grades?

References

Bartlett, M. (1937). Properties of sufficiency and statistical tests. *Proceedings of the Royal Society* A160: 268–282.

Brown, M., and Forsythe, A. (1974). Robust tests for the equality of variances. *Journal of the American Statistical Association* 69: 364–67.

Levene, H. (1960). *Contributions to Probability and Statistics: Essays in Honor of Harold Hotelling*, ed. I. Olkin, 278–292. Stanford, CA: Stanford University Press.

Ryan, T., and Joiner, B. (1976). Normal probability plots and tests for normality. Technical report. University Park, PA: The Pennsylvania State University.

11

Nonparametric Statistics

11.1 Introduction

All the topics that we have studied up to now are considered *parametric* methods of statistical inference. This is because many of these techniques rely on a specific collection of model assumptions. For instance, you may recall that in order to make reasonable inferences from an analysis of variance, we are assuming that the error component is normally distributed, each factor level has constant variance, and the observations are independent of each other. Also, in many of the analyses we have seen thus far, the presence of outliers can often have a profound impact on any inferences that we make.

Nonparametric statistics refers to a set of statistical methods and techniques that do not rely on such strong underlying assumptions, and such methods also tend to be more resilient to outliers. Nonparametric hypothesis tests are sometimes referred to as *distribution-free* tests because they can be performed without reliance on a specific shape of the distribution, such as the shape of a normal curve. However, nonparametric tests do rely on some basic assumptions, but such assumptions tend to be more relaxed than the parametric tests.

You may think that since nonparametric methods do not rely on strict distributional assumptions, these methods would obviously be preferred over the standard parametric methods of inference. However, this is not usually the case because nonparametric methods do rely on some basic distributional assumptions. Furthermore, nonparametric methods are not as powerful as parametric methods when all the model assumptions have been met. So, in practice, nonparametric methods only tend to be used when there are gross violations of the model assumptions, as is often the case when there are a significant number of outliers that could impact the analysis or when smaller sample sizes are used.

11.2 Wilcoxon Signed-Rank Test

The first nonparametric test that we will be considering is called the *Wilcoxon signed-rank test*. The Wilcoxon signed-rank test is similar to a *t*-test for comparing a population mean against a specific hypothesized value. However, the difference between the Wilcoxon signed-rank test and a one-sample *t*-test is that while the population *median* is tested against some hypothesized value in the former, the population *mean* is tested against some hypothesized value in the latter. The possible null and alternative hypotheses for a Wilcoxon signed-rank test are as follows:

$$H_0 : \eta = \eta_0$$
$$H_1 : \eta \neq \eta_0$$

$$H_0 : \eta = \eta_0$$
$$H_1 : \eta < \eta_0$$

$$H_0 : \eta = \eta_0$$
$$H_1 : \eta > \eta_0$$

Where η is the symbol for the population median and η_0 is the median being tested under the null hypothesis.

The Wilcoxon signed-rank test relies on the assumption that the sample is drawn from a population that has a *symmetric distribution*, but no specific shape of the distribution is required. Recall in Chapter 4 that for a one-sample t-test with a small sample size (usually when $n < 30$), we assumed that the population we sampled from was approximately normally distributed. If the population sampled from does not come from a normal distribution, then we may want to consider using the Wilcoxon signed-rank test as an alternative to a one-sample t-test provided that the underlying distribution is symmetric.

Example 11.1

Consider the data set provided in Table 11.1, which consists of a sample of the final examination grades for 14 students selected at random from the population of students in an introductory statistics course.

Because we have such a small sample size, we may be unsure whether the population being sampled from has the shape of a normal distribution. We can thus use the Wilcoxon signed-rank test as an alternative to a one-sample t-test.

Suppose we were interested in determining whether the true population median final exam grade is greater than 70. The difference between using a one-sample t-test and the Wilcoxon signed-rank test is that we will be looking at whether the unknown population median (versus the population mean) is greater than 70. Thus, the appropriate null and alternative hypotheses would be specified as follows:

$$H_0 : \eta = 70$$

$$H_1 : \eta > 70$$

where $\eta = 70$ is the unknown population median that we are testing under the null hypothesis.

To conduct the Wilcoxon signed-rank test, we need to rank each observation in the data set according to the distance the observation is from the median that we are testing under the null hypothesis (which in this case is $\eta = 70$). To do this, we take the difference between each observation and the hypothesized median (Column 3 of Table 11.2), take the absolute value of this difference (Column 5 of

TABLE 11.1

Final Examination Grades for a Sample of 14 Students in an Introductory Statistics Course

80	90	71	79	75	64	75
68	83	89	66	58	82	88

TABLE 11.2

Ranks and Signed Ranks for the Data in Table 11.1

Observation	Grade	Difference d	Sign of d	Absolute Value of d $|d|$	Rank of $|d|$	Signed Rank
1	80	$80 - 70 = 10$	+	10	8	8
2	90	$90 - 70 = 20$	+	20	14	14
3	71	$71 - 70 = 1$	+	1	1	1
4	79	$79 - 70 = 9$	+	9	7	7
5	75	$75 - 70 = 5$	+	5	4.5	4.5
6	64	$64 - 70 = -6$	−	6	6	−6
7	75	$75 - 71 = 5$	+	5	4.5	4.5
8	68	$68 - 70 = -2$	−	2	2	−2
9	83	$83 - 70 = 13$	+	13	11	11
10	89	$89 - 70 = 19$	+	19	13	13
11	66	$66 - 70 = -4$	−	4	3	−3
12	58	$58 - 70 = -12$	−	12	9.5	−9.5
13	82	$82 - 70 = 12$	+	12	9.5	9.5
14	88	$88 - 70 = 18$	+	18	12	12

Table 11.2), rank the absolute value of this difference (Column 6 of Table 11.2), and then attach the sign of the difference to this rank (Column 7 of Table 11.2).

Notice that when there are ties for the ranks, the rank for each of the tied values is assigned the mean of the ranks. For instance, there are two observations that have $|d| = 5$, hence they are each assigned the mean rank of 4.5 since they are tied for the fourth and fifth positions. This also holds for the two observations assigned the mean rank of 9.5 because both have $|d| = 12$ which corresponds to being tied for the ninth and tenth positions.

We can then take the sum of the absolute values of the ranks of the differences in Column 6 of Table 11.2 to get:

$$8 + 14 + 1 + 7 + 4.5 + 6 + 4.5 + 2 + 11 + 13 + 3 + 9.5 + 9.5 + 12 = 105$$

The basic idea behind the Wilcoxon signed-rank test is to see if the sum of the positive ranks is significantly more than the sum of the negative ranks. So for the current example, since the sum of the absolute value of all the ranks is 105, and if the sum of the positive ranks is significantly greater than half of this, which is 52.5 (with some predetermined level of significance), then this would provide evidence to reject the null hypothesis and accept the alternative hypothesis, which would imply that the median final examination score for the population is significantly greater than 70 ($\eta > 70$).

From Column 7 in Table 11.2, we can see that the sum of the positive ranks is as follows:

$$8 + 14 + 1 + 7 + 4.5 + 4.5 + 11 + 13 + 9.5 + 12 = 84.5$$

This value of the sum of the positive ranks, $R^+ = 84.5$, is then used to calculate the test statistic that represents the sum of the positive ranks as described:

$$Z_W^+ = \frac{R^+ - \dfrac{n(n+1)}{4} - \dfrac{1}{2}}{\sqrt{\dfrac{n(n+1)(2n+1)}{24}}}$$

where R^+ is the sum of the positive ranks and n is the sample size. The sampling distribution of this test statistic is approximately normally distributed.

So for $R^+ = 84.5$ and $n = 14$, we get:

$$Z_R^+ = \frac{R^+ - \dfrac{n(n+1)}{4} - \dfrac{1}{2}}{\sqrt{\dfrac{n(n+1)(2n+1)}{24}}} = \frac{84.5 - \dfrac{14(14+1)}{4} - \dfrac{1}{2}}{\sqrt{\dfrac{14(14+1)(2 \cdot 14+1)}{24}}} \approx 1.98$$

In order to accept the alternative hypothesis and reject the null hypothesis, the value of the test statistic Z_R^+ would have to be greater than the value of z that defines the rejection region for a given level of significance and the specified direction of the alternative hypothesis on the standard normal distribution. For a significance level of $\alpha = 0.05$ for a right-tailed test, we would compare $Z_R^+ = 1.98$ to the value $z = 1.645$, as illustrated in Figure 11.1.

Since $Z_R^+ = 1.98$ falls in the rejection region as illustrated in Figure 11.1, we can accept the alternative hypothesis and reject the null hypothesis. In other words, we can claim that the population median final examination grade is significantly greater than 70.

Because the sampling distribution of the test statistic for a Wilcoxon signed-rank test is approximately normally distributed, we can easily find the corresponding *p*-value, as illustrated in Figure 11.2, by finding the probability that $z > Z_R^+$.

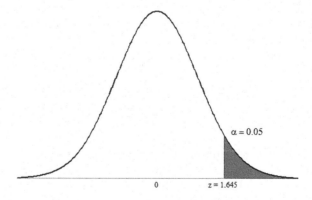

FIGURE 11.1
Rejection region for a one-sided test of whether the population median is greater than 70 for the final examination data in Table 11.1.

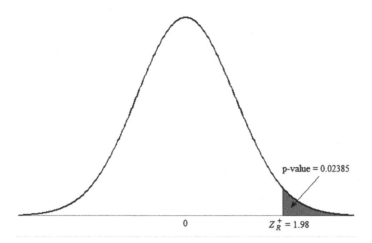

FIGURE 11.2
Corresponding p-value for the test statistic $Z_R^+ = 1.98$.

Similarly, since the p-value ($p \approx 0.024$) is less than our predetermined level of significance ($\alpha = 0.05$), we can infer that the median final examination grade is significantly greater than 70.

If we were to test a two-tailed alternative hypothesis (H_A: $\eta \neq 70$), the p-value would be $2(0.024) \approx 0.048$.

If we wanted to test a left-tailed alternative hypothesis (H_A: $\eta < 70$), we could sum up the *negative* ranks as follows:

$$R^- = 6 + 2 + 3 + 9.5 = 20.5$$

The test statistic would be calculated as follows:

$$Z_R^- = \frac{R^- - \dfrac{n(n+1)}{4} - \dfrac{1}{2}}{\sqrt{\dfrac{n(n+1)(2n+1)}{24}}} = \frac{20.5 - \dfrac{14(14+1)}{4} - \dfrac{1}{2}}{\sqrt{\dfrac{14(14+1)(2 \cdot 14+1)}{24}}} \approx -2.04$$

The corresponding p-value would be 0.9793. Notice that calculating the p-value for a left-tailed test requires finding the probability that $z > Z_R^-$. The reason for this is because the test statistic uses either the sum of the positive or the sum of the negative ranks.

11.3 Using Minitab for the Wilcoxon Signed-Rank Test

We can use Minitab to perform the Wilcoxon signed-rank test for the data in Table 11.1 by selecting **Stat < Nonparametrics < 1-Sample Wilcoxon**, as illustrated in Figure 11.3.

We can test whether the median score is greater than 70 using the Wilcoxon signed-rank test dialog box, as illustrated in Figure 11.4 ($\alpha = 0.05$). This gives the Minitab printout in Figure 11.5.

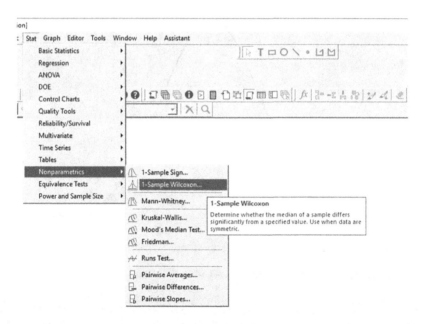

FIGURE 11.3
Minitab commands for the Wilcoxon signed-rank test.

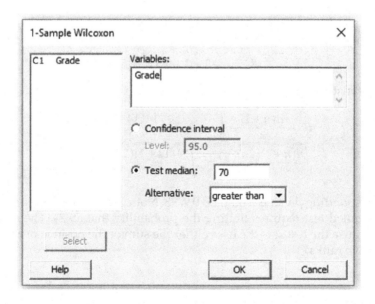

FIGURE 11.4
Minitab dialog box for the Wilcoxon signed-rank test.

Similar to what we found when doing the calculations manually, the *p*-value given in Figure 11.5 is equal to 0.024, which is less than 0.05, and therefore, we can claim that the median grade (for the population) is greater than 70.

It is important to note that when performing a left-tailed test where the sum of the negative ranks is used, Minitab will always report the value of the Wilcoxon statistic as the

Wilcoxon Signed Rank Test: Grade

Method

η: median of Grade

Descriptive Statistics

Sample	N	Median
Grade	14	77

Test

Null hypothesis	H_0: η = 70
Alternative hypothesis	H_1: η > 70

Sample	N for Test	Wilcoxon Statistic	P-Value
Grade	14	84.50	0.024

FIGURE 11.5
Minitab printout for the Wilcoxon signed-rank test of whether the population median final examination grade is greater than 70.

sum of the positive ranks. However, Minitab calculates the *p*-value based on the sum of the negative ranks.

Even though the Wilcoxon signed-rank test is a nonparametric test, this does not mean that there are absolutely no underlying assumptions. In fact, for this particular test, it is assumed that the population being sampled from has a symmetric distribution but no specific shape of the distribution is required. We can run an exploratory check on this by plotting a histogram of the data, as in Figure 11.6.

And as we have seen before, with small sample sizes it can be very difficult to assess by eye the shape of the underlying distribution. When this is the case, we can also use the Ryan–Joiner test as a formal check on the normality of the distribution of the data, as illustrated in Figure 11.7.

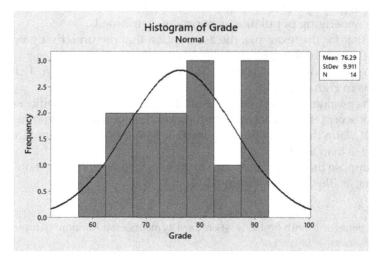

FIGURE 11.6
Histogram of scores on a statistics final examination for a random sample of 14 students.

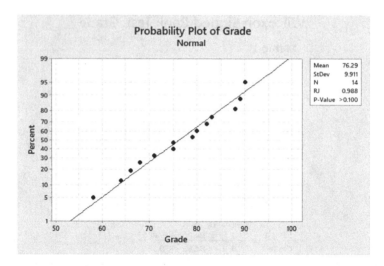

FIGURE 11.7
Ryan–Joiner test comparing the distribution of the sample data from Table 11.1 with a normal distribution.

The Ryan–Joiner test can be used to compare the distribution of the sample data to a normal distribution, and since the p-value is greater than 0.100, we have reason to believe that the data comes from a population that is normally distributed. We can also calculate the skewness to help assess if the distribution is symmetric (see Exercise 21 in Chapter 3).

Example 11.2

The data set in Table 11.3 gives the amount spent (in dollars) per month on monthly car payments for a random sample of 15 customers. Suppose we are interested in determining whether the (population) median amount that individual consumers spend on their monthly car payment is less than $400 ($\alpha = 0.05$).

We can first graph an exploratory histogram, as in Figure 11.8, to see if the data are roughly symmetrical, and run a Ryan-Joiner test, as in Figure 11.9, to see if the underlying population is normally distributed.

Notice that for this example, the assumption that the underlying population is normally distributed appears to hold true.

By running the Wilcoxon signed-rank test in Minitab in Figure 11.10, we get the results in Figure 11.11.

Since the p-value is not less than the predetermined level of significance of 0.05, we cannot accept the alternative hypothesis and reject the null hypothesis. Thus, we cannot claim that the population median amount spent on monthly car payments is less than $400.

Now suppose that we look at what happens when we run a t-test on the same set of data, as illustrated in Figure 11.12.

TABLE 11.3

Amount Spent Per Month (in Dollars) on Car Payments for a Random Sample of 15 Consumers

187	425	445	200	384	290	255	420
488	460	260	377	300	425	290	

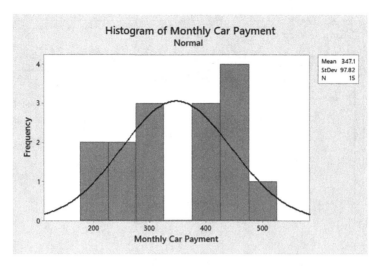

FIGURE 11.8
Histogram of monthly car payments for a random sample of 15 consumers.

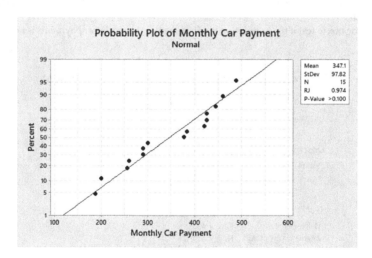

FIGURE 11.9
Ryan–Joiner test of the car payment data in Table 11.3.

Notice for the one-sample t-test that the p-value in Figure 11.12 is less than 0.05, whereas the p-value for the Wilcoxon signed-rank test presented in Figure 11.11 is not less than 0.05. Aside from the fact that the Wilcoxon signed-rank test is for testing *medians* and the t-test is for testing *means*, one reason why parametric methods tend to be preferred over nonparametric methods is that the former are often more powerful than the latter, and thus you are more likely to find an effect if such an effect exists. For this example, the Ryan–Joiner test in Figure 11.9 indicates that the assumption that the sample comes from a population that has a normal distribution does not appear to have been violated. But because parametric tests are more powerful, we would be more likely to find an effect (if one actually exists) when using a parametric test versus a nonparametric test.

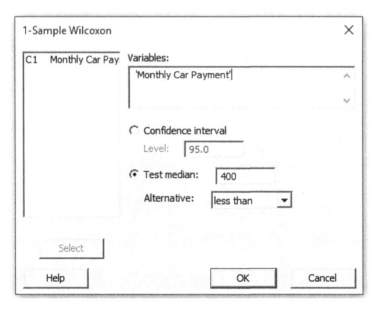

FIGURE 11.10
Minitab dialog box to test if the population median spent on monthly car payments is less than $400.

Wilcoxon Signed Rank Test: Monthly Car Payment

Method

η: median of Monthly Car Payment

Descriptive Statistics

Sample	N	Median
Monthly Car Payment	15	347

Test

Null hypothesis	H_0: η = 400
Alternative hypothesis	H_1: η < 400

Sample	N for Test	Wilcoxon Statistic	P-Value
Monthly Car Payment	15	32.00	0.059

FIGURE 11.11
Minitab output for the Wilcoxon signed-rank test of whether the true population median spent on monthly car payments is less than $400.

If you have reason to believe that the underlying assumption of normality has not been violated, you may prefer to use parametric methods versus nonparametric methods. Even though an exploratory analysis of your data may suggest that the underlying distributions are not exactly what they should be, using nonparametric methods may not provide any significant results because these methods have less power than their equivalent parametric statistical methods when the assumptions have been reasonably met. In other words, nonparametric methods are less likely to detect an effect if such an effect were to exist.

One-Sample T: Monthly Car Payment

Descriptive Statistics

N	Mean	StDev	SE Mean	95% Upper Bound for μ
15	347.1	97.8	25.3	391.6

μ: mean of Monthly Car Payment

Test

Null hypothesis	H_0: μ = 400
Alternative hypothesis	H_1: μ < 400

T-Value	P-Value
-2.10	0.027

FIGURE 11.12
One-sample t-test of whether the true population mean amount spent on monthly car payments is less than \$400.

11.4 The Mann–Whitney Test

The Mann–Whitney test is a nonparametric alternative to a two-sample t-test. Recall that a two-sample t-test can be used to compare the population means from two *independent* samples. And similar to the Wilcoxon signed-rank test, the Mann–Whitney test relies on comparing *medians* versus *means*. The three different null and alternative hypotheses for the Mann–Whitney test are as follows:

$$H_0 : \eta_1 = \eta_2 \quad\quad H_0 : \eta_1 = \eta_2 \quad\quad H_0 : \eta_1 = \eta_2$$

$$H_1 : \eta_1 \neq \eta_2 \quad\quad H_1 : \eta_1 > \eta_2 \quad\quad H_1 : \eta_1 < \eta_2$$

Where η_1 is the median from the first population and η_2 is the median from the second population.

The assumptions for the Mann–Whitney test rely on the populations being independent, and the scale for each of the populations can either be continuous or ordinal.

Example 11.3

Suppose we want to use the Mann–Whitney test to determine if there is a difference in the median time it takes for two different pumps to dispense three gallons of gasoline ($\alpha = 0.05$).

The data in Table 11.4 gives the time (in seconds) to dispense three gallons of gasoline for two different brands of pumps.

Since we are looking to see whether there is a difference in the median time it takes to pump three gallons of gasoline, we could consider a two-tailed test where the null and alternative hypothesis would be stated as follows:

$$H_0 : \eta_1 = \eta_2$$

$$H_1 : \eta_1 \neq \eta_2$$

TABLE 11.4

Time (in Seconds) to Dispense Three Gallons of
Gasoline for Two Different Brands of Pumps

Pump 1	Pump 2
14	11
16	17
18	12
19	15
26	21
23	—

The test statistic for the Mann–Whitney test follows a standard normal distribution. For a two-tailed test, the test statistic is as follows:

$$Z_W = \frac{\left| W - \frac{n_1(n_1 + n_2 + 1)}{2} \right| - 0.50}{\sqrt{\frac{n_1 \cdot n_2 (n_1 + n_2 + 1)}{12}}}$$

where W is the sum of the ranks for the first group, n_1 is the sample size of the first group, and n_2 is the sample size of the second group.

So for our example, we would first need to rank order our combined data as can be seen in Table 11.5. If there are ties for some of the ranks, then the mean rank for the tied values would be used.

To calculate W we need to sum the ranks of the first sample as follows:

$$W = 3 + 5 + 7 + 8 + 11 + 10 = 44$$

The test statistic would then be:

TABLE 11.5

Rank Order for the Combined Data from Table 11.4

	Combined Data	Combined Rank
Brand 1	14	3
	16	5
	18	7
	19	8
	26	11
	23	10
Brand 2	11	1
	17	6
	12	2
	15	4
	21	9

$$Z_W = \frac{\left| W - \dfrac{n_1(n_1 + n_2 + 1)}{2} \right| - 0.50}{\sqrt{\dfrac{n_1 \cdot n_2(n_1 + n_2 + 1)}{12}}} = \frac{\left| 44 - \dfrac{6(6 + 5 + 1)}{2} \right| - 0.50}{\sqrt{\dfrac{6 \cdot 5(6 + 5 + 1)}{12}}} \approx 1.369$$

We could then use this test statistic to find the corresponding p-value as is illustrated in Figure 11.13.

Since this is a two-tailed test, the p-value would be:

$$2(0.08550) \approx 0.171$$

Since the p-value is greater than $\alpha = 0.05$, this suggests that there is not a significant difference in the median amount of time it takes for the two brands of pumps to dispense three gallons of gasoline.

For a right-tailed test, the test statistic would be:

$$Z_W = \frac{W - \dfrac{n_1(n_1 + n_2 + 1)}{2} - 0.50}{\sqrt{\dfrac{n_1 \cdot n_2(n_1 + n_2 + 1)}{12}}}$$

For a left-tailed test, the test statistic would be:

$$Z_W = \frac{S - \dfrac{n_1(n_1 + n_2 + 1)}{2} + 0.50}{\sqrt{\dfrac{n_1 \cdot n_2(n_1 + n_2 + 1)}{12}}}$$

where $S = W - n_1(n_1 + n_2 + 1)$.

For both a right- and left-tailed test, W would correspond to the sum of the ranks from the first sample, n_1 is the sample size from the first sample, and n_2 is the sample size from the second sample.

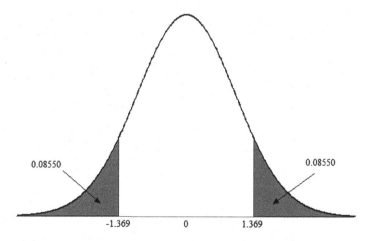

FIGURE 11.13
Standard normal distribution showing the corresponding p-value for the test statistic of $Z_W = 1.369$.

If there are ties in the ranks, the denominator of the above test statistics will need to be adjusted as follows:

$$\sqrt{\frac{n_1 \cdot n_2}{12}\left[(n_1 + n_2 + 1) - \frac{\sum_{i=1}^{j}\left(t_i^3 - t_i\right)}{(n_1 + n_2)(n_1 + n_2 - 1)}\right]}$$

where j is the number of ties, and t_i is the number of tied values in the ith set of ties.

Example 11.4

Suppose we want to use the Mann–Whitney test to determine if tire Brand A has a higher median number of miles until the tire is considered worn-out as compared to the tire Brand B. A sample of seven tires from Brand A and four tires from Brand B were tested and the number of miles until they are considered worn-out is given in Table 11.6.

This would be a one-tailed test and the null and alternative hypothesis would be:

$$H_0 : \eta_A = \eta_B$$

$$H_1 : \eta_A > \eta_B$$

We first need to rank the combined data as is illustrated in Table 11.7. Whenever there are ties, we assign the mean rank to each of the tied values.

To calculate W, we need to sum the ranks of the first sample as follows:

$$W = 6 + 2.5 + 7 + 9 + 8 + 10 + 11 = 53.5$$

Since this is a right-tailed test and there are ties, the denominator of the test statistic would need to be adjusted as follows:

$$Z_W = \frac{W - \frac{n_1(n_1 + n_2 + 1)}{2} - 0.50}{\sqrt{\frac{n_1 \cdot n_2}{12}\left[(n_1 + n_2 + 1) - \frac{\sum_{i=1}^{j}\left(t_i^3 - t_i\right)}{(n_1 + n_2)(n_1 + n_2 - 1)}\right]}}$$

TABLE 11.6

Number of Miles Until the Tire is Considered Worn-out for Two Different Brands of Tires

Brand A	Brand B
39,820	33,850
33,850	38,900
42,900	32,600
46,200	34,500
44,500	—
46,350	—
48,290	—

TABLE 11.7

Rank Order for the Combined Data from Table 11.6

	Combined Data	Combined Rank
Brand A	39,820	6
	33,850	2.5
	42,900	7
	46,200	9
	44,500	8
	46,350	10
	48,290	11
Brand B	33,850	2.5
	38,900	5
	32,600	1
	34,500	4

Since there is only one tie, then $j = 1$, and $t_1 = 2$ since there are two values that are tied. The test statistic will thus be:

$$Z_W = \frac{W - \frac{n_1(n_1 + n_2 + 1)}{2} - 0.50}{\sqrt{\frac{n_1 \cdot n_2}{12}\left[(n_1 + n_2 + 1) - \frac{t_1^3 - t_1}{(n_1 + n_2)(n_1 + n_2 - 1)}\right]}} = \frac{53.5 - \frac{7(7 + 4 + 1)}{2} - 0.50}{\sqrt{\frac{7 \cdot 4}{12}(7 + 4 + 1) - \frac{2^3 - 2}{(7 + 4)(7 + 4 - 1)}}}$$

$$\approx \frac{11}{\sqrt{28 - 0.05455}} \approx \frac{11}{5.27} \approx 2.08$$

The *p*-value would be the area to the right of $Z_W = 2.08$, as can be seen in Figure 11.14.

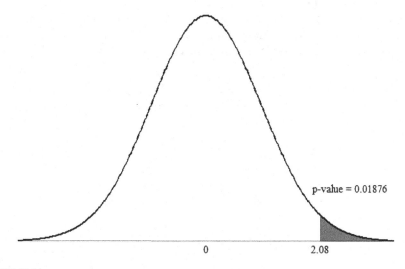

FIGURE 11.14

Standard normal distribution of a right-tailed Mann–Whitney test for the test statistic $Z_W = 2.08$ with a corresponding *p*-value of 0.01876.

Since the *p*-value is less than $\alpha = 0.05$, we can accept the alternative hypothesis and reject the null hypothesis. This suggests that the median number of miles for Brand A tires is significantly higher than the median number of miles for Brand B tires.

11.5 Using Minitab for the Mann–Whitney Test

For Minitab to conduct a Mann–Whitney test for the data provided in Table 11.6, the data needs to be entered where the two samples are entered in separate columns as can be seen in Figure 11.15.

Then select **Stat > Nonparametrics > Mann–Whitney**, as is illustrated in Figure 11.16. This brings up the dialog box that is shown in Figure 11.17.

All that has to be specified is Brand A as the first sample, Brand B as the second sample, the desired confidence level, and the direction of the alternative hypothesis. This gives the results that are presented in Figure 11.18.

Notice that the sum of the positive ranks $W = 53.5$ as shown in Figure 11.18 is the same as we found and the *p*-value that was adjusted for ties is also similar to what we found (the slight difference is due to the round-off error).

11.6 Kruskal–Wallis Test

The Kruskal–Wallis test is a nonparametric alternative to a one-way analysis of variance (ANOVA). Recall that a one-way ANOVA relies on the assumption that the levels of the populations we are sampling from are all normally distributed (see Section 10.4). The Kruskal–Wallis test can be used for continuous data collected from populations that have the same shape. However, unlike the Wilcoxon signed-rank test, the Kruskal–Wallis test does not require the distributions to be normal or even symmetric; the only requirement is that the distributions for all of the populations sampled from have the same shape. Just like the Wilcoxon signed-rank test, the test statistic for the Kruskal–Wallis test is also based on ranks.

↓	C1	C2
	Brand A	Brand B
1	39820	33850
2	33850	38900
3	42900	32600
4	46200	34500
5	44500	*
6	46350	*
7	48290	*
-		

FIGURE 11.15
Minitab worksheet for the Mann–Whitney test.

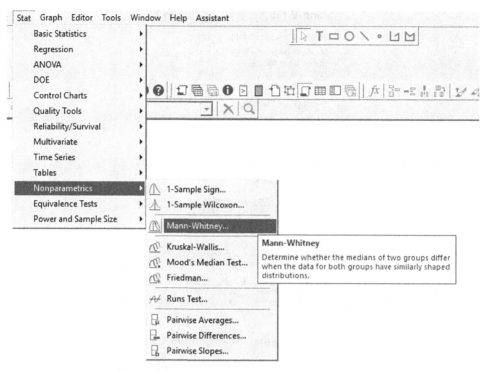

FIGURE 11.16
Minitab commands for the Mann–Whitney test.

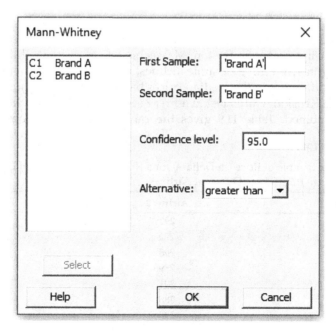

FIGURE 11.17
Minitab dialog box for the Mann–Whitney test.

Mann-Whitney: Brand A, Brand B

Method

η_1: median of Brand A
η_2: median of Brand B
Difference: $\eta_1 - \eta_2$

Descriptive Statistics

Sample	N	Median
Brand A	7	44500
Brand B	4	34175

Estimation for Difference

Difference	Lower Bound for Difference	Achieved Confidence
9220	1250	96.37%

Test

Null hypothesis	H_0: $\eta_1 - \eta_2 = 0$
Alternative hypothesis	H_1: $\eta_1 - \eta_2 > 0$

Method	W-Value	P-Value
Not adjusted for ties	53.50	0.019
Adjusted for ties	53.50	0.019

FIGURE 11.18
Results of the Mann–Whitney test for comparing the median wear times for the two different brands of tires.

The null and alternative hypotheses for the Kruskal–Wallis test are as follows:

H_0 : The median for all the levels of the populations are the same.

H_1 : At least two levels of the populations have different medians.

Example 11.5

Consider the data set provided in Table 11.8, which shows the price (in dollars) for a round-trip ticket from Boston, Massachusetts to Chicago, Illinois for nine different flights from three separate airlines. We are interested in determining if there is a difference in the median prices based on the different airlines.

To use the Kruskal–Wallis test, we first need to rank the data from all three samples combined. Table 11.9 gives the rank-ordered data from Table 11.8.

TABLE 11.8

Random Sample of Price (in Dollars) for a Round-Trip Ticket from Boston to Chicago from Three Different Airlines

Airline 1	Airline 2	Airline 3
285	292	288
297	272	269
299	298	275
290	289	285
291	291	279
299	288	280
302	290	281
295	279	289
293	283	273

Notice in Table 11.9 that if two or more observations have the same rank, then we need to find the mean of the rank and assign it to each of the tied values. For example, the highlighted portion in Table 11.9 illustrates that for a round-trip ticket price of $279 we have a tie because there are two observations that share this same price. Instead of ranking one with a five and the other with a six, the ranks assigned to each of these observations is the mean of 5.5.

Once we obtain the ranks for each of the individual observations, we can then find the total of the ranks and the mean of the ranks for each of the k different samples, as presented in Table 11.10. The total of the ranks is found by adding up the ranks separately for each of the three different airlines, and the mean rank is found by dividing this value by the number of observations collected from each airline.

If there are no ties among the ranks, the Kruskal–Wallis test is based on the following test statistic that follows a χ^2 distribution with $k - 1$ degrees of freedom. This test statistic represents the variability in the rank positions between the k different samples:

$$H = \frac{12}{n(n+1)} \sum_{i=1}^{k} \frac{R_i^2}{n_i} - 3(n+1)$$

where n is the total sample size, n_i is the sample size for group i, and R_i^2 is the squared rank for the ith group.

If there are ties, the test statistic gets adjusted as follows:

$$H_{adj} = \frac{H}{1 - \frac{\sum_{i=1}^{j} \left(t_i^3 - t_i\right)}{\left(n^3 - n\right)}}$$

TABLE 11.9

Rank-Ordered Data Where the Highlighted Data Represent Ties

Airline 1	Rank	Airline 2	Rank	Airline 3	Rank
285	10.5	292	20	288	12.5
297	23	272	2	269	1
299	25.5	298	24	275	4
290	16.5	289	14.5	285	10.5
291	18.5	291	18.5	279	5.5
299	25.5	288	12.5	280	7
302	27	290	16.5	281	8
295	22	279	5.5	289	14.5
293	21	283	9	273	3
Sum of the ranks	**189.5**	—	**122.5**	—	**66**

TABLE 11.10

Total and Average Ranks for Each of the Three Airlines

	Total Rank	Average Rank
Airline 1	189.5	189.5/9 = 21.06
Airline 2	122.5	122.5/9 = 13.61
Airline 3	66	66/9 = 7.33

where j is the number of ranks, t_i is the number of ties in the ith rank. The test statistic adjusted for ties also follows the χ^2 distribution with $k-1$ degrees of freedom.

For our example, the value of H can be calculated as follows:

$$H = \frac{12}{27(28)}\left[\frac{189.5^2}{9} + \frac{122.5^2}{9} + \frac{66^29}{9}\right] - 3(27+1) \approx 13.48$$

To calculate the test statistic adjusted for ties, we would have to find $\sum_{i=1}^{j}\left(t_i^3 - t_i\right)$ for all of the ties in the sample. This value is found by counting up the number of ties for each rank. For our example, since there are 0 ties for the first rank, the value $\left(t_1^3 - t_1\right) = \left(0^3 - 0\right) = 0$. The same can be said for the second, third, and fourth ranks. However, notice that there are two values that are tied for the fifth rank; thus $\left(t_5^3 - t_5\right) = \left(2^3 - 2\right) = 6$. Similarly, there are two values tied for the tenth, twelfth, fourteenth, sixteenth, eighteenth, and twenty-fifth positions. By summing up all the values of $\left(t_i^3 - t_i\right)$ across all of the different ranks, we get:

$$\sum_{i=1}^{j}\left(t_i^3 - t_i\right) = 6+6+6+6+6+6+6 = 42$$

Thus, the value of H_{adj} would be:

$$H_{adj} = \frac{H}{1 - \dfrac{\sum_{i=1}^{j}\left(t_i^3 - t_i\right)}{\left(n^3 - n\right)}} = \frac{13.48}{1 - \dfrac{42}{\left(27^3 - 27\right)}} \approx 13.51$$

Now, by comparing this value to the value of the χ^2 distribution with $3 - 1 = 2$ degrees of freedom $\left(\chi^2_{0.05} = 5.991\right)$, we see that H_{adj} falls in the rejection region, as illustrated in Figure 11.19.

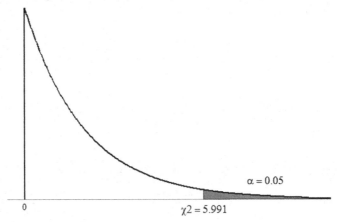

α = 0.05

0   χ2 = 5.991

FIGURE 11.19
χ^2 distribution with 2 degrees of freedom, where $H_{adj} = 13.51$ falls in the rejection region.

Since the test statistic falls in the rejection region, we can reject the null hypothesis and accept the alternative hypothesis. In other words, there is a difference in the median ticket prices based on the different airlines.

A typical rule of thumb for the Kruskal–Wallis test is to make sure that there are at least five observations sampled from each population. If less than five observations are obtained from any of the different populations being sampled from, then the value of the test statistic H or H_{adj} may not follow the χ^2 distribution, and thus any inference could be suspect.

11.7 Using Minitab for the Kruskal–Wallis Test

To use Minitab for the Kruskal–Wallis test, we need to enter the data in stacked form. This means there are two columns—one column for the response variable (which, for this example, would be the price of the round-trip tickets) and the other column for the factor (which, for this example, describes the airline). Figure 11.20 shows what a portion of this data set would look like in Minitab.

The Minitab commands for a Kruskal–Wallis test are given in Figure 11.21.

This brings up the dialog box for a Kruskal–Wallis test in Figure 11.22.

Notice that the response represents the price of the airline tickets and the factor is the airline. The Minitab printout is given in Figure 11.23.

Notice that Figure 11.23 gives the value of the test statistic H, the test statistic adjusted for ties H_{adj}, along with the corresponding p-values. Also notice in the Minitab printout in Figure 11.23 that the value of the test statistic H is only slightly different when it is adjusted for ties, H_{adj}, as is similar to the calculations obtained manually. Recall that ties occur when the observations are ranked such that there are two or more observations that have the same rank and if there are only a few ties, then H and H_{adj} will not be much different.

↓	C1	C2
	Price	Airline
1	285	1
2	297	1
3	299	1
4	290	1
5	291	1
6	299	1
7	302	1
8	295	1
9	293	1
10	292	2
11	272	2
12	298	2
13	289	2
14	291	2

FIGURE 11.20
Data for round-trip tickets from Boston to Chicago by price and airline.

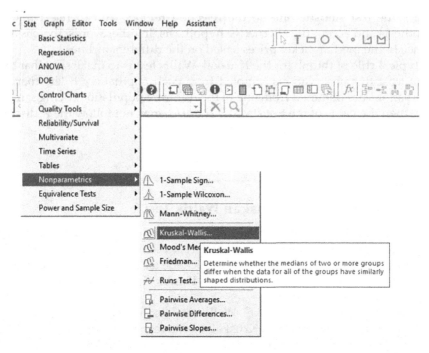

FIGURE 11.21
Minitab commands to run a Kruskal–Wallis test.

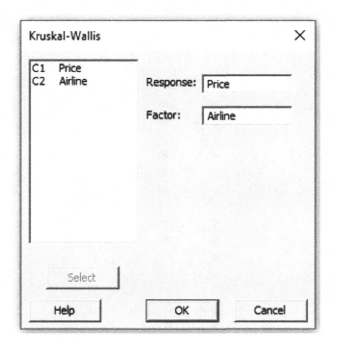

FIGURE 11.22
Minitab dialog box for the Kruskal–Wallis test.

Kruskal–Wallis Test: Price versus Airline

Descriptive Statistics

Airline	N	Median	Mean Rank	Z-Value
1	9	295	21.1	3.27
2	9	289	13.6	-0.18
3	9	280	7.3	-3.09
Overall	27		14.0	

Test

Null hypothesis H_0: All medians are equal
Alternative hypothesis H_1: At least one median is different

Method	DF	H-Value	P-Value
Not adjusted for ties	2	13.48	0.001
Adjusted for ties	2	13.51	0.001

FIGURE 11.23
Minitab printout for the Kruskal–Wallis test of whether the price for a round-trip ticket from Boston to Chicago is different by the airline.

The p-value of 0.001 suggests that we can reject the null hypothesis and accept the alternative hypothesis. Thus, we can claim that the median price for a round-trip ticket from Boston to Chicago are significantly different for at least two of the different airlines.

One reason why the Kruskal–Wallis test may be preferred over a one-way ANOVA is because the only assumptions for the Kruskal–Wallis test are that the data are sampled from populations that have a similar distribution shape. Thus, the distributions of the underlying populations do not have to be symmetric or normal; the only requirement is that they are similar.

Example 11.6

Table 11.11 gives a data set that provides the year-to-date rate of return for a random sample of low-, medium-, and high-risk mutual funds.

If we were to run a one-way ANOVA as in Figure 11.24, based on the p-value, we would find that there is not enough evidence to suggest a difference in the mean year-to-date rate of return for the mutual funds based on the degree of risk.

TABLE 11.11

Year-to-Date Returns for a Sample of 30 Different Mutual Funds Based on Risk

Low Risk	Medium Risk	High Risk
3.47	5.17	18.24
2.84	4.69	17.63
6.78	11.52	10.22
5.24	7.63	54.24
4.98	5.68	18.72
5.22	7.23	−15.24
3.81	8.82	10.54
4.43	7.44	−32.45
5.19	6.78	19.25
8.54	9.54	10.20

One-way ANOVA: Rate versus Risk

Method

Null hypothesis	All means are equal
Alternative hypothesis	Not all means are equal
Significance level	$\alpha = 0.05$

Equal variances were assumed for the analysis.

Factor Information

Factor	Levels	Values
Risk	3	High, Low, Medium

Analysis of Variance

Source	DF	Adj SS	Adj MS	F-Value	P-Value
Risk	2	187.9	93.94	0.54	0.591
Error	27	4735.3	175.38		
Total	29	4923.2			

Model Summary

S	R-sq	R-sq(adj)	R-sq(pred)
13.2432	3.82%	0.00%	0.00%

Means

Risk	N	Mean	StDev	95% CI
High	10	11.14	22.78	(2.54, 19.73)
Low	10	5.050	1.652	(-3.543, 13.643)
Medium	10	7.450	2.089	(-1.143, 16.043)

Pooled StDev = 13.2432

Interval Plot of Rate vs Risk

Residual Plots for Rate

FIGURE 11.24
One-way ANOVA results for the year-to-date rate of return versus risk for mutual funds.

However, upon further investigation, the distributional assumptions of normality and constant variance for a one-way ANOVA could likely be found to have been violated, as illustrated in Figures 11.25–11.27.

Because the one-way ANOVA assumptions appear to have been violated, we can run the Kruskal–Wallis test to see if there is a difference in the median year-to-date rate of returns based on the degree of risk. The results for the Kruskal–Wallis test are given in Figure 11.28.

The results from Figure 11.28 suggest that at least two populations have different medians, and thus we can conclude that the medians for the year-to-date rates of return are not all the same. This is a different finding from what we obtained when using a one-way ANOVA. The key difference being that a one-way ANOVA tests if there is a difference in the *mean rates* of return, whereas the Kruskal–Wallis test explores if there is a difference between the *median rates*.

However, even with the Kruskal–Wallis test, the populations we are sampling from are assumed to have a similar shape. Looking at the boxplot in Figure 11.29,

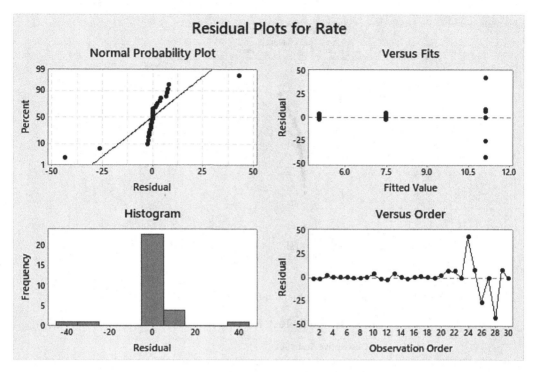

FIGURE 11.25
Four-in-one residual plots for a one-way ANOVA for the mutual fund data in Table 11.11.

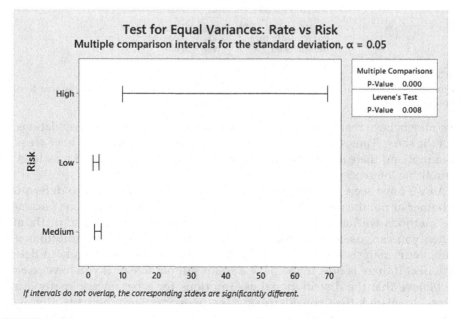

FIGURE 11.26
Levene's test for equal variances for the mutual fund data in Table 11.11.

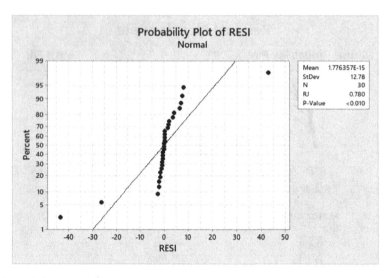

FIGURE 11.27
Ryan–Joiner test for normality for the mutual fund data in Table 11.11.

Kruskal–Wallis Test: Rate versus Risk

Descriptive Statistics

Risk	N	Median	Mean Rank	Z-Value
High	10	14.085	21.2	2.51
Low	10	5.085	9.3	-2.75
Medium	10	7.335	16.1	0.24
Overall	30		15.5	

Test

Null hypothesis	H_0: All medians are equal
Alternative hypothesis	H_1: At least one median is different

Method	DF	H-Value	P-Value
Not adjusted for ties	2	9.27	0.010
Adjusted for ties	2	9.27	0.010

FIGURE 11.28
Minitab printout for the Kruskal–Wallis test for the difference in ranks between the year-to-date rates of return by risk.

we may believe that the shape of the distributions between the populations is not the same. Thus, even when using nonparametric methods that have weaker assumptions, there are still some basic assumptions that if violated may lend to unreliable inferences.

As we have seen with small sample sizes, it can be difficult to determine whether or not the basic distributional assumptions for the various parametric methods we have studied have been met. As a general rule of thumb, when you can, use parametric methods. When the model assumptions are met, your analysis will have more power and you are more likely to detect an effect if there is one. If you have a small sample size or if you have reason to believe that the distributional assumptions for a parametric method are grossly violated, then consider using the nonparametric methods. But even when using nonparametric methods, always be aware that there are still some assumptions that need to hold true in order to make meaningful inferences.

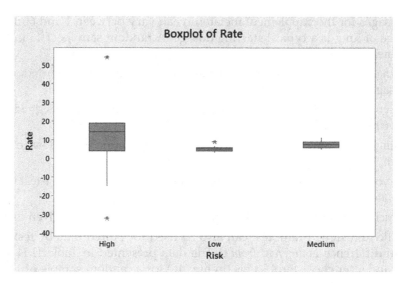

FIGURE 11.29
Boxplot of the year-to-date rate of return for mutual funds based on the degree of risk.

Exercises

Unless otherwise specified, let $\alpha = 0.05$.

1. Using the data given in Table 11.3, perform a Wilcoxon signed-rank test manually by completing Table 11.12.

TABLE 11.12

Cell Phone Bill, Difference, Sign of Difference, Absolute Value of Difference, Rank of Difference, and Signed Rank

Cell Phone Bill	Difference d	Sign of d	Absolute Value of d $\lvert d \rvert$	Rank of $\lvert d \rvert$	Signed Rank
187					
425					
445					
200					
384					
290					
255					
420					
488					
460					
260					
377					
300					
425					
290					

2. The dosages for the supplement melatonin can vary between 1 and 60 mg, where a dosage of 5 mg is a typical starting dosage. A random sample of 15 individuals using melatonin was collected; the dosage is provided in Table 11.13.

   a. Using the Wilcoxon Signed-Rank test, determine if the population median dosage is greater than 5 mg.

   b. Do you believe that the assumptions underlying the Wilcoxon signed-rank test have been violated?

   c. Using a one-sample *t*-test, determine if the population mean dosage is greater than 5 mg.

   d. Do you believe that the assumptions underlying a one-sample *t*-test have been violated?

   e. Which analysis do you think is the most appropriate for the given data?

3. The Wilcoxon signed-rank test can also be used for paired data by testing if the median difference is *different from 0*. The data presented in Table 11.14 gives the before and after cholesterol levels (in mg/dL) for a random sample of 11 patients who were given a cholesterol-lowering drug.

   a. Using the Wilcoxon signed-rank test, determine if the cholesterol drug was effective in reducing cholesterol.

   b. Do you believe that the assumptions underlying the Wilcoxon signed-rank test have been violated?

4. An educational researcher is interested in whether students do better in online classes as compared to on-ground classes. A random sample of 50 students was

TABLE 11.13

Dosages of Melatonin for a Random Sample of 15 Individuals

40	30	10	15	20	5	2	1	15	20	10	5	4	4	5

TABLE 11.14

Before and After Cholesterol Levels for a Random Sample of 11 Patients Given a Cholesterol-Lowering Drug

Observation	Cholesterol Level (Before)	Cholesterol Level (After)	Difference (Before – After)
1	261	227	34
2	256	223	33
3	250	245	5
4	244	206	38
5	269	219	50
6	260	205	55
7	249	224	25
8	241	242	−1
9	264	246	18
10	270	243	27
11	260	225	35

TABLE 11.15

Final Examination Scores for Students
Randomly Assigned to an Online or
On-ground Statistics Class

Online	On-Ground
90	97
83	99
91	80
63	80
86	98
71	93
71	74
92	97
73	79
74	97
62	77
91	99
90	91
73	84
60	92
71	71
83	93
65	73
86	71
87	97
93	77
91	97
92	74
85	77
—	82
	97

randomly assigned to participate in either an online or an on-ground statistics class. The final examination was the same for both classes, the grades of which are given in Table 11.15.

a. Using the Mann–Whitney test, determine if there is a difference in the population median examination scores between students who take online classes as compared to those who take on-ground classes.

b. Do you believe that the assumptions underlying the Mann–Whitney test have been violated?

c. Using an independent two-sample *t*-test, determine if there is a difference in the population mean examination scores between students who take online classes as compared to those who take on-ground classes.

d. Do you believe that the assumptions underlying the two-sample *t*-test have been violated?

e. Which analysis do you think is the most appropriate for the given data?

5. A real estate developer wants to estimate if there is a difference in the price per unit for multifamily apartments based on the where the property is located. A random sample of 12 per unit prices for multifamily apartments in one location and a random sample of 14 per unit prices for multifamily apartments in another location are presented in Table 11.16.

   a. Is there a significant difference in the median price per unit based on the location?

   b. Is there a significant difference in the mean price per unit based on the location?

   c. Check any relevant model assumptions for both part a. and part b.

   d. Which analysis do you believe is the most appropriate for the data that was collected?

6. Is there a difference in the yearly cost to service an automobile based on whether it is a luxury car, economy car, sports utility vehicle, or truck? The data given in Table 11.17 represents a random sample of the yearly price for service based on the type of vehicle.

   a. Is there a difference in the price paid for yearly service by type of vehicle? Run both a Kruskal–Wallis test and a one-way ANOVA and check any relevant model assumptions.

   b. On the basis of results from part a., which method do you think best represents the underlying data?

7. Is there a difference in the time spent on social media between males and females? The data given in Table 11.18 gives a random sample of 16 males and 16 females who were asked how much time they spend on social media each day.

   a. Using both parametric and nonparametric tests, determine if there is a difference in the time spent on social media between males and females.

   b. Test any relevant model assumptions.

   c. Decide which method is preferred for the underlying data.

TABLE 11.16

Price Per Unit (in Thousands of Dollars) for Multi-family Apartments from Two Different Locations

Location A	Location B
44	55
35	41
50	55
38	41
41	41
40	44
47	50
36	48
43	45
35	54
36	46
46	45
—	48
—	41

TABLE 11.17

Yearly Service Cost for Luxury Cars, Economy Cars, Sport Utility Vehicles, or Trucks

Luxury Cars	Economy Cars	Sport Utility Vehicles	Trucks
852	630	744	794
874	693	692	801
846	577	625	758
844	639	639	764
725	566	605	958
845	562	825	961
745	641	808	803
937	493	807	826
981	677	612	985
924	479	885	752
978	465	856	923

TABLE 11.18

Time (in Hours) Spent Using Social Media Per Day Based on Gender

Males	Females
6	11
11	10
10	16
8	10
10	8
6	14
9	16
10	9
12	11
6	7
7	12
6	15
4	15
4	7
5	8
6	8

8. A study was conducted where 36 patients were randomly assigned to use one of the three over-the-counter pain medications and were asked how long the medicine provided pain relief (rounded to the nearest hour). The data is provided in Table 11.19.

    Using a nonparametric test, determine if there is a difference in the number of hours of pain relief provided by the type of over-the-counter medication.

9. Can your diet make you happier? A researcher wants to know if the food you eat has an effect on your happiness. A random sample of 41 participants was asked

TABLE 11.19

Number of Hours of Pain Relief by the Type of Over-the-
Counter Medication

Aspirin	Acetaminophen	Ibuprofen
4	6	6
5	5	8
3	7	9
5	8	7
4	4	8
6	5	9
4	7	6
5	8	8
4	4	7
5	7	7
6	4	8
8	4	9

TABLE 11.20

Responses to Happiness Survey by the Type of Diet

No Specific Diet	Vegetarian	Vegan	Low Carb	Low Fat
70	92	78	74	58
55	77	74	73	94
56	84	94	72	97
50	92	62	69	73
53	74	61	69	55
75	76	86	—	86
71	81	90	—	67
50	86	66	—	—
53	72	73	—	—
66	—	58	—	—

to complete a survey on happiness and a questionnaire on their diet. The happiness survey is scored on a scale of 50–100, where higher scores are associated with someone who has a higher degree of happiness and satisfaction with their life. The data from the survey and the type of diet are provided in Table 11.20.

Using a nonparametric test, determine if there is a difference in the responses to the happiness survey based on the type of diet.

12

Two-Way Analysis of Variance and Basic Time Series

12.1 Two-Way Analysis of Variance

Recall that a one-way analysis of variance (ANOVA) compares the means of a continuous variable that is classified by *only one categorical factor*. However, there may be situations where you may want to compare the means of a variable that is classified by *two categorical factors*. To do this, we could conduct a *two-way ANOVA*.

Example 12.1

Suppose the manager of a manufacturing company is interested in estimating whether there is a difference in the mean number of defective products that are produced at three different manufacturing plants using five different types of machines. The data provided in Table 12.1 consists of a random sample of the number of defective products that were produced by the plant and machine at two different collection times.

We are interested in testing whether there is a difference in the mean number of defective products that are produced by the machine and by the plant. We can describe one factor that represents the five different types of machines (this is referred to as the *row factor*) and another factor that represents the three different plants (this is referred to as the *column factor*).

In a fashion similar to that of a one-way ANOVA, let δ describe the effect of the different levels of the machine (the row factor), and let β describe the effect

TABLE 12.1

Number of Defective Products Produced by Machine and by Plant

	Plant 1	Plant 2	Plant 3
Machine 1	12	18	19
	15	13	17
Machine 2	16	22	21
	19	15	10
Machine 3	12	10	8
	12	9	8
Machine 4	25	20	20
	24	21	19
Machine 5	18	19	26
	10	8	9

of the different levels of the plant (column factor). We can then model the average number of defective items for the ith row and the jth column by using the following regression equation:

$$y_{i,j,k} = \mu + \delta_i + \beta_j + (\delta\beta)_{i,j} + \varepsilon_{i,j,k}$$

where $y_{i,j,k}$ is the kth observation at level δ_i and level β_j, μ represents the mean common to all populations, $(\delta\beta)_{i,j}$ represents the interaction between the row factor δ at level i and the column factor β at level j, and $\varepsilon_{i,j,k}$ represents the error component that corresponds to the kth observation of the row factor at level i and the column factor at level j.

A two-way ANOVA can be used to test whether there is a significant difference between the mean number of defective products based on the type of machine or the plant where it was manufactured. A two-way ANOVA can also test whether or not there is an interaction or dependency between the row and column factors. For our example, this would represent a dependency between the machine and the plant. For instance, an interaction can occur if there is a particular machine at a certain plant that is producing a larger number of defects than can be explained by the type of machine and the plant on their own.

The appropriate null and alternative hypotheses for the row effect, the column effect, and the interaction effect corresponding to the five machines and three plants would be as follows:

$H_0 : \delta_1 = \delta_2 = \delta_3 = \delta_4 = \delta_5$

$H_1 :$ At least two population machine (row) effects are not equal.

$H_0 : \beta_1 = \beta_2 = \beta_3$

$H_1 :$ At least two population plant (column) effects are not equal.

$H_0 :$ There is no interaction between the population row and column factors.

$H_1 :$ There is an interaction between the population row and column factors.

A two-way ANOVA also requires using the F-distribution with the following test statistics for each of the row, column, and interaction effects, along with the respective degrees of freedom for the numerator and denominator for each:

$$F = \frac{\text{MSRow}}{\text{MSE}}, \text{df} = \left[r - 1, rc(m - 1) \right]$$

$$F = \frac{\text{MSColumn}}{\text{MSE}}, \text{df} = \left[c - 1, rc(m - 1) \right]$$

$$F = \frac{\text{MSInteraction}}{\text{MSE}}, \text{df} = \left[(r - 1)(c - 1), rc(m - 1) \right]$$

TABLE 12.2

Average of the Observations in Each Cell from the Data in Table 12.1

	Plant 1	Plant 2	Plant 3	Row Mean $\left(\bar{R}_i\right)$
Machine 1	$\dfrac{12+15}{2}=13.5$	$\dfrac{18+13}{2}=15.5$	$\dfrac{19+17}{2}=18.0$	$\dfrac{13.5+15.5+18}{3}\approx 15.6667$
Machine 2	17.5	18.5	15.5	17.1667
Machine 3	12.0	9.5	8.0	9.8333
Machine 4	24.5	20.5	19.5	21.5
Machine 5	14.0	13.5	17.5	15.0
Column mean $\left(\bar{C}_j\right)$	$\dfrac{13.5+17.5+12.0+24.5+14.0}{5}=16.30$	15.50	15.70	15.333

where r is the number of row factors, c is the number of column factors, and m is the number of replicates in each cell.

In order to calculate the mean squares for the row, column, and interaction effects, first we need to find the appropriate sum of squares and mean squares for the row, column, and interaction. In Table 12.1, notice that there are two observations (or replicates) for each machine and plant. In order to find the sum of squares for the rows and the columns, it is usually easier to first take the average of the observations in each cell, as illustrated in Table 12.2.

To find the sum of squares for the row factor, we can use the following formula:

$$\text{SSRow} = mc \sum_{i=1}^{r} \left(\bar{R}_i - \bar{y}_G\right)^2$$

where m is the number of observations (or replicates) in each cell (in this example $m = 2$), c is the number of column factors (in this example, $c = 3$ because there are three plants), and r is the number of row factors (in this example, $r = 5$ because there are five types of machines). The statistic $\bar{R}_i$ is the mean of the ith row, and $\bar{y}_G$ is the grand mean of all of the observations.

To find the sum of squares for the row factor, we first need to find the grand mean of all of the observations in the data set. For the data in Table 12.2, the grand mean is calculated by adding up all the values and dividing by 30. We can also find the grand mean by first summing up each column, adding them together, and then dividing by 30, as follows:

$$\bar{y}_G = \frac{163+155+157}{30} \approx 15.8333$$

The squared difference between each of the five row means and the total mean of all the observations is:

$$\left(\bar{R}_1 - \bar{y}_G\right)^2 = (15.6667 - 15.8333)^2 \approx 0.0278$$

$$\left(\bar{R}_2 - \bar{y}_G\right)^2 = (17.1667 - 15.8333)^2 \approx 1.7780$$

$$\left(\bar{R}_3 - \bar{y}_G\right)^2 = (9.8333 - 15.8333)^2 \approx 36.0$$

$$\left(\bar{R}_4 - \bar{y}_G\right)^2 = (21.50 - 15.8333)^2 \approx 32.1115$$

$$\left(\bar{R}_5 - \bar{y}_G\right)^2 = (15.0 - 15.8333)^2 \approx 0.6944$$

Then summing all of these squared differences gives

$$\sum_{i=1}^{5}\left(\bar{R}_i - \bar{y}_G\right)^2 = 0.0278 + 1.7780 + 36.0 + 32.1115 + 0.6944 \approx 70.6117$$

Therefore:

$$\text{SSRow} = mc\sum_{i=1}^{5}\left(\bar{R}_i - \bar{y}\right)^2 \approx (2)(3)(70.6117) \approx 423.6702$$

And:

$$\text{MSRow} = \frac{\text{SSRow}}{r-1} = \frac{423.6702}{5-1} \approx 105.9176$$

Similarly, to find the sum of squares and the mean square for the column factor:

$$\text{SSColumn} = mr\sum_{i=1}^{c}\left(\bar{C}_i - \bar{y}_G\right)^2 = (2)(5)\left[(16.3 - 15.8333)^2 + (15.5 - 15.8333)^2 + (15.7 - 15.8333)^2\right]$$

$$\approx 3.4667$$

where m is the number of observations in each cell, r is the number of rows, c is the number of columns, $\bar{C}_i$ is the mean of the ith column, and $\bar{y}_G$ is the grand mean of all of the observations. Hence,

$$\text{MSColumn} = \frac{\text{SSColumn}}{c-1} = \frac{3.4667}{3-1} \approx 1.7334$$

In order to test for an interaction between the row and the column factor, we need to find the sum of squares for the interaction term. This is done by summing the squares of the individual observations in each of the cells, dividing by the number of observations in each cell, and then subtracting the sum of squares for both the row and column factors, and finally subtracting the sum of

TABLE 12.3

Squared Sum of the Observations for Each Cell From the Data in Table 12.1

	Plant 1	Plant 2	Plant 3
Machine 1	$(12+15)^2 = 729$	$(18+13)^2 = 961$	$(19+17)^2 = 1296$
Machine 2	$(16+19)^2 = 1225$	$(22+15)^2 = 1369$	$(21+10)^2 = 961$
Machine 3	$(12+12)^2 = 576$	$(10+9)^2 = 361$	$(8+8)^2 = 256$
Machine 4	$(25+24)^2 = 2401$	$(20+21)^2 = 1681$	$(20+19)^2 = 1521$
Machine 5	$(18+10)^2 = 784$	$(19+8)^2 = 729$	$(26+9)^2 = 1225$

all the values squared divided by the total sample size. This can be represented by the following formula:

$$\text{SSInteraction} = \frac{\text{Sum of squares of all the cell values}}{m} - \text{SSRow} - \text{SSColumn} - \frac{\left(\sum_{i=1}^{n} y_i\right)^2}{n}$$

For our example, we first sum up the values of the observations in each cell, and then square them, as illustrated in Table 12.3.

Then, summing all of these values in Table 12.3, we get a total of 16,075. Dividing this value by $m = 2$ gives 8037.50.

To find the SSInteraction, we subtract from this value SSRow and SSColumn, and since

$$\left(\sum_{i=1}^{30} y_i\right)^2 = (475)^2 = 225,625$$

$$\frac{\left(\sum_{i=1}^{30} y_i\right)^2}{30} \approx 7,520.8333$$

Then:

$$\text{SSInteraction} = 8,037.5 - 423.6702 - 3.4670 - 7,520.8333 \approx 89.5295$$

$$\text{MSInteraction} = \frac{\text{SSInteraction}}{(c-1)(r-1)} = \frac{89.5295}{(3-1)(5-1)} \approx 11.1912$$

To find SSE, we need to subtract the means from the value of the observations in each cell, square them, and then add them all together, as illustrated in Table 12.4.

Then:

$$\text{SSE} = \sum_{i=1}^{n}(y_i - \text{cell mean})^2 = 347.50$$

TABLE 12.4

Square of the Difference Between the Each of the Observations and their
Respective Cell Means

	Plant 1	Plant 2	Plant 3
Machine 1	$(12-13.5)^2 = 2.25$	$(18-15.5)^2 = 6.25$	$(19-18)^2 = 1$
	$(15-13.5)^2 = 2.25$	$(13-15.5)^2 = 6.25$	$(17-18)^2 = 1$
Machine 2	$(16-17.5)^2 = 2.25$	$(22-18.5)^2 = 12.25$	$(21-15.5)^2 = 30.25$
	$(19-17.5)^2 = 2.25$	$(15-18.5)^2 = 12.25$	$(10-15.5)^2 = 30.25$
Machine 3	$(12-12)^2 = 0$	$(10-9.5)^2 = 0.25$	$(8-8)^2 = 0$
	$(12-12)^2 = 0$	$(9-9.5)^2 = 0.25$	$(8-8)^2 = 0$
Machine 4	$(25-24.5)^2 = 0.25$	$(20-20.5)^2 = 0.25$	$(20-19.5)^2 = 0.25$
	$(24-24.5)^2 = 0.25$	$(21-20.5)^2 = 0.25$	$(19-19.5)^2 = 0.25$
Machine 5	$(18-14)^2 = 16$	$(19-13.5)^2 = 30.25$	$(26-17.5)^2 = 72.25$
	$(10-14)^2 = 16$	$(8-13.5)^2 = 30.25$	$(9-17.5)^2 = 72.25$

And:

$$MSE = \frac{SSE}{(c \cdot r)(m-1)} = \frac{347.50}{(3 \cdot 5)(2-1)} \approx 23.1667$$

Thus, the test statistics for the row, column, and interaction effect can be determined as follows:

$$F = \frac{MSRow}{MSE} = \frac{105.9176}{23.1667} \approx 4.5720$$

$$F = \frac{MSColumn}{MSE} = \frac{1.7335}{23.1667} \approx 0.0748$$

$$F = \frac{MSInteraction}{MSE} = \frac{11.1913}{23.1667} \approx 0.4831$$

We then need to compare each of these test statistics to the respective F-distributions with the desired level of significance and appropriate degrees of freedom.

For a significance level of $\alpha = 0.05$, the row factor will have $r - 1 = 5 - 1 = 4$ degrees of freedom for the numerator and $r \cdot c(m-1) = (5 \cdot 2)(2-1) = 15$ degrees of freedom for the denominator. Then we compare the test statistic $F = 4.5720$ to the value $f(4,15) = 3.06$ that defines the rejection region, as is illustrated in Figure 12.1.

As Figure 12.1 illustrates, since $F = 4.52 > f = 3.06$, we can accept the alternative hypothesis that at least two population machine (row) effects are different from each other.

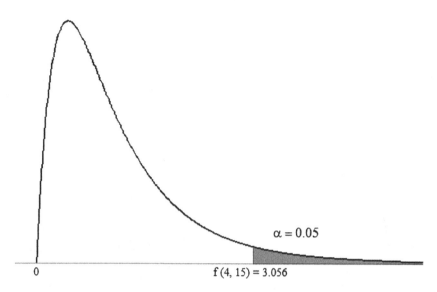

FIGURE 12.1
Rejection region for the row factor for a two-way ANOVA.

Similarly, for the column factor, with $k - 1 = 3 - 1 = 2$ degrees of freedom for the numerator and 15 degrees of freedom for the denominator, we can compare the value of $f(2,15) = 3.68$ to the value of the test statistic $F = 0.0748$, as in Figure 12.2.

From Figure 12.2, since $F = 0.0748 < f = 3.68$, we do not have enough evidence to suggest that the plant (column) effects are different from each other.

And finally, for the interaction with $(r - 1)(c - 1) = (4)(2) = 8$ degrees of freedom for the numerator and 15 degrees of freedom for the denominator, we could compare the value that defines the rejection region, $f(8,15) = 2.64$ to the test statistic $F = 0.4831$, as in Figure 12.3.

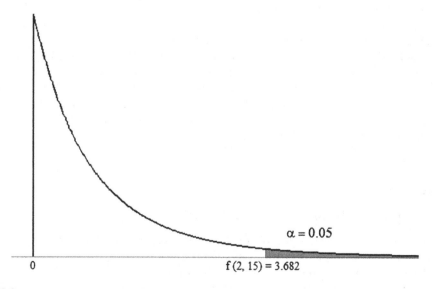

FIGURE 12.2
Rejection region for the column factor for a two-way ANOVA.

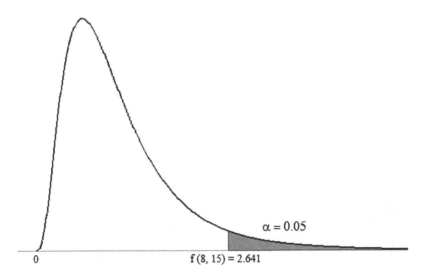

FIGURE 12.3
Rejection region for the interaction for a two-way ANOVA.

Similarly, from Figure 12.3, since $F = 0.4831 < f = 2.64$, we do not have enough evidence to suggest that there is an interaction between the two factors. Thus, our two-way ANOVA reveals that the only significant differences are in the mean number of defective products produced by machine, and not by plant or the interaction between plant and machine.

One thing to keep in mind when running a two-way ANOVA is that in addition to the model assumptions (which happen to be the same as for a one-way ANOVA), the model is constructed such that there are only two factors that can impact the number of defects. Similar to a one-way ANOVA, we are assuming that there are no other factors that could impact the number of defects that are produced. If you have reason to believe that there are more than two factors, such as the temperature or the time at which the defects were obtained, you may want to reconsider using a two-way ANOVA.

The analysis just described is called a *balanced two-way ANOVA*. This is because the same number of replicates is contained in each cell. Furthermore, in order to test for an interaction, there must be at least two observations in each cell. There are more sophisticated methods that can be used to handle unbalanced two-way ANOVAs, but they are beyond the scope of this book and can be found in some more advanced texts.

12.2 Using Minitab for a Two-Way ANOVA

Using Minitab for a two-way ANOVA for the data given in Table 12.1 requires that the data be entered in three separate columns: one column representing the response (the number of defects), one column representing the row factor (machine), and one column representing the column factor (plant). A portion of how the data are to be entered in Minitab is illustrated in Figure 12.4.

The Minitab commands for running a two-way ANOVA are found under the **General Linear Model** portion of the **ANOVA** command from the **Stat** menu bar, as illustrated in Figure 12.5.

This gives the dialog box presented in Figure 12.6, where the response, row factor, and the column factor need to be specified.

In order to find the interaction term, we need to select the **Model** tab so we can specify the interaction term. This is done by first highlighting the Factors and then adding their product to the model, as is illustrated in Figure 12.7.

↓	C1	C2	C3
	Number of Defects	Plant	Machine
1	12	1	1
2	15	1	1
3	16	1	2
4	19	1	2
5	12	1	3
6	12	1	3
7	25	1	4
8	24	1	4
9	18	1	5
10	10	1	5
11	18	2	1

Worksheet 1 ***

FIGURE 12.4
Format for entering data into Minitab in order to run a two-way ANOVA.

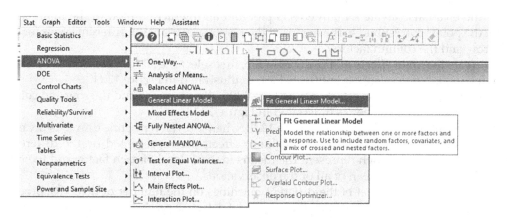

FIGURE 12.5
Minitab commands to run a two-way ANOVA.

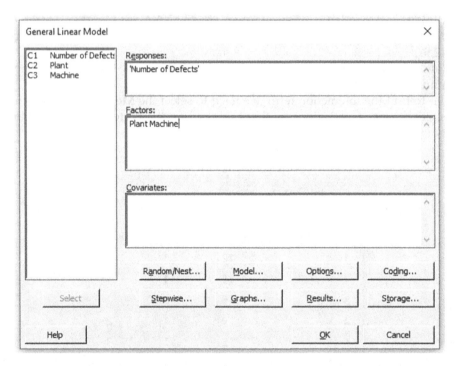

FIGURE 12.6
Minitab dialog box to run a two-way ANOVA.

This gives the Minitab printout that is illustrated in Figure 12.8 (the regression equation was omitted for simplicity). Notice that the values for SS, MS, and F match those that were calculated manually for the row, column, and interaction factors. Also, in interpreting the highlighted p-values for the row factor, column factor, and interaction, we see that the only significant effect is the row factor (machine). This is because the p-value is approximately 0.013 and this is less than our predetermined level of significance of $\alpha = 0.05$.

We can also use Minitab to check the usual regression model assumptions by selecting **Graphs** on the General Linear Model dialog box to get the dialog box illustrated in Figure 12.9.

The four-in-one residual plot is given in Figure 12.10.

Minitab can also create a main effects plot and an interaction plot. In the main effects plot, the points correspond to the means of the response variable at different levels for both the row and the column factors.

To create the main effects plot using Minitab, first select the ANOVA command from the Stat menu and then select Main Effects Plot, as illustrated in Figure 12.11.

This brings up the main effects dialog box, where the response and the row and column factors need to be specified, as illustrated in Figure 12.12.

The main effects plot is given in Figure 12.13, where the reference line drawn is the mean of all the observations. The row and column effects are represented by the differences between the points, which represent the means for each level of each factor and the reference line, which is the mean of all of the observations.

In Figure 12.13, the effect of the different machines on the number of defective products produced is quite large compared to the number of defective products produced at the different plants. Notice that there is more variability from the mean of all the observations in the means in the main effects plot by machine versus the variability in the means by the plant.

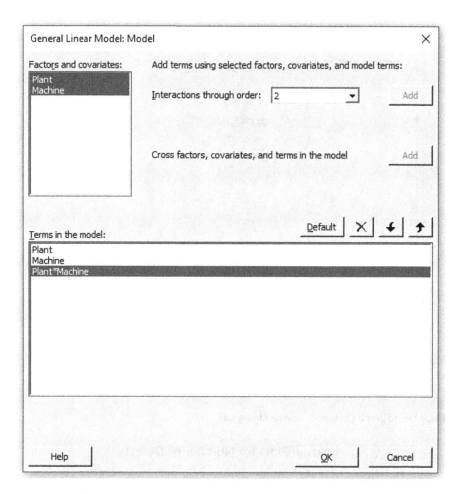

FIGURE 12.7
Minitab dialog box to add the interaction term of plant*machine.

(a)
Analysis of Variance

Source	DF	Adj SS	Adj MS	F-Value	P-Value
Plant	2	3.467	1.733	0.07	0.928
Machine	4	423.667	105.917	4.57	0.013
Plant*Machine	8	89.533	11.192	0.48	0.850
Error	15	347.500	23.167		
Total	29	864.167			

(b)
Model Summary

S	R-sq	R-sq(adj)	R-sq(pred)
4.81318	59.79%	22.26%	0.00%

FIGURE 12.8
Portion of the Minitab printout for the two-way ANOVA testing to check whether there is a difference in the mean number of defective products that are produced by the machine and by the plant.

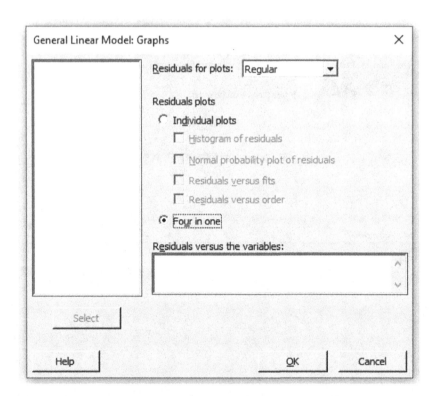

FIGURE 12.9
Minitab dialog box to graph the four-in-one residual plots.

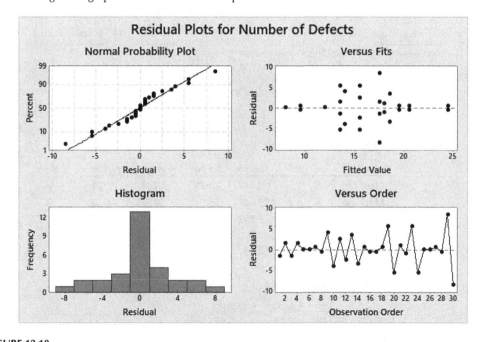

FIGURE 12.10
Four-in-one plot for the two-way ANOVA testing to assess whether there is a difference in the mean number of defective products that are produced by the machine and by the plant.

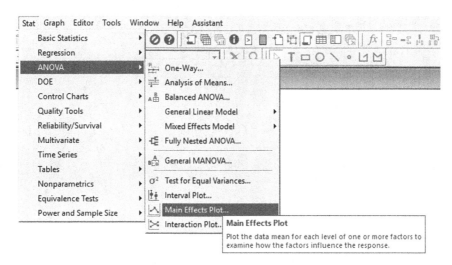

FIGURE 12.11
Minitab commands to create a main effects plot.

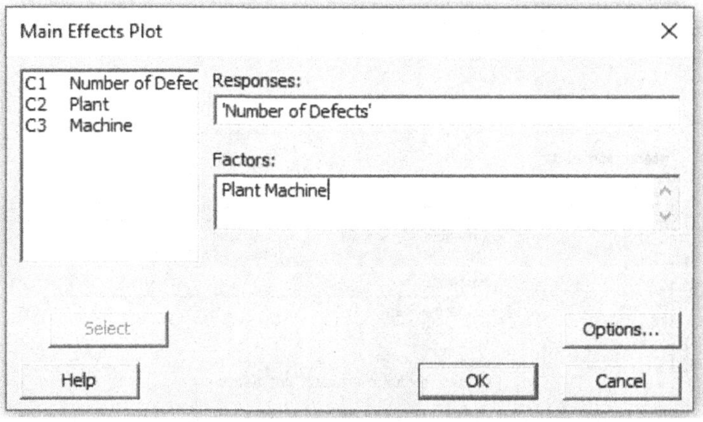

FIGURE 12.12
Minitab dialog box for a main effects plot.

Similarly, Minitab can also graph an interaction plot. An interaction plot consists of the mean values for each level of a row factor with respect to the column factor. Recall that an interaction occurs when the row and column factors affect each other. An interaction plot depicts potential interactions as nonparallel lines. The dialog box for an interaction plot is given in Figure 12.14.

Figure 12.15 gives the interaction plot for the data in Table 12.1. This plot shows the mean number of defects versus the plant for each of the five different types of machines.

An interaction would occur if the number of defects produced on any given machine depends on a given plant (or plants). If such an interaction were present, it would show up in the interaction plot as nonparallel lines. When there is no interactive effect, then the interaction plot would primarily consist of parallel lines because the row and column factors do not depend on each other. If an interaction is present, then it does not make sense

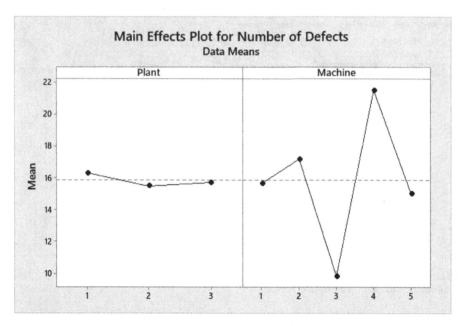

FIGURE 12.13
Main effects plot for the two-way ANOVA using the data in Table 12.1.

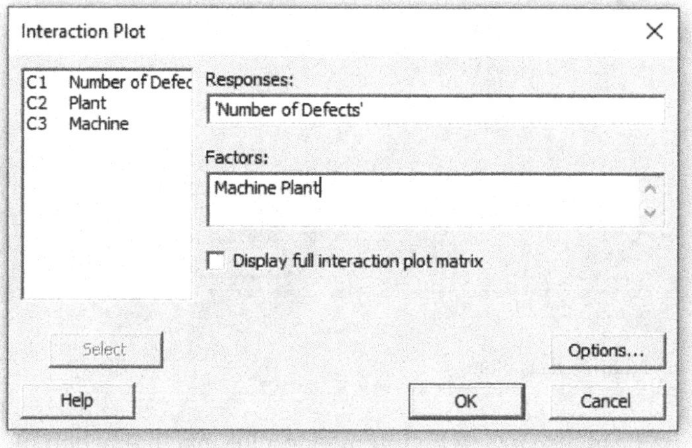

FIGURE 12.14
Minitab dialog box for an interaction plot.

to conduct individual tests for the row and column effects because an interaction suggests that differences in the row factor depend on the levels of the column factor (or vice versa). If the row and column variables are dependent, then testing them individually will not be appropriate. Notice that the interaction plot in Figure 12.15 does not reveal any obvious nonparallel lines among machines by the plant. This supports the formal analysis, which suggests that there may not be a significant interaction between the row (machine) and column (plant) factors.

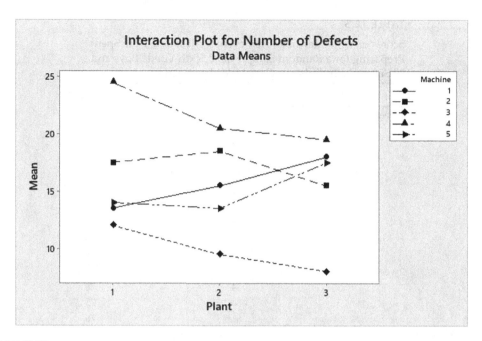

FIGURE 12.15
Interaction plot for the two-way ANOVA using the data in Table 12.1.

Example 12.2

The data set in Table 12.5 presents the scores on a standardized achievement test for a random sample of 30 sixth-grade boys and 30 sixth-grade girls and the length of time they spent preparing for the achievement test.

We want to know whether there is a difference in the population mean score received on the standardized test by gender and by the amount of time that was spent preparing for the test. We would use a two-way ANOVA if we had reason to believe that these are the only two factors along with a potential interaction effect that could impact the scores received on the standardized examination.

Thus, our null and alternative hypotheses could be stated as follows:

$H_0 : \alpha_1 = \alpha_2 = \alpha_3$

H_1 : There is a difference in the population mean test score by time preparing.

$H_0 : \beta_1 = \beta_2$

H_1 : There is a difference in the population mean test score based on gender.

H_0 : There is no interaction between time preparing and gender.

H_1 : There is an interaction between the time preparing and gender.

TABLE 12.5

Scores on a Standardized Achievement Test and Time Spent
Preparing for a Random Sample of 30 Sixth-Grade Boys and
30 Sixth-Grade Girls at a Certain Middle School

	Boys	Girls
1 week	65	78
	78	65
	84	81
	82	55
	76	72
	69	70
	81	68
	95	84
	87	80
	90	75
2 weeks	78	62
	84	78
	86	75
	81	76
	69	81
	79	83
	82	84
	98	89
	90	90
	86	70
3 weeks	85	85
	89	78
	91	82
	94	86
	82	85
	78	81
	90	83
	85	89
	87	91
	91	90

Using Minitab to run this analysis gives a portion of the results for the two-way
ANOVA that is presented in Figure 12.16.

There does not appear to be an interaction effect between time preparing for
the examination and gender ($p = 0.482$). Therefore, it makes sense to look at
the individual row and column effects. On the basis of the p-values from the
Minitab printout in Figure 12.16, there does appear to be a significant difference
in the mean examination score based on the amount of time spent preparing
for the examination ($p = 0.001$) and by gender ($p = 0.014$). These row and column
effects can also be seen by creating a simple boxplot with groups (**Graphs <
Boxplot < One Y < With Groups**), as in Figure 12.17.

By visually inspecting the graph in Figure 12.17, it appears that there may be
a difference by gender because for each of these sets of boxplots, boys appear to

Factor Information

Factor	Type	Levels	Values
Time	Fixed	3	One-Week, Three-Weeks, Two-Weeks
Gender	Fixed	2	Boys, Girls

Analysis of Variance

Source	DF	Adj SS	Adj MS	F-Value	P-Value
Time	2	876.10	438.05	7.88	0.001
Gender	1	355.27	355.27	6.39	0.014
Time*Gender	2	82.23	41.12	0.74	0.482
Error	54	3001.00	55.57		
Total	59	4314.60			

Model Summary

S	R-sq	R-sq(adj)	R-sq(pred)
7.45480	30.45%	24.01%	14.13%

FIGURE 12.16
Portion of the Minitab printout for the two-way ANOVA testing to check if there is a difference in the mean examination score based on time for preparation (row factor) and gender (column factor).

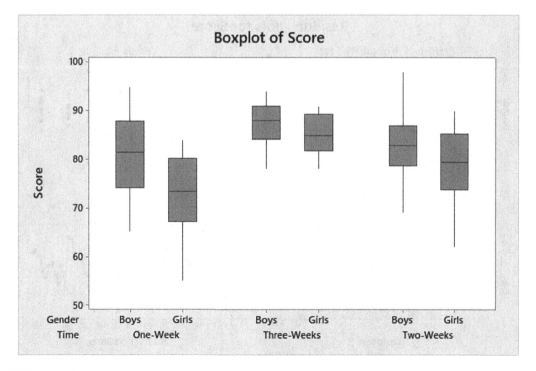

FIGURE 12.17
Boxplot of exam score based on time preparing for the exam and gender.

score higher on the examination than girls. Also, there appears to be a difference in the scores based on the amount of time that was spent in preparing for the examinations because the scores are higher with more time spent studying. To check the two-way ANOVA model assumptions, a four-in-one residual plot is given in Figure 12.18, and the results of the Ryan–Joiner test are presented in Figure 12.19.

The graphs of the residual plots in Figure 12.18 suggest that the model assumptions for a two-way ANOVA appear to hold true. The results of the Ryan–Joiner test presented in Figure 12.19 also support the claim that the assumption of normality of the error component does not appear to have been violated.

The main effects plot and the interaction plot also support these conclusions, as represented in Figures 12.20 and 12.21.

The main effects plot in Figure 12.20 suggests that there could be both a row (time preparing) and column (gender) effect, whereas the interaction plot in Figure 12.21 suggests that there does not appear to be an interaction between preparation and gender because the mean values by gender do not appear to be affected by the time spent preparing for the examination.

Example 12.3

The data set provided in Table 12.6 presents a sample of the flight delay times in minutes by airline and by whether the delay was caused by the airline carrier or the weather. We want to determine if there is a difference in the population

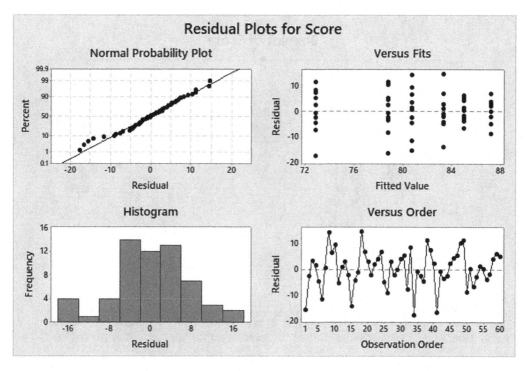

FIGURE 12.18
Residual plots for the two-way ANOVA from Figure 12.16.

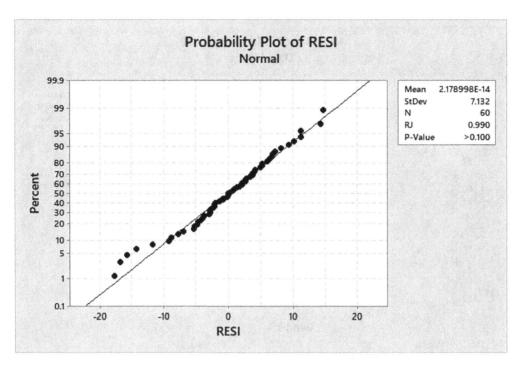

FIGURE 12.19
Ryan–Joiner test of the assumption that the residuals are normally distributed.

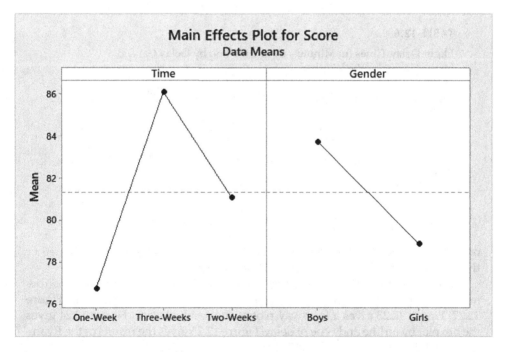

FIGURE 12.20
Main effects plot for scores on a standardized achievement test based on gender and amount of time preparing for the test.

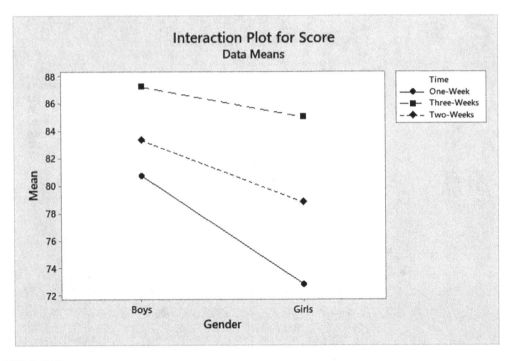

FIGURE 12.21
Interaction plot for scores on a standardized achievement test based on gender and amount of time preparing for the test.

TABLE 12.6

Flight Delay Times (in Minutes) by Airline and by Delay Caused by the Carrier or the Weather

	Airline 1	Airline 2	Airline 3	Airline 4
Airline carrier	257	3,850	1,260	1,650
	229	1,542	3,300	208
	451	3,501	1,521	710
	164	1,980	205	198
Weather	285	125	102	180
	270	250	76	12
	480	167	109	59
	195	102	111	102

mean delay time based on the airline carrier and the type of delay and whether there is an interaction between the airline carrier and the type of delay.

Running a two-way ANOVA using Minitab, with the type of delay as the row factor and the airline carrier as the column factor, gives the printout in Figure 12.22. Figure 12.23 gives a four-way plot of the residuals, and Figure 12.24 gives the boxplot by airline and type of delay. Figure 12.25 gives the results of the Ryan–Joiner test. Figures 12.26 and 12.27 give the main effects and interaction plots.

The results of the two-way ANOVA in Figure 12.22 suggest a significant interaction effect. The interaction plot in Figure 12.27 illustrates the interaction between

Analysis of Variance

Source	DF	Adj SS	Adj MS	F-Value	P-Value
Airline	3	6579552	2193184	5.11	0.007
Carrier/Weather	1	10581150	10581150	24.67	0.000
Airline*Carrier/Weather	3	7561374	2520458	5.88	0.004
Error	24	10291982	428833		
Total	31	35014059			

Model Summary

S	R-sq	R-sq(adj)	R-sq(pred)
654.853	70.61%	62.03%	47.74%

FIGURE 12.22
Portion of the Minitab printout for a two-way ANOVA for the flight delay data from Table 12.6.

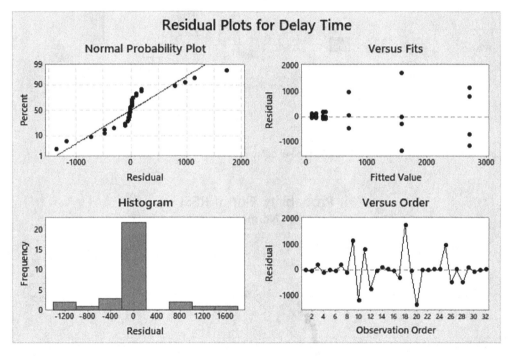

FIGURE 12.23
Four-way residual plots for the airline data from Table 12.6.

the type of delay and airline. An interaction may be present because some of the lines in Figure 12.27 that represent the different airlines are nonparallel to each other. For instance, the graph representing the mean delay time for Airline 1 is perpendicular to the graph representing the mean delay times for Airlines 2 and 3. The interaction plot suggests an interaction between the airline carrier and the type of delay. In other words, the cause of the delay depends upon the different airline carriers. Because there is an interaction effect, it is not appropriate to interpret the individual row and column effects. Even though the *p*-values and the main effects plot in Figure 12.26 suggest a row effect (type of delay) and column

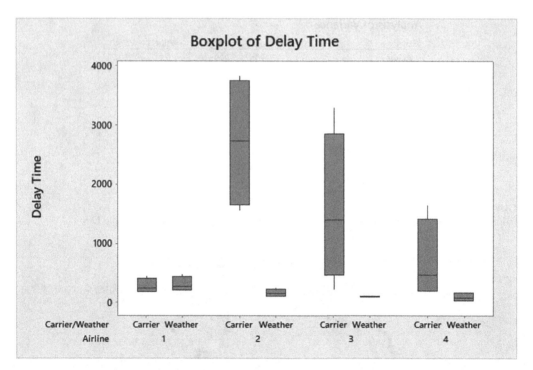

FIGURE 12.24
Boxplot of airline delay data by airline and type of delay.

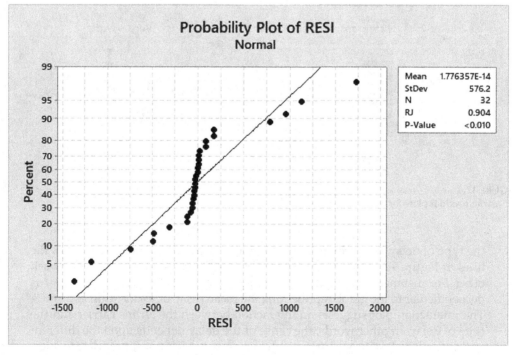

FIGURE 12.25
Ryan–Joiner test of the normality assumption for the airline data given in Table 12.6.

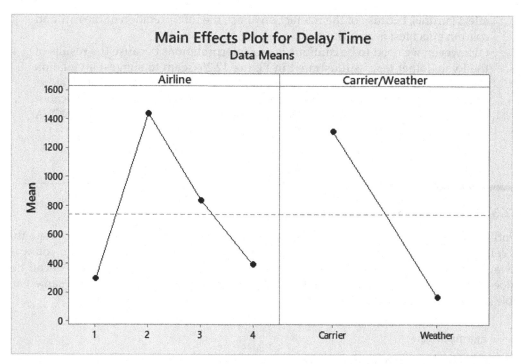

FIGURE 12.26
Main effects plot for the flight delay data by airline and type of delay.

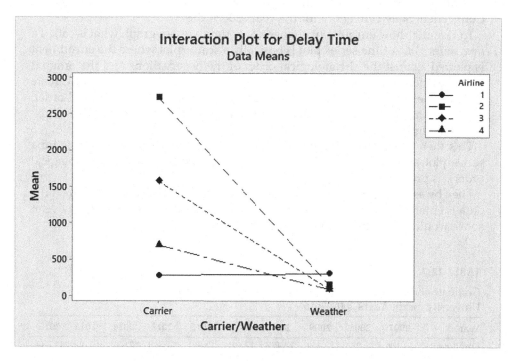

FIGURE 12.27
Interaction plot for the flight delay data by airline and the type of delay.

effect (airline), because of the interaction effect, the interpretation of the row and column effect test is meaningless.

However, we need to be cautious of our interpretations because the results of the Ryan–Joiner test, as illustrated in Figure 12.25, seem to suggest a violation of the assumption that the error component is normally distributed. Therefore, any inferences that we made based on our analysis could be suspect because the assumption of normality of the error component for the two-way ANOVA may not have been met.

12.3 Basic Time Series Analysis

Until now, we have primarily discussed statistical methods that can be used for data that is collected at a single point in time. However, there may be circumstances when you want to analyze data that is collected on the same variable over a period of time. We call this type of data *time series data* because it consists of data that is measured on the same variable sequentially over time.

Example 12.4

Suppose we are interested in how the enrollment patterns at a given university change over time. Such data would be considered time series data because the same variable, namely, enrollment, is measured sequentially over time. Table 12.7 gives the total graduate and undergraduate fall enrollments for Central Connecticut State University for the years 2007–2016.

To visualize how enrollment changes over time, we can graph what is called a *time series plot*. A time series plot is basically a scatterplot where the enrollment is plotted against the chronological order of the observations and the ordered pairs are then connected by straight lines. Minitab can easily be used to create a time series plot by first entering the enrollment data in chronological order, then selecting **Time Series > Time Series Plot** under the **Stat** menu, as illustrated in Figure 12.28.

This then gives the dialog box in Figure 12.29. We will select a simple time series plot and then select the variable (which needs to be in chronological order), as presented in Figure 12.30.

Then by selecting the **Time/Scale** tab, we can specify the format in which the data is entered. In our example, the data were entered chronologically by year, so we would select the **Calendar** as **Year**, as illustrated in Figure 12.31.

TABLE 12.7

Graduate and Undergraduate Fall Enrollments for Central Connecticut State University for the Years 2007–2016

Year	2007	2008	2009	2010	2011	2012	2013	2014	2015	2016
Enrollment	12,106	12,233	12,461	12,477	12,521	12,091	11,865	12,037	12,086	11,784

Source: http://docs.ccsu.edu/oira/institutionalData/factbook/enrollments/headcount/Fall_Enrollment_Historical.pdf.

FIGURE 12.28
Minitab commands to draw a time series plot.

Also notice that the years start at 2007. This gives the time series plot of the enrollment data by year, as illustrated in Figure 12.32.

One way to describe time series data is by *trends*. One such type of trend is called a *long-term trend*. A long-term trend would suggest whether the series tends to increase or decrease over a period of time. We can also determine if there is a *seasonal trend*. A seasonal trend could be described by a repeating pattern that seems to reoccur during a certain interval of time. Another trend that could be observed is a *cyclical trend*. A cyclical trend would be a pattern that oscillates about a long-term trend. Because we cannot describe every possible type of trend, there is also a *random trend*, which consists of any trend that may not have been observed. It is what is left over after the long-term, seasonal, and cyclical trends have been accounted for. Notice in Figure 12.32 that there appears to be a decreasing long-term trend. In other words, it appears that the enrollments at CCSU appear to be decreasing over time.

There are some relatively simple forecasting techniques that can be used to analyze time series data. These techniques require modeling observable trends or patterns in the data and then using such patterns to predict future observations. This is very similar to what we described when we covered regression analysis.

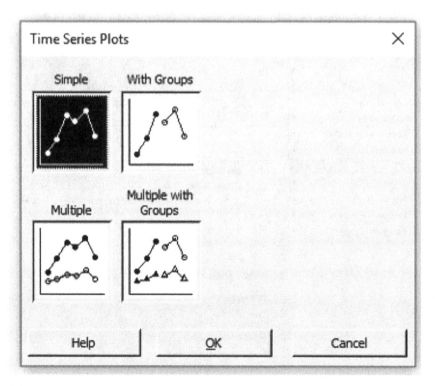

FIGURE 12.29
Minitab dialog box to draw a simple time series plot.

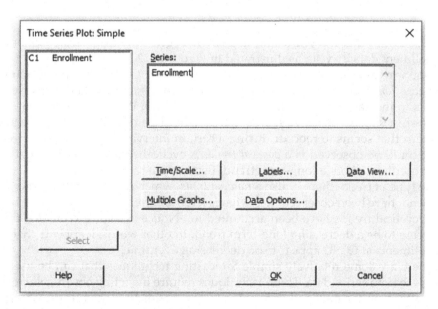

FIGURE 12.30
Minitab dialog box with Enrollment selected for the time series plot.

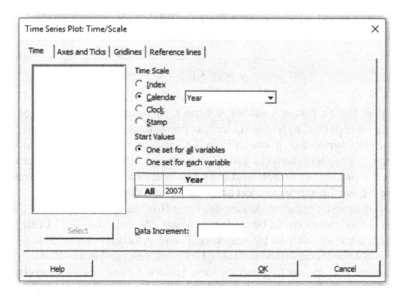

FIGURE 12.31
Minitab dialog box to specify the chronological ordering of the data.

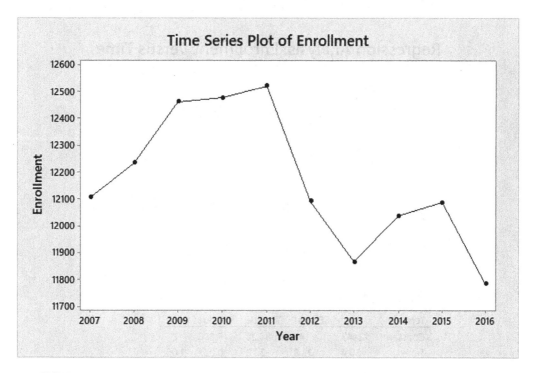

FIGURE 12.32
Time series plot of enrollment by year for the data in Table 12.7.

One way to see if there is a long-term trend is to do a *trend analysis*. A trend analysis entails fitting a regression model to time series data and then using such a model to extrapolate for future time periods.

A linear forecast model uses a linear model to represent the relationship between the response variables of interest at time t, as follows:

$$y_t = \beta_0 + \beta_1 t_t + \varepsilon_t$$

where y_t is the response variable at time t, β_1 represents the average change from one time period to the next, and t represents a given time period.

For the time series data given in Table 12.7, we can perform a linear forecast by running a regression analysis. This is done by creating time periods and using them as the predictor variable and the enrollments as the response variable and running a basic linear regression analysis, as seen in Figure 12.33.

We can then use the estimated model from this regression analysis to forecast what the enrollments could be for future years. We can also use Minitab to perform such a trend analysis by selecting **Trend Analysis** under **Stat** and **Time Series** to get the trend analysis dialog box presented in Figure 12.34.

We can use this model to generate forecasts for future observations by checking the **Generate forecasts** box and then indicating the **Number of forecasts**. The number of forecasts will be the same scale as the time series data. Since in our example the time is in years, we will use the model to forecast the enrollments for the next 3 years. If we select the **Time Scale** tab, the **Calendar** slot can

Regression Analysis: Enrollment versus Time

Analysis of Variance

Source	DF	Adj SS	Adj MS	F-Value	P-Value
Regression	1	209412	209412	4.48	0.067
Time	1	209412	209412	4.48	0.067
Error	8	373639	46705		
Total	9	583051			

Model Summary

S	R-sq	R-sq(adj)	R-sq(pred)
216.113	35.92%	27.91%	0.00%

Coefficients

Term	Coef	SE Coef	T-Value	P-Value	VIF
Constant	12443	148	84.28	0.000	
Time	-50.4	23.8	-2.12	0.067	1.00

Regression Equation

Enrollment = 12443 - 50.4 Time

FIGURE 12.33
Regression analysis of enrollments over time for the enrollment data in Table 12.7.

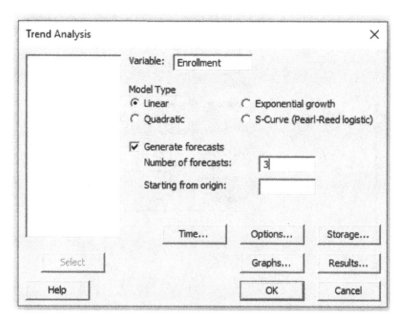

FIGURE 12.34
Minitab dialog box to perform a trend analysis for the enrollment data in Table 12.7.

be selected as the **Year** and the start value of 2007 can be provided, as can be seen in Figure 12.35.

This will generate the trend analysis graph presented in Figure 12.36.

Notice in Figure 12.36 that the linear trend model is the same as the linear regression model that is presented in Figure 12.33. The trend analysis plot not only provides the time series plot but also provides the linear trend equation for the forecast (or estimated) values. Also, you will notice that there is a set of accuracy measures in the box on the right-hand side of Figure 12.36. These accuracy measures can be used to assess how well the model fits the data as they compare the observed values of the time series data with the forecast values and they can be very useful for comparing different models.

The first measure, MAPE, stands for the *mean absolute percentage error*. This measure gives the average percentage error by finding the average of the absolute value of the difference between each observation and the fitted value multiplied by 100. The smaller this value is, the better the fit of the forecast model. It can be calculated by using the following equation:

$$\text{MAPE} = \frac{\sum_{t=1}^{n} \left| \frac{y_t - \hat{y}_t}{y_t} \right|}{n} \cdot 100$$

where y_t is the value of the observation at time t, $\hat{y}_t$ is the fitted value at time t, and n is the sample size.

Table 12.8 gives the actual observations along with the corresponding fitted values for the data in Table 12.7.

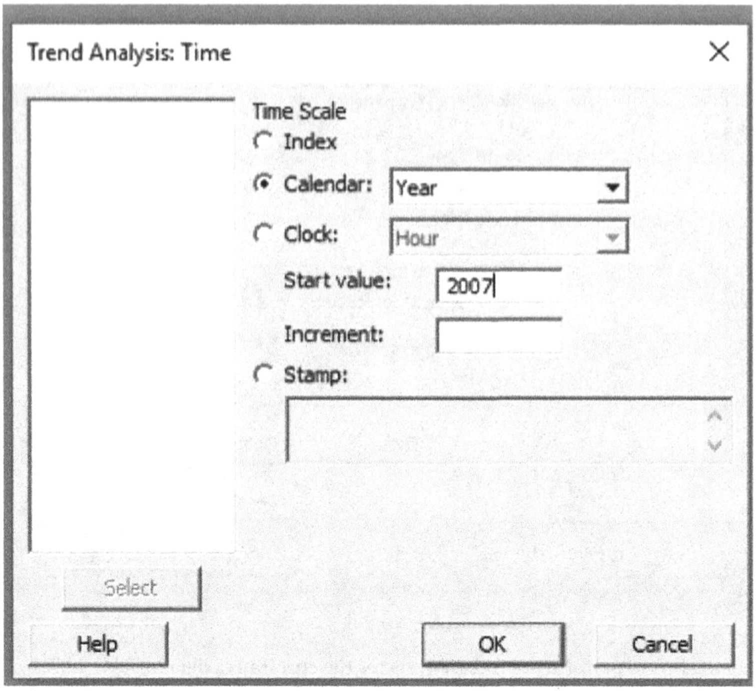

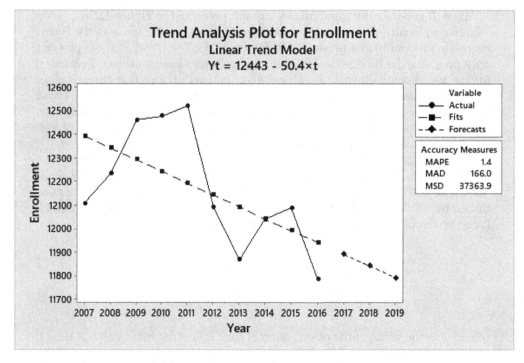

FIGURE 12.35
Minitab dialog box to specify the chronological ordering of the data.

FIGURE 12.36
Minitab graph of a linear trend analysis for the change in enrollments over time.

TABLE 12.8

Enrollment and Fitted Values for the Data Given in Table 12.7

Time (in Years) t	Enrollment y_t	Fitted (or Estimated) Enrollment $\hat{y}_t$	$y_t - \hat{y}_t$	$(y_t - \hat{y}_t)^2$	$\lvert y_t - \hat{y}_t \rvert$	$\left\lvert \dfrac{y_t - \hat{y}_t}{y_t} \right\rvert$
1	12,106	12,392.60	−286.60	82,139.56	286.6	0.02367
2	12,233	12,342.20	−109.20	11,924.64	109.2	0.00893
3	12,461	12,291.80	169.20	28,628.64	169.2	0.01358
4	12,477	12,241.40	235.60	55,507.36	235.6	0.01888
5	12,521	12,191.00	330.00	108,900	330	0.02636
6	12,091	12,140.60	−49.60	2,460.16	49.6	0.00410
7	11,865	12,090.20	−225.20	50,715.04	225.2	0.01898
8	12,037	12,039.80	−2.80	7.84	2.8	0.00023
9	12,086	11,989.40	96.60	9,331.56	96.6	0.00799
10	11,784	11,939.00	−155.00	24,025	155	0.01315
Total				373,639.8	1,659.80	0.13587

To calculate MAPE, we add up the seventh column from Table 12.8, divide by the sample size, then multiply by 100 as follows:

$$\text{MAPE} = \frac{\sum_{t=1}^{n} \left\lvert \dfrac{y_t - \hat{y}_t}{y_t} \right\rvert}{n} \cdot 100 = \frac{0.13587}{10} \cdot 100 \approx 1.3587 \approx 1.4$$

Another accuracy measure, the *mean absolute deviation*, MAD, measures the accuracy of the fitted model by finding the average difference between the observations and the fitted value, and is calculated as follows:

$$\text{MAD} = \frac{\sum_{t=1}^{n} \lvert y_t - \hat{y}_t \rvert}{n}$$

where y_t is the value of the observation at time t, $\hat{y}_t$ is the fitted value at time t, and n is the sample size.

Using the calculations in Table 12.8, we can calculate the MAD by adding up the sixth column and dividing by the sample size as follows:

$$\text{MAD} = \frac{\sum_{t=1}^{n} \lvert y_t - \hat{y}_t \rvert}{n} = \frac{1,659.80}{10} = 165.98 \approx 166.0$$

Similar to the MAPE, the smaller the value of the MAD, the better the fit of the estimated model.

Another accuracy measure is the MSD or the *mean squared deviation*. This measure gives the average of the squared differences between the observed and the fitted values, and it is found using the following formula:

$$MSD = \frac{\sum_{i=1}^{n}(y_t - \hat{y}_t)^2}{n}$$

where y_t is the value of the observation at time t, $\hat{y}_t$ is the fitted value at time t, and n is the sample size.

Using the calculations in Table 12.8, we can calculate the MSD by adding up the fifth column and dividing by the sample size as follows:

$$MSD = \frac{\sum_{i=1}^{n}(y_t - \hat{y}_t)^2}{n} = \frac{373,639.8}{10} = 37,363.98$$

Because the long-term trend may be nonlinear, we can also use Minitab to fit a quadratic model and forecast the enrollments for the next 3 years, as can be seen in Figure 12.37.

This gives the quadratic trend model presented in Figure 12.38.

Notice that the accuracy measures in Figure 12.38 for the quadratic model are smaller than those for the linear model in Figure 12.36. This suggests that the quadratic model fits the data better than the linear model because the deviations

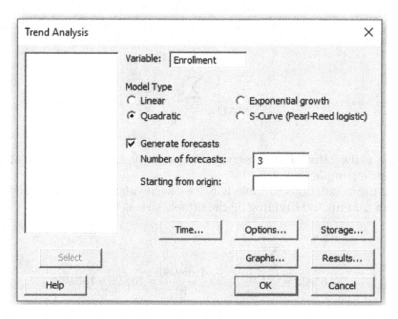

FIGURE 12.37
Minitab dialog box for a quadratic trend analysis.

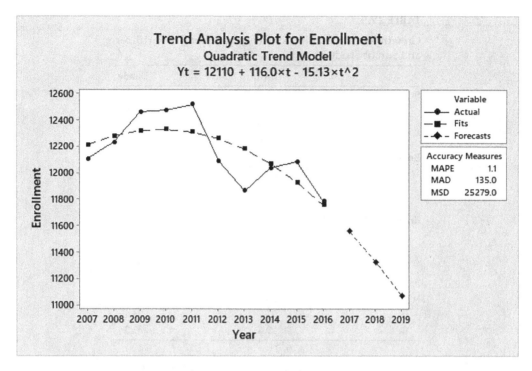

FIGURE 12.38
Minitab quadratic trend analysis for the enrollment data in Table 12.7.

between the observed and the fitted values are smaller for the quadratic model. Also, using this model, we can forecast the enrollments for the next 3 years.

Exercises

Unless otherwise specified, let $\alpha = 0.05$.

1. An experiment was conducted to estimate if there is a difference in the mean growth rate of grass based on four brands of fertilizer and whether the conditions are full sun or shade. A randomized block design was used to randomly assign the different fertilizers to two parcels of grass, one in the sun and one in the shade. The growth rate after two weeks (in centimeters) is provided in Table 12.9.

   a. Which is the row factor and which is the column factor?

   b. How many levels are there for each factor?

   c. How many replicates are in each cell?

   d. Using Minitab, determine if there is a difference in the mean growth rate based on the brand of fertilizer and whether the grass was in full sun or in the shade.

   e. Is there an interaction between the brand of fertilizer and whether the grass was in the sun or the shade?

TABLE 12.9

Growth Rate (in Centimeters) Based on Brand of Fertilizer and Sun or Shade

	Sun	Shade
Brand 1	25.3	26.4
	26.4	27.9
	27.9	29.4
	26.8	30.5
Brand 2	22.4	23.6
	24.3	22.6
	18.7	25.8
	26.3	23.4
Brand 3	29.7	31.5
	28.6	33.6
	30.4	33.7
	33.9	36.2
Brand 4	22.5	26.5
	28.6	30.2
	23.7	26.4
	29.4	33.3

    f. Create a main effects plot and an interaction plot and comment on whether these plots support your findings.

    g. Check any relevant model assumptions and describe any violations that you see.

2. The data presented in Table 12.10 gives the lifetime (in hours) of cell phone batteries based on the age of the battery and the capacity of the battery (in milliamps per hour).

    a. Which is the row factor and which is the column factor?

    b. How many levels are there for each factor?

    c. How many replicates are in each cell?

    d. Using Minitab, determine if there is a difference in the mean lifetime of cell phone batteries based on age and capacity.

TABLE 12.10

Lifetime (in mAh, Milliamp Hours) for a Random Sample of Cell Phone Batteries Based on Age of Battery and Hours

	1,400 mAh	1,600 mAh	1,800 mAh
Less than 1 year old	24.3	26.4	33.5
	22.1	18.5	30.4
	18.6	27.4	19.5
	27.2	29.9	27.7
	15.4	30.3	33.5
More than 1 year old	22.4	25.4	30.5
	23.1	29.1	31.7
	25.3	26.5	33.8
	16.8	20.1	32.1
	15.2	19.9	25.4

TABLE 12.11

Mutual Fund Prices for a 2-Year Period

	Jan	Feb	Mar	Apr	May	Jun	Jul	Aug	Sep	Oct	Nov	Dec
2017	69.96	68.38	72.49	69.24	69.17	69.08	67.74	68.70	70.80	66.10	69.23	67.50
2016	60.23	62.94	64.17	67.56	67.92	66.03	67.34	70.64	66.25	68.43	67.15	67.76

    e. Is there an interaction between the age of the cell phone and the capacity?

    f. Create a main effects plot and an interaction plot and comment on whether these plots support your findings.

    g. Check any relevant model assumptions and describe any violations that you see.

3. An investor wants to invest in a mutual fund. She is hoping that the long-term trend for the fund will be increasing. She has data on the mutual fund price collected each month for the years 2016 and 2017. The date and the price are presented in Table 12.11.

    a. Run a linear trend model on the price of the mutual fund for 2016.

    b. Run a quadratic trend model on the price of the mutual fund for 2016.

    c. Run a linear trend model on the price of the mutual fund for 2017.

    d. Run a quadratic trend model on the price of the mutual fund for 2017.

    e. Run a linear trend model on the price of the mutual fund for both years combined.

    f. Run a quadratic trend model on the price of the mutual fund for both 2016 and 2017 combined.

    g. Which of these models best fit the data?

    h. Using the best fitting model, forecast the price for the 3 months of 2018.

4. A random sample of new home sales (in dollars) for a four-bedroom house with 2.5 bathrooms and a two-car garage based on the age of the home and the region of the United States is given in Table 12.12.

    a. Which is the row factor and which is the column factor?

    b. How many levels are there for each factor?

    c. How many replicates are in each cell?

    d. Using Minitab, determine if there is a difference in the mean selling price of homes based on age and region of the country.

    e. Is there an interaction between the age and the region of the country?

    f. Create a main effects plot and an interaction plot and comment on whether these plots support your findings.

    g. Check any relevant model assumptions and describe any violations that you see.

5. An educational researcher wants to investigate the tuition patterns at Central Connecticut State University (CCSU). The data in Table 12.13 gives the tuition (in dollars) for full-time undergraduate students for each academic year from 2008 to 2017.

TABLE 12.12

New Home Sales (in Thousands of Dollars) Based on Age and Region of the United States

	Northeast	Southeast	Midwest	Western
Less than 40 years	450,080	354,045	257,390	410,230
	373,510	477,370	374,145	433,280
	287,045	484,410	204,565	433,870
	360,755	355,235	380,435	463,980
	488,089	489,400	362,290	484,550
	331,570	432,430	245,670	431,080
	256,510	387,825	352,505	465,635
40–60 years	390,785	350,620	287,960	350,955
	262,280	424,640	363,815	460,405
	246,000	309,395	221,580	377,900
	286,090	338,590	365,745	349,800
	286,645	467,995	262,150	224,865
	471,905	431,280	311,815	255,650
	487,390	323,410	440,180	415,105
60+ years	311,955	454,300	321,560	371,155
	244,820	455,985	238,560	377,270
	409,700	461,815	326,615	465,835
	296,580	368,535	263,170	453,800
	218,885	356,010	275,850	388,180
	480,440	382,455	378,700	418,620
	286,075	336,320	438,410	384,560

TABLE 12.13

Tuition (in Dollars) for Full-Time Undergraduate Students from 2008–2017 at Central Connecticut State University

Year	2008–2009	2009–2010	2010–2011	2011–2012	2012–2013	2013–2014	2014–2015	2015–2016	2016–2017	2017–2018
Tuition	7,042	7,414	7,861	8,055	8,321	8,706	8,877	9,300	9,741	10,225

a. Run a linear trend model on the tuition at CCSU.

b. Run a quadratic trend model on the tuition at CCSU.

c. Which of these models best fit the data?

d. Using the best fitting model, forecast the tuition at CCSU for 2020 and 2021.

Appendix

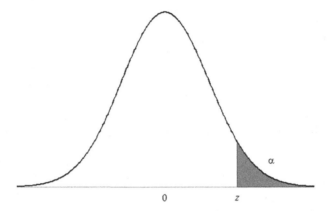

TABLE A.1

Standard Normal Distribution

z	0.00	0.01	0.02	0.03	0.04	0.05	0.06	0.07	0.08	0.09
0.00	0.5000	0.4960	0.4920	0.4880	0.4840	0.4801	0.4761	0.4721	0.4681	0.4641
0.10	0.4602	0.4562	0.4522	0.4483	0.4443	0.4404	0.4364	0.4325	0.4286	0.4247
0.20	0.4207	0.4168	0.4129	0.4090	0.4052	0.4013	0.3974	0.3936	0.3897	0.3859
0.30	0.3821	0.3783	0.3745	0.3707	0.3669	0.3632	0.3594	0.3557	0.3520	0.3483
0.40	0.3446	0.3409	0.3372	0.3336	0.3300	0.3264	0.3228	0.3192	0.3156	0.3121
0.50	0.3085	0.3050	0.3015	0.2981	0.2946	0.2912	0.2877	0.2843	0.2810	0.2776
0.60	0.2743	0.2709	0.2676	0.2643	0.2611	0.2578	0.2546	0.2514	0.2483	0.2451
0.70	0.2420	0.2389	0.2358	0.2327	0.2296	0.2266	0.2236	0.2206	0.2177	0.2148
0.80	0.2119	0.2090	0.2061	0.2033	0.2005	0.1977	0.1949	0.1922	0.1894	0.1867
0.90	0.1841	0.1814	0.1788	0.1762	0.1736	0.1711	0.1685	0.1660	0.1635	0.1611
1.00	0.1587	0.1562	0.1539	0.1515	0.1492	0.1469	0.1446	0.1423	0.1401	0.1379
1.10	0.1357	0.1335	0.1314	0.1292	0.1271	0.1251	0.1230	0.1210	0.1190	0.1170
1.20	0.1151	0.1131	0.1112	0.1093	0.1075	0.1056	0.1038	0.1020	0.1003	0.0985
1.30	0.0968	0.0951	0.0934	0.0918	0.0901	0.0885	0.0869	0.0853	0.0838	0.0823
1.40	0.0808	0.0793	0.0778	0.0764	0.0749	0.0735	0.0721	0.0708	0.0694	0.0681
1.50	0.0668	0.0655	0.0643	0.0630	0.0618	0.0606	0.0594	0.0582	0.0571	0.0559
1.60	0.0548	0.0537	0.0526	0.0516	0.0505	0.0495	0.0485	0.0475	0.0465	0.0455
1.70	0.0446	0.0436	0.0427	0.0418	0.0409	0.0401	0.0392	0.0384	0.0375	0.0367
1.80	0.0359	0.0351	0.0344	0.0336	0.0329	0.0322	0.0314	0.0307	0.0301	0.0294
1.90	0.0287	0.0281	0.0274	0.0268	0.0262	0.0256	0.0250	0.0244	0.0239	0.0233
2.00	0.0228	0.0222	0.0217	0.0212	0.0207	0.0202	0.0197	0.0192	0.0188	0.0183
2.10	0.0179	0.0174	0.0170	0.0166	0.0162	0.0158	0.0154	0.0150	0.0146	0.0143
2.20	0.0139	0.0136	0.0132	0.0129	0.0125	0.0122	0.0119	0.0116	0.0113	0.0110

(Continued)

TABLE A.1 (*Continued*)

Standard Normal Distribution

z	0.00	0.01	0.02	0.03	0.04	0.05	0.06	0.07	0.08	0.09
2.30	0.0107	0.0104	0.0102	0.0099	0.0096	0.0094	0.0091	0.0089	0.0087	0.0084
2.40	0.0082	0.0080	0.0078	0.0075	0.0073	0.0071	0.0069	0.0068	0.0066	0.0064
2.50	0.0062	0.0060	0.0059	0.0057	0.0055	0.0054	0.0052	0.0051	0.0049	0.0048
2.60	0.0047	0.0045	0.0044	0.0043	0.0041	0.0040	0.0039	0.0038	0.0037	0.0036
2.70	0.0035	0.0034	0.0033	0.0032	0.0031	0.0030	0.0029	0.0028	0.0027	0.0026
2.80	0.0026	0.0025	0.0024	0.0023	0.0023	0.0022	0.0021	0.0021	0.0020	0.0019
2.90	0.0019	0.0018	0.0018	0.0017	0.0016	0.0016	0.0015	0.0015	0.0014	0.0014
3.00	0.0013	0.0013	0.0013	0.0012	0.0012	0.0011	0.0011	0.0011	0.0010	0.0010
3.10	0.0010	0.0009	0.0090	0.0009	0.0008	0.0008	0.0008	0.0008	0.0007	0.0007
3.20	0.0007	0.0007	0.0006	0.0060	0.0006	0.0006	0.0006	0.0005	0.0005	0.0005
3.30	0.0005	0.0005	0.0005	0.0004	0.0004	0.0004	0.0004	0.0004	0.0004	0.0003
3.40	0.0003	0.0003	0.0003	0.0003	0.0003	0.0003	0.0003	0.0003	0.0003	0.0002
3.50	0.0002	0.0002	0.0002	0.0002	0.0002	0.0002	0.0002	0.0002	0.0002	0.0002
3.60	0.0002	0.0002	0.0001	0.0001	0.0001	0.0001	0.0001	0.0001	0.0001	0.0001
3.70	0.0001	0.0001	0.0001	0.0001	0.0001	0.0001	0.0001	0.0001	0.0001	0.0001
3.80	0.0001	0.0001	0.0001	0.0001	0.0001	0.0001	0.0001	0.0001	0.0001	0.0001
3.90	0.0000	0.0000	0.0000	0.0000	0.0000	0.0000	0.0000	0.0000	0.0000	0.0000
4.00	0.0000	0.0000	0.0000	0.0000	0.0000	0.0000	0.0000	0.0000	0.0000	0.0000

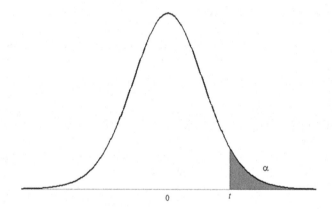

TABLE A.2

T-Distribution

	$\alpha = 0.10$	$\alpha = 0.05$	$\alpha = 0.025$	$\alpha = 0.01$	$\alpha = 0.005$
$df = 1$	3.078	6.314	12.706	31.821	63.657
2	1.886	2.920	4.303	6.965	9.925
3	1.638	2.353	3.182	4.541	5.841
4	1.533	2.132	2.776	3.747	4.604
5	1.476	2.015	2.571	3.365	4.032
6	1.440	1.943	2.447	3.143	3.707

(*Continued*)

TABLE A.2 (*Continued*)

T-Distribution

	$\alpha = 0.10$	$\alpha = 0.05$	$\alpha = 0.025$	$\alpha = 0.01$	$\alpha = 0.005$
7	1.415	1.895	2.365	2.998	3.499
8	1.397	1.860	2.306	2.896	3.355
9	1.383	1.833	2.262	2.821	3.250
10	1.372	1.812	2.228	2.764	3.169
11	1.363	1.796	2.201	2.718	3.106
12	1.356	1.782	2.179	2.681	3.055
13	1.350	1.771	2.160	2.650	3.012
14	1.345	1.761	2.145	2.624	2.977
15	1.341	1.753	2.131	2.602	2.947
16	1.337	1.746	2.120	2.583	2.921
17	1.333	1.740	2.110	2.567	2.898
18	1.330	1.734	2.101	2.552	2.878
19	1.328	1.729	2.093	2.539	2.861
20	1.325	1.725	2.086	2.528	2.845
21	1.323	1.721	2.080	2.518	2.831
22	1.321	1.717	2.074	2.508	2.819
23	1.319	1.714	2.069	2.500	2.807
24	1.318	1.711	2.064	2.492	2.797
25	1.316	1.708	2.060	2.485	2.787
26	1.315	1.706	2.056	2.479	2.779
27	1.314	1.703	2.052	2.473	2.771
28	1.313	1.701	2.048	2.467	2.763
29	1.311	1.699	2.045	2.462	2.756
30	1.310	1.697	2.042	2.457	2.750
35	1.306	1.690	2.030	2.438	2.724
40	1.303	1.684	2.021	2.423	2.704
45	1.301	1.679	2.014	2.412	2.690
50	1.299	1.676	2.009	2.403	2.678
100	1.290	1.660	1.984	2.364	2.626
∞	1.282	1.645	1.960	2.326	2.576

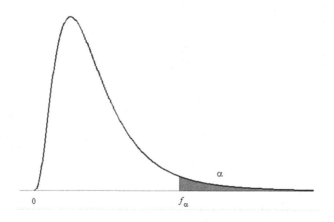

TABLE A.3

F-Distribution

($\alpha = 0.10$) $df_{denom} = 1$	$df_{num} = 1$	2	3	4	5	6	7	8	9	10	25	50	100	∞
1	39.863	49.500	53.593	55.833	57.240	58.204	58.906	59.439	59.858	60.195	62.055	62.688	63.007	63.328
2	8.526	9.000	9.162	9.243	9.293	9.326	9.349	9.367	9.381	9.392	9.451	9.471	9.481	9.491
3	5.538	5.462	5.391	5.343	5.309	5.285	5.266	5.252	5.240	5.230	5.175	5.155	5.144	5.134
4	4.545	4.325	4.191	4.107	4.051	4.010	3.979	3.955	3.936	3.920	3.828	3.795	3.778	3.761
5	4.060	3.780	3.619	3.520	3.453	3.405	3.368	3.339	3.316	3.297	3.187	3.147	3.126	3.105
6	3.776	3.463	3.289	3.181	3.108	3.055	3.014	2.983	2.958	2.937	2.815	2.770	2.746	2.722
7	3.589	3.257	3.074	2.961	2.883	2.827	2.785	2.752	2.725	2.703	2.571	2.523	2.497	2.471
8	3.458	3.113	2.924	2.806	2.726	2.668	2.624	2.589	2.561	2.538	2.400	2.348	2.321	2.293
9	3.360	3.006	2.813	2.693	2.611	2.551	2.505	2.469	2.440	2.416	2.272	2.218	2.189	2.159
10	3.285	2.924	2.728	2.605	2.522	2.461	2.414	2.377	2.347	2.323	2.174	2.117	2.087	2.055
11	3.225	2.860	2.660	2.536	2.451	2.389	2.342	2.304	2.274	2.248	2.095	2.036	2.005	1.972
12	3.177	2.807	2.606	2.480	2.394	2.331	2.283	2.245	2.214	2.188	2.031	1.970	1.938	1.904
13	3.136	2.763	2.560	2.434	2.347	2.283	2.234	2.195	2.164	2.138	1.978	1.915	1.882	1.846
14	3.102	2.726	2.522	2.395	2.307	2.243	2.193	2.154	2.122	2.095	1.933	1.869	1.834	1.797
15	3.073	2.695	2.490	2.361	2.273	2.208	2.158	2.119	2.086	2.059	1.894	1.828	1.793	1.755
16	3.048	2.668	2.462	2.333	2.244	2.178	2.128	2.088	2.055	2.028	1.860	1.793	1.757	1.718
17	3.026	2.645	2.437	2.308	2.218	2.152	2.102	2.061	2.028	2.001	1.831	1.763	1.726	1.686
18	3.007	2.624	2.416	2.286	2.196	2.130	2.079	2.038	2.005	1.977	1.805	1.736	1.698	1.657
19	2.990	2.606	2.397	2.266	2.176	2.109	2.058	2.017	1.984	1.956	1.782	1.711	1.673	1.631
20	2.975	2.589	2.380	2.249	2.158	2.091	2.040	1.999	1.965	1.937	1.761	1.690	1.650	1.607
25	2.918	2.528	2.317	2.184	2.092	2.024	1.971	1.929	1.895	1.866	1.683	1.607	1.565	1.518
50	2.809	2.412	2.197	2.061	1.966	1.895	1.840	1.796	1.760	1.729	1.529	1.441	1.388	1.327
100	2.756	2.356	2.139	2.002	1.906	1.834	1.778	1.732	1.695	1.663	1.453	1.355	1.293	1.214
∞	2.706	2.303	2.084	1.945	1.847	1.774	1.717	1.670	1.632	1.599	1.375	1.263	1.185	1.000

(*Continued*)

TABLE A.3 (Continued)

F-Distribution

($\alpha = 0.05$)	$df_{num} = 1$	2	3	4	5	6	7	8	9	10	25	50	100	∞
$df_{denom} = 1$	161.448	199.500	215.707	224.583	230.162	233.986	236.768	238.883	240.543	241.882	249.260	251.774	253.041	254.314
2	18.513	19.000	19.164	19.247	19.296	19.330	19.353	19.371	19.385	19.396	19.456	19.476	19.486	19.496
3	10.128	9.552	9.277	9.117	9.013	8.941	8.887	8.845	8.812	8.786	8.634	8.581	8.554	8.526
4	7.709	6.944	6.591	6.388	6.256	6.163	6.094	6.041	5.999	5.964	5.769	5.699	5.664	5.628
5	6.608	5.786	5.409	5.192	5.050	4.950	4.876	4.818	4.772	4.735	4.521	4.444	4.405	4.365
6	5.987	5.143	4.757	4.534	4.387	4.284	4.207	4.147	4.099	4.060	3.835	3.754	3.712	3.669
7	5.591	4.737	4.347	4.120	3.972	3.866	3.787	3.726	3.677	3.637	3.404	3.319	3.275	3.230
8	5.318	4.459	4.066	3.838	3.687	3.581	3.500	3.438	3.388	3.347	3.108	3.020	2.975	2.928
9	5.117	4.256	3.863	3.633	3.482	3.374	3.293	3.230	3.179	3.137	2.893	2.803	2.756	2.707
10	4.965	4.103	3.708	3.478	3.326	3.217	3.135	3.072	3.020	2.978	2.730	2.637	2.588	2.538
11	4.844	3.982	3.587	3.357	3.204	3.095	3.012	2.948	2.896	2.854	2.601	2.507	2.457	2.404
12	4.747	3.885	3.490	3.259	3.106	2.996	2.913	2.849	2.796	2.753	2.498	2.401	2.350	2.296
13	4.667	3.806	3.411	3.179	3.025	2.915	2.832	2.767	2.714	2.671	2.412	2.314	2.261	2.206
14	4.600	3.739	3.344	3.112	2.958	2.848	2.764	2.699	2.646	2.602	2.341	2.241	2.187	2.131
15	4.543	3.682	3.287	3.056	2.901	2.790	2.707	2.641	2.588	2.544	2.280	2.178	2.123	2.066
16	4.494	3.634	3.239	3.007	2.852	2.741	2.657	2.591	2.538	2.494	2.227	2.124	2.068	2.010
17	4.451	3.592	3.197	2.965	2.810	2.699	2.614	2.548	2.494	2.450	2.181	2.077	2.020	1.960
18	4.414	3.555	3.160	2.928	2.773	2.661	2.577	2.510	2.456	2.412	2.141	2.035	1.978	1.917
19	4.381	3.522	3.127	2.895	2.740	2.628	2.544	2.477	2.423	2.378	2.106	1.999	1.940	1.878
20	4.351	3.493	3.098	2.866	2.711	2.599	2.514	2.447	2.393	2.348	2.074	1.966	1.907	1.843
25	4.242	3.385	2.991	2.759	2.603	2.490	2.405	2.337	2.282	2.236	1.955	1.842	1.779	1.711
50	4.034	3.183	2.790	2.557	2.400	2.286	2.199	2.130	2.073	2.026	1.727	1.599	1.525	1.438
100	3.936	3.087	2.696	2.463	2.305	2.191	2.103	2.032	1.975	1.927	1.616	1.477	1.392	1.283
∞	3.841	2.996	2.605	2.372	2.214	2.099	2.010	1.938	1.880	1.831	1.506	1.350	1.243	1.000

(Continued)

TABLE A.3 (Continued)

F-Distribution

($\alpha = 0.01$) $df_{denom} = 1$	$df_{num} = 1$	2	3	4	5	6	7	8	9	10	25	50	100	∞
	4052.181	4999.500	5403.352	5624.583	5763.650	5858.986	5928.356	5981.070	6022.473	6055.847	6239.825	6302.517	6334.110	6365.864
2	98.503	99.000	99.166	99.249	99.299	99.333	99.356	99.374	99.388	99.399	99.459	99.479	99.489	99.499
3	34.116	30.817	29.457	28.710	28.237	27.911	27.672	27.489	27.345	27.229	26.579	26.354	26.240	26.125
4	21.198	18.000	16.694	15.977	15.522	15.207	14.976	14.799	14.659	14.546	13.911	13.690	13.577	13.463
5	16.258	13.274	12.060	11.392	10.967	10.672	10.456	10.289	10.158	10.051	9.449	9.238	9.130	9.020
6	13.745	10.925	9.780	9.148	8.746	8.466	8.260	8.102	7.976	7.874	7.296	7.091	6.987	6.880
7	12.246	9.547	8.451	7.847	7.460	7.191	6.993	6.840	6.719	6.620	6.058	5.858	5.755	5.650
8	11.259	8.649	7.591	7.006	6.632	6.371	6.178	6.029	5.911	5.814	5.263	5.065	4.963	4.859
9	10.561	8.022	6.992	6.422	6.057	5.802	5.613	5.467	5.351	5.257	4.713	4.517	4.415	4.311
10	10.044	7.559	6.552	5.994	5.636	5.386	5.200	5.057	4.942	4.849	4.311	4.115	4.014	3.909
11	9.646	7.206	6.217	5.668	5.316	5.069	4.886	4.744	4.632	4.539	4.005	3.810	3.708	3.602
12	9.330	6.927	5.953	5.412	5.064	4.821	4.640	4.499	4.388	4.296	3.765	3.569	3.467	3.361
13	9.074	6.701	5.739	5.205	4.862	4.620	4.441	4.302	4.191	4.100	3.571	3.375	3.272	3.165
14	8.862	6.515	5.564	5.035	4.695	4.456	4.278	4.140	4.030	3.939	3.412	3.215	3.112	3.004
15	8.683	6.359	5.417	4.893	4.556	4.318	4.142	4.004	3.895	3.805	3.278	3.081	2.977	2.868
16	8.531	6.226	5.292	4.773	4.437	4.202	4.026	3.890	3.780	3.691	3.165	2.967	2.863	2.753
17	8.400	6.112	5.185	4.669	4.336	4.102	3.927	3.791	3.682	3.593	3.068	2.869	2.764	2.653
18	8.285	6.013	5.092	4.579	4.248	4.015	3.841	3.705	3.597	3.508	2.983	2.784	2.678	2.566
19	8.185	5.926	5.010	4.500	4.171	3.939	3.765	3.631	3.523	3.434	2.909	2.709	2.602	2.489
20	8.096	5.849	4.938	4.431	4.103	3.871	3.699	3.564	3.457	3.368	2.843	2.643	2.535	2.421
25	7.770	5.568	4.675	4.177	3.855	3.627	3.457	3.324	3.217	3.129	2.604	2.400	2.289	2.169
50	7.171	5.057	4.199	3.720	3.408	3.186	3.020	2.890	2.785	2.698	2.167	1.949	1.825	1.683
100	6.895	4.824	3.984	3.513	3.206	2.988	2.823	2.694	2.590	2.503	1.965	1.735	1.598	1.427
∞	6.635	4.605	3.782	3.319	3.017	2.802	2.639	2.511	2.407	2.321	1.773	1.523	1.358	1.000

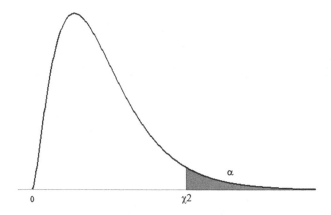

TABLE A.4

Chi-Square Distribution

df	$\alpha = 0.10$	$\alpha = 0.05$	$\alpha = 0.025$	$\alpha = 0.01$	$\alpha = 0.005$
1	2.7055	3.8415	5.0239	6.6349	7.8794
2	4.6052	5.9915	7.3778	9.2103	10.5966
3	6.2514	7.8147	9.3484	11.3449	12.8382
4	7.7794	9.4877	11.1433	13.2767	14.8603
5	9.2364	11.0705	12.8325	15.0863	16.7496
6	10.6446	12.5916	14.4494	16.8119	18.5476
7	12.0170	14.0671	16.0128	18.4753	20.2777
8	13.3616	15.5073	17.5345	20.0902	21.9550
9	14.6837	16.9190	19.0228	21.6660	23.5894
10	15.9872	18.3070	20.4832	23.2093	25.1882
11	17.2750	19.6751	21.9200	24.7250	26.7568
12	18.5493	21.0261	23.3367	26.2170	28.2995
13	19.8119	22.3620	24.7356	27.6882	29.8195
14	21.0641	23.6848	26.1189	29.1412	31.3193
15	22.3071	24.9958	27.4884	30.5779	32.8013
16	23.5418	26.2962	28.8454	31.9999	34.2672
17	24.7690	27.5871	30.1910	33.4087	35.7185
18	25.9894	28.8693	31.5264	34.8053	37.1565
19	27.2036	30.1435	32.8523	36.1909	38.5823
20	28.4120	31.4104	34.1696	37.5662	39.9968
21	29.6151	32.6706	35.4789	38.9322	41.4011
22	30.8133	33.9244	36.7807	40.2894	42.7957
23	32.0069	35.1725	38.0756	41.6384	44.1813
24	33.1962	36.4150	39.3641	42.9798	45.5585
25	34.3816	37.6525	40.6465	44.3141	46.9279
26	35.5632	38.8851	41.9232	45.6417	48.2899
27	36.7412	40.1133	43.1945	46.9629	49.6449
28	37.9159	41.3371	44.4608	48.2782	50.9934
29	39.0875	42.5570	45.7223	49.5879	52.3356

(Continued)

TABLE A.4 (*Continued*)

Chi-Square Distribution

df	$\alpha = 0.10$	$\alpha = 0.05$	$\alpha = 0.025$	$\alpha = 0.01$	$\alpha = 0.005$
30	40.2560	43.7730	46.9792	50.8922	53.6720
35	46.0588	49.8018	53.2033	57.3421	60.2748
40	51.8051	55.7585	59.3417	63.6907	66.7660
45	57.5053	61.6562	65.4102	69.9568	73.1661
50	63.1671	67.5048	71.4202	76.1539	79.4900
55	68.7962	73.3115	77.3805	82.2921	85.7490
60	74.3970	79.0819	83.2977	88.3794	91.9517
65	79.9730	84.8206	89.1771	94.4221	98.1051
70	85.5270	90.5312	95.0232	100.4252	104.2149
75	91.0615	96.2167	100.8393	106.3929	110.2856
80	96.5782	101.8795	106.6286	112.3288	116.3211
85	102.0789	107.5217	112.3934	118.2357	122.3246
90	107.5650	113.1453	118.1359	124.1163	128.2989
95	113.0377	118.7516	123.8580	129.9727	134.2465
100	118.4980	124.3421	129.5612	135.8067	140.1695

TABLE A.5

Approximate Rejection Regions for the Ryan–Joiner Test of Normality

n	$\alpha = 0.10$	$\alpha = 0.05$	$\alpha = 0.01$
4	0.8951	0.8734	0.8318
5	0.9033	0.8804	0.8320
10	0.9347	0.9180	0.8804
15	0.9506	0.9383	0.9110
20	0.9600	0.9503	0.9290
25	0.9662	0.9582	0.9408
30	0.9707	0.9639	0.9490
40	0.9767	0.9715	0.9597
50	0.9807	0.9764	0.9644
60	0.9835	0.9799	0.9710
75	0.9865	0.9835	0.9757

Source: Minitab website
(https://www.minitab.com/uploadedFiles/Content/News/Published_Articles/normal_probability_plots.pdf).

Index

A

Alternative hypothesis, 101, 149
   correlation inference, 255, 256
   Kruskal–Wallis test, 400
   Mann–Whitney test, 400
   normality assumption, 265
   one-sample t-test, 101, 103
   one-way ANOVA, 344
   population regression parameters, 228–229
Analyses, types of, 9–10
Analysis of variance (ANOVA), 9–10, 341–383
   assumption of constant variance, 350–355
      Bartlett's test, 350
      χ^2 distribution, 351
      example, 352
      Levene's test, 352
      rejection of null hypothesis, 351
      test statistic, 351
   balanced two-way, 424
   basic experimental design, 341–342
      one-way analysis of variance, 342
      random assignment of brands, 342
      randomized block design, 342
      randomized design, 341
   exercises, 378–383
   F-distribution, 345
   MINITAB use for multiple comparisons, 373–376
   MINITAB use for one-way ANOVAs, 357–370
      commands, 362
      dialog box, 358, 360, 363
      example, 366
      histogram of residuals, 361, 368
      interval plot, 360, 368
      normal probability plot, 362, 368
      output, 365, 369–370
      printout, 367
      residual *versus* fitted values, 361, 368
      residual *versus* order plot, 361
      Ryan–Joiner test, 361, 362, 369
      worksheet, 400
   model assumptions, 349–350
   multiple comparison techniques, 370–373
      confidence interval, interpretation of, 373
      example, 371
      Fisher's Least Significant Difference (LSD), 371
      MINITAB use, 373–376
      t-distribution, 371
   normality assumption, 355–357
      example, 356
      rejection of null hypothesis, 357
   one-way ANOVA, 342–349
      alternative hypothesis, 344
      balanced, 343
      degrees of freedom for the denominator, 344
      degrees of freedom for the numerator, 344
      example, 345
      factor, 343
      F-distribution, 344
      fixed-effects ANOVA model, 343
      mean square error, 347
      null hypothesis, 344
      samples, 343
      table, 349
   power analysis and one-way ANOVA, 374–378
      example, 376
      MINITAB printout, 377
      rejection of null hypothesis, 374
      sample size estimation, 375
   randomized block design, 342
Analysis of variance table, 292–296
   degrees of freedom for the denominator, 292–293
   degrees of freedom for the numerator, 292–293
   example, 293
   F-distribution, 292–293, 294
   F-test, 292
   printout, 294
ANOVA, *see* Analysis of variance
Automatic setting, in MINITAB, 18

B

Balanced two-way ANOVA, 424
Bar charts, 24
   discrete data, 24

Bar charts (*cont.*)
    MINTAB use, 24–27
    variables, 24
Bartlett's test, 350
Basic statistical inference, 9
Basic time series, *see also* Time series analysis,
        basic
    exercises, 449–452
Best practices, 11–12, 201–203
    graphs and charts, 38, 40, 41
Best subsets regression, 324–329
    example, 325
    full regression model, 324
    Mallow's C_p statistic, 324
    MINITAB use, 329–331
    model fitting, 331
    predictor variables, 324
    regression analysis including variables, 326,
        327, 328
    statistical modeling using, 325
Binary variables, 314
Binomial distribution, 74–75
Box plots, 27–31
    construction, 27
    example, 28–31
    general form, 28
    Kruskal–Wallis test (MINITAB), 410
    median, 29
    quartiles, 27
    two-way ANOVA, 434, 430
    upper and lower limits, 27–28
    whiskers, 27, 30

C

Cartesian plane, 33
Categorical variables, 24
Central limit theorem, 64
Chi-square distribution, 78–79, 130, 131
Coefficient of correlation, 250–253
    example, 252
    formula, 250
    linear relationship, 253
    linear relationship between variables, 250
    negative relationship, 250
    no relationship, 250
    positive relationship, 250
    sample standard deviation, 251
    scatter plot, 253
Coefficient of determination, 247–249
    example, 248
    MINITAB use, 249
    predictor variable, 247

    sample mean, 247
    SAT–GPA data set, 247
    scatter plot, 248
Coefficient of variation (COV), 88
Column factor, 417
Columns, of data set, 1
Conceptual populations, 5
Confidence interval, 93–99
    calculation, 95–96
    degrees of freedom, 94
    for difference between two means, 157–160
        calculation, 157
        degrees of freedom, 159
        example, 158
        hypothesis tests, 157
        MINITAB calculation, 160–162
        population mean lifetimes, 159
        unequal variances, 158
    example, 96–97
    hypothesis tests for proportions and,
        120–124
        distribution of sample proportion, 121
        example, 121
        population proportion, 120–121
        p-value, 123
        rejection region, 123
        standard normal tables, 123
        test statistic, 122, 124
    MINITAB calculation, 99–100
        commands, 99
        dialog box, 100
        printout, 100
    for one-sample count variable, 140–142
    for one-sample variance, 129–132
        example, 132–133
    point estimate, 93
    t-distribution, 94, 98
    theory for population mean, 94
    and t-test, 172–176
    two-count variables, and hypothesis tests
        for, 195–197
        example, 195
    two variances, hypothesis tests for, 184–191
    unknown population mean, 95
    use, 93
Continuous random variable, 63
Continuous variables, 6, 24
Control group, 5
Control variables, 285
Cook's distance, 274–275
Correlation analysis, use of MINITAB for,
    257–258
Correlation inference, 254–257

example, 256, 257
negative linear correlation, 254
null and alternative hypotheses, 255, 256
population coefficient of correlation, 254
positive linear correlation, 254
rejection region, 255, 256, 257
sampling distribution, 254
test statistic, 255
true population coefficient of correlation, 255
Count variable, 140
COV, *see* Coefficient of variation
Cyclical trend, time series analysis, 441

D

Data, 1
descriptive representations of, *see*
Descriptive statistics
qualitative, 1
quantitative, 1
residual, 262
Data sets, 1
box plots, 32
kurtosis of, 91
SAT–GPA, 247
skewness of, 91
Degrees of freedom, 94
for the denominator, 292–293, 344
for the numerator, 292–293, 344
Dependent populations, 172
Dependent variable, 32, 33
Descriptive representations, of data, 2, 3–4
Descriptive statistics, 47–91, 188
binomial distribution, 74–75
chi-square distribution, 78–79
coefficient of variation, 88
definition of, 47
discrete random variables, 58–60
probability distribution, 58
representation, 59
summarized, 59
exercises, 85–91
F-distribution, 79
kurtosis of data set, 91
measures of central tendency, 48–52
definition of, 48
example, 48, 50–52
median of numeric variable, 50
median position, 50
mode, 51
population mean, 49
sample average, 48
summation notation, 48

measures of variability, 52–55
example, 52
interquartile range, 53
population standard deviation, 55
population variance, 55
range, 52
sample standard deviation, 53
sample variance, 53
MINITAB, 55
dialog box, 56
output, 57
statistic properties, 57
mode, 51
Poisson distribution, 75–76
probability distribution, 73
purpose of calculating, 47
range, 52
sampling distributions, 61–64, 74
area under the curve, 65
central limit theorem, 64
continuous random variable, 64
example, 61
graph, 63
nonstandard normal distribution, 69–73
normal distribution, 65–69
population parameter, 63
probability distribution, 60, 62
sample mean, 62
standard normal distribution, 65–69
standard normal table, 67
skewness of data set, 91
standard error, 63
standard normal distribution, 65–69
t-distribution, 77–78
types, 52
weighted mean, 89
Discrete variables, 6
random, 58–60
Distribution-free tests, 385

E

Equation of line of best fit, finding, 221–224
formulas, 221
mean square error, 223
population parameters, 221
residuals, 221–223
root mean square error, 223
statistics, 221
unknown population parameters, 222
Error(s)
component, 216, 223
mean square error, 223

Error(s) (*cont.*)
   observed values, 219
   residual, mean square error due to, 292
   root mean square error, 223
   round-off, 216
   standard, 63
   Type I, 103
Exercises
   analysis of variance, 378–383
   descriptive statistics, 85–91
   graphs and charts, 41–44
   introduction, 12–13
   multiple regression analysis, 306–311,
      334–339
   nonparametric statistics, 411–416
   simple linear regression, 277–283
   simple regression, 242–246
   statistical inference, 151–155
      for two-sample data, 203–211
   two-way analysis of variance and basic time
      series, 449–452
Experimental studies, 5

F

F-distribution, 79, 292, 344, 345, 418
Fisher's Least Significant Difference (LSD), 371
Fitted line plot, 235, 236, 237, 241, 260
Fixed-effects ANOVA model, 343
Four-way residual plots, 409, 437
Frequency distribution, and histogram, 15
F-tables, 187
F-test, 292, 296
Full regression model, 324

G

GPAs, *see* Grade point averages
Grade point averages (GPAs), 47
Graphical displays, 2
Graphs and charts, 15–46
   bar charts, 24
      discrete data, 24
      MINITAB, 24–25
      variables, 24
   box plots, 27–31
      construction, 27, 28
      example, 28–31
      general form, 28
      median, 29
      MINITAB, 31
      multiple, 31
      quartiles, 27

      upper and lower limits, 27–28
      whiskers, 27, 30
   exercises, 41–44
   frequency distribution, histogram drawn
      from, 15
   histograms, 15–17
      construction, 16
      frequency distribution, histogram drawn
         from, 15
      MINITAB, 17–21
      purpose of drawing, 16
      software use, 17
   marginal plots, 33–35
   matrix plots, 36–38
   scatter plots, 32–33
      Cartesian plane, 32
      data set, 33
      example, 32
      MINITAB, 33
      predictor, 32
      response, 32
      variables, 32
   stem-and-leaf plots, 21–22
      example, 21
      MINITAB, 22–24
      purpose, 21
      stem, 23

H

Histogram(s), 15–17
   construction, 16
   frequency distribution, histogram drawn
      from, 15
   marginal plot with, 36, 214
   MINITAB, 17
   purpose of drawing, 16
   residuals
      one-way ANOVAs, 361, 368
      two-way analysis of variance, 428
      values, 262, 305
   software use, 17
   Wilcoxon signed-rank test, 390, 392
Hypothesis, 175
   alternative, 101
   analysis of variance, 351, 359
   correlation inference, 255, 256
   Kruskal–Wallis test, 400
   Mann–Whitney test, 395
   normality assumption, 265
   null, 101
   for one-sample count variable, 142–144
   one-sample t-test, 101, 103

for one-sample variance, 132–133
one-way ANOVA, 344–345
population regression parameters, 228
population slope parameter, 231
test, 101, 102, 157
two-count variables, and confidence
 intervals for, 195–197
 example, 195
two variances, confidence interval for,
 184–191

I

Independent variable, 32, 33
Indicator variables, 314
Inferential statistics, 2, 4
Interquartile range (IQR), 52
Interval plots, one-way ANOVAs, 360
Interval variable, 8
IQR, *see* Interquartile range
*i*th residual, 217

K

Kruskal–Wallis test, 400–405
 χ^2 distribution, 404
 example, 402
 MINITAB, 405–411
  box plot, 411
  commands, 405
  data entry, 405
  dialog box, 406
  example, 407
  four-way residual plots, 409
  Levene's tests, 409
  one-way ANOVA, 409
  printout, 405, 410
  p-value, 405
  Ryan–Joiner test for normality, 410
 null and alternative hypotheses, 402
 ranking of data, 400
 rule of thumb, 405
 test statistic, 403, 404
Kurtosis of data set, 91

L

Least Significant Difference (LSD), 371
Least squares line, 219
Left-tailed test, 103
Level of significance, 102
Levene's test, 352, 409
Leverage value, 269

Linear forecast model, time series analysis, 444
Linear regression, *see* Simple linear regression
Line of best fit, 219
Long-term trend, time series analysis, 441
Lowess smoother, 288
LSD, *see* Least Significant Difference

M

MAD, *see* Mean absolute deviation
Mallow's C_p statistic, 324
Mann–Whitney test, 395–400
 example, 395
 MINITAB, 400–402
  commands, 401
  dialog box, 401
  results, 402
  worksheet, 400
 ranking of data, 403
 test statistic, 396–397
MAPE, *see* Mean absolute percentage error
Marginal plots, 33–35
Matrix plots, 36–38, 287
Mean absolute deviation (MAD), 447
Mean absolute percentage error (MAPE), 445
Mean squared deviation (MSD), 448
Mean square error, 223
Mean square error due to the regression (MSR),
 292–294
Mean square error due to the residual error
 (MSE), 292–294
Mean square for the treatment (MSTR), 346
Measures of central tendency, 48–52
 definition of, 48
 example, 48, 50–52
 median of numeric variable, 50
 median position, 50
 mode, 51
 population mean, 49
 sample average, 48
 summation notation, 48
Measures of variability, 52–55
 example, 52, 53
 interquartile range, 53
 population standard deviation, 55
 population variance, 55
 range, 52
 sample standard deviation, 53
 sample variance, 53
MINITAB
 bar chart, 24–27
  categorical variable, 24, 25, 27
  commands, 18, 32

MINITAB (*cont.*)
    dialog box, 27
    example, 28
    type selection, 19
    worksheet, 24
    best subsets regression, 329–331
        dialog box, 329
        printout, 330
    box plot
        dialog box, 31
        example, 28–29
        multiple box plots, 31
    categorical predictor variables, 314–323
        dialog box, 316, 319
        example, 316
        printout, 317
        regression dialog box, 320
        worksheet, 319
    coefficient of determination, 249
        error sum of squares, 249
        example, 249
        printout, 249
        total sum of squares, 249
    confidence and prediction intervals, finding
        of, 235–242
        data set, 236
        dialog box, 236
        example, 235–239
        fitted line plot, 235, 237, 241
        options box, 236
        printout, 239
        regression options tab, 237
        scatter plot, 240
    confidence and prediction intervals
        for multiple regression analysis,
        calculation of, 331–333
        options dialog box, 332
        printout, 332
    confidence interval for difference between
        two means, calculation of, 160–162
        commands, 160, 161
        dialog box, 161
        options dialog box, 161
        printout, 162
    confidence interval for one-sample count
        variable, 142–144
    confidence interval for population mean,
        99–100
        commands, 99
        dialog box, 100
        printout, 100
    Cook's distance, 274–275
    correlation analysis, 257–258

        dialog box, 258
        printout, 258
    descriptive statistics, calculation of, 55
        dialog box, 56
        output, 57
        statistic properties, 57
    difference between two means, testing of,
        166–167
        dialog box, 169
        options box, 171
        printout, 167, 172
    entering data into, 10–11
    histogram creation using, 17–21
        automatic setting in, 18
        commands for drawing, 18
        dialog box, 18, 19
        sample histogram, 19, 20
        worksheet, 17
    hypothesis tests for one-sample count
        variable, 146–147
    individual regression parameters, 299
    interval plot, 167–170
        dialog box, 169
        example, 170
    Kruskal–Wallis test, 405
        box plot, 411
        commands, 405
        data entry, 405
        dialog box, 406
        example, 407
        four-way residual plots, 409
        Levene's tests, 409
        one-way ANOVA, 408
        printout, 405, 410
        p-value, 405
        Ryan–Joiner test for normality, 410
    leverage values, calculation of, 269–272
        example, 271
        printout, 271–272
        regression analysis, 269
        storage dialog box, 270
        storage of leverage values, 270
    lowess smoother, 288
    Mann–Whitney test, 400–402
        commands, 401
        dialog box, 401
        results, 402
        worksheet, 400
    marginal plot, 33–35
        commands, 35
        dialog box, 35
        with histograms, 36
    matrix plot, 287–289

commands, 35
dialog box, 38, 288
lowess smoother, 288
smoother, 289
multiple comparisons (ANOVA), 373–376
multiple regression, 289–290
printout, 291
regression dialog box, 290
multiple regression model assumptions, 304
error component, 305–306
normal probability plot of residuals, 306
normal probability plot, Ryan–Joiner test statistic, and *p*-value, 306
residual *versus* fitted values plot, 305
for one-sample Poisson rate, 147–149
dialog box, 148
printout, 148
one-sample proportion, 124–127, 129
commands, 125
dialog box, 125, 128
options box, 126, 129
printout, 126, 129
one-sample *t*-test, 106–114
dialog box, 107
example, 109–113
options box, 107
printout, 108
p-value, 107
rejection region, 108, 111
sample mean, 109
sample standard deviation, 110
unknown population, 110
value of test statistic, 111
one-sample *t*-test, power analysis for, 116–120
commands, 117
dialog box, 117
mean difference, 116
options tab, 119
printout, 118
results, 120
sample size, 119
for one-sample variance, 134–136
one-way ANOVAs, 357–370
commands, 362
dialog box, 358, 360, 363
example, 366
histogram of residuals, 361, 368
interval plot, 360, 368
normal probability plot, 362, 369
output, 363, 369–370
printout, 367
residual *versus* fitted values, 361, 368

residual *versus* order plot, 361
Ryan–Joiner test, 361, 362, 369
worksheet, 358
population slope parameter, testing of, 230–232
assumptions, 231
confidence intervals, calculation of, 231
hypothesis test, 232
intercept of true population equation, 231
null hypothesis, 231–232
printout, 231
p-value, 231
standard error, 231
test statistic, 231
power analysis for one-sample variance, 138, 139
probability distributions
dialog box, 80
graphing, 79–82
residuals, exploratory plots of, 259–264
area under standard normal curve, 262–263
dialog box, 260
fitted line plot, 260
histogram of residuals, 260–262, 264
normal probability plot, 262
regression dialog box, 261
regression graphs box, 261
regression storage options, 259
residual data, 262
scatter plot, 260
Ryan–Joiner test, 267–268
dialog box, 267
normality plot, 268
saving of project or worksheet in, 11
scatter plot, 33
dialog box, 33
variables, 33
worksheet, 33
session and worksheet window, 11
simple linear regression, 224–226
dialog box, 224
fitted line plot, 225
printout, 226
regression dialog box, 225
scatter plot, 224
standardized residuals, calculation of, 272–273
printout, 274
regression dialog box, 273
storage of residuals, 274
statistical inference, 56–57
stem-and-leaf plot, 21–24

MINITAB (*cont.*)
    dialog box, 23
    median, 24
    time series analysis
        commands, 441
        dialog box, 442, 443, 445, 448
        output, 446
        quadratic trend analysis, 448
    for two-sample Poisson, 198–199
        commands, 198
        dialog box, 199
        option box, 198
        printout, 200
    two-sample proportion confidence intervals
        and hypothesis tests, 182–184
        commands, 183
        dialog box, 182
        options box, 182, 186
        printout, 185, 186
    two-sample *t*-test, power analysis for,
        170–172
        dialog box, 169, 171
        options box, 171
        printout, 172
    two-sample variances, power analysis for,
        193–195
        dialog box, 194
        printout, 195
    two-way ANOVA, 424
        box plot, 432, 438
        commands, 425, 429
        dialog box, 425, 429
        example, 431
        four-way residual plots, 437
        histogram of residual plots, 428
        interaction plot, 426, 436, 439
        main effects plot, 435, 439
        normal probability plot of residuals, 428
        printout, 432, 436
        residual plots, 434
        residual *versus* fitted values, 428
        Ryan–Joiner test, 434, 435, 438
    use for testing two sample variances,
        191–193
        dialog box, 192
        options box, 192
        printout, 193
    variance inflation factors, calculation of,
        303–304
        printout, 303
        VIF values, 304
    Wilcoxon signed-rank test, 389

        commands, 390
        dialog box, 390
        example, 392
        histogram, 391
        one-sample *t*-test, 393
        printout, 391, 394
        results, 394
        Ryan–Joiner test, 392, 392
    worksheet illustrating date, text, and
        numeric forms of data, 11
Mode, 51
MSD, *see* Mean squared deviation
MSE, *see* Mean square error due to the residual
    error
MSR, *see* Mean square error due to the
    regression
MSTR, *see* Mean square for the treatment
Multiple regression analysis, 285–311, 313–339
    adjusted R^2, 321–322
    analysis of variance table, 292–296
        example, 293
        F-distribution, 292–293, 294
        F-test, 292
        printout, 297
    basics, 285–287
        example, 286
        MINITAB use, 287–289
        model development, 286
        model fit, 285
        population linear multiple regression
            equation, 286
        predictor variables, 286
        random error component, 286
        variables, 285
    best subsets regression, 324–329
        example, 325
        full regression model, 324
        Mallow's C_p statistic, 324
        MINITAB use, 329–331
        model fitting, 331
        predictor variables, 324
        regression analysis including variables,
            326, 327, 328
        statistical modeling using, 325
    binary variables, 314
    categorical predictor variables, using,
        313–314
        characteristic of interest, 313
        example, 313–314
        MINITAB use, 315–321
    coefficient of determination for multiple
        regression, 291

confidence and prediction intervals for multiple regression, 331
   calculations, 331
   MINITAB use, 331–333
   specific value, 331
control variables, 285
degrees of freedom, 292–293
exercises, 306–311, 334–339
indicator variables, 314
lowess smoother, 288
matrix plot, 287
MINITAB use for best subsets regression, 329–331
   dialog box, 329
   printout, 330
MINITAB use for categorical predictor variables, 315–321
   categorical predictor variable, 314, 321
   dialog box, 316, 319
   example, 316
   printout, 315
   regression dialog box, 320
   worksheet, 319
MINITAB use for multiple regression, 289–290
   printout, 291
   regression dialog box, 290
MINITAB use to calculate confidence and prediction intervals for, 331–333
   options dialog box, 332
   printout, 332
MINITAB use to calculate variance inflation factors, 303–304
   printout, 303
   VIF values, 303
MINITAB use to check multiple regression model assumptions, 304
   error component, 305–306
   histogram of residual values, 305
   normal probability plot of residuals, 306
   normal probability plot, Ryan–Joiner test statistic, and p-value, 306
   residual *versus* fitted values plot, 305
MINITAB use to create matrix plot, 287–289
   dialog box, 288
   lowess smoother, 288
   matrix plot and smoother, 289
MINITAB use to test individual regression parameters, 299
multicollinearity, 300–302
   example, 300
   predictor variables, 300
   regression parameter estimates, 300

multiple regression model assumptions, 304
outliers, assessment of, 333–334
random error component, 286
testing individual population regression parameters, 296–299
   confidence interval, interpretation of, 298
   example, 297, 299
   population predictor variables, 297
   response variable, 297
   t-distribution, 298
variance inflation factors, 302–303
   calculation, 302
   example, 302
   predictor variables, 302
   printout, 299
   use of MINITAB to calculate, 303–304
   value too high, 304

N

Nominal variables, 7
Nonparametric statistics, 385–416
   exercises, 411–416
Nonstandard normal distribution, 69–73
Normality assumption, formal test of, 264–266
   error component, 264
   null and alternative hypotheses, 265
   Ryan–Joiner test, 266–268
   test statistic value, 266
   violated assumption of normality, 268
Normal probability plot, 262
Null hypothesis, 101, 150
   analysis of variance, 351, 359
   correlation inference, 255, 256
   Kruskal–Wallis test, 400
   Mann–Whitney test, 402
   normality assumption, 265
   one-sample t-test, 101, 103
   one-way ANOVA, 344
   population regression parameters, 228
   population slope parameter, 230–232

O

Observational studies, 5, 6
One-sample count variable
   confidence intervals for, 140–142
   MINITAB
      commands, 142
      dialog box, 143
      hypothesis test, 146–147
      printout, 144, 147

One-sample Poisson, 199
  power analysis for, 147–149
    dialog box, 148
    printout, 148
One-sample *t*-test, 100–106
  alternative hypothesis, 101, 103
  decision process, 102
  error types, 103
  graph, 102
  left-tailed test, 103
  level of significance, 102
  mean of random sample, 101
  null hypothesis, 101, 103
  power analysis for, 115–116
    conclusions, 115–116
    null hypothesis, 115
    power of test, 115
    test statistic, 114
  rejection region, 102, 105–106
  right-tailed test, 103
  sampling error, 101
  significance level, 103
  two-tailed test, 103
  value of test statistic, 105–106
One-sample variance
  confidence intervals for, 129–131
    example, 132–133
  hypothesis tests for, 132–133
  MINITAB, 134–136
One-tailed hypothesis test, 149–150
One-way ANOVA, 342–349
  alternative hypothesis, 344
  balanced, 343
  degrees of freedom for the denominator, 344
  degrees of freedom for the numerator, 344
  example, 345
  factor, 343
  F-distribution, 344
  fixed-effects ANOVA model, 343
  Kruskal–Wallis test, 410
  mean square error, 347
  null hypothesis, 344
  samples, 343
  table, 349
Ordinal variables, 7–8
Outliers
  assessment of, 268
    Cook's distance, 274–275
    leverage values, 269–272
    multiple regression analysis, 331–333
    standardized residuals, 272–273
    unusual observation, 268
  dealing with, 276–277

    example, 276
    MINITAB printout, 277
    unusual observation, 276

P

Paired confidence interval
  MINITAB, 176–178
    option box, 177, 178
    printout, 178, 179
  and *t*-test, 176–178
Paired *t*-test, 172
Parameter, 4
Parametric methods of statistical inference, 385
Physical populations, 5
Plots, *see* Variables, graphing of
Point estimate, 93
Poisson distribution, 75–76
Population mean count, 140
Populations, 4, 5
  coefficient of correlation, 254
  mean, 49
    confidence intervals, 94
    one-sample *t*-test for, 100–106
    unknown, 95
  regression parameters, inferences about,
      227–230
    confidence intervals, calculation of, 227
    example, 229
    null and alternative hypotheses, 229
    predictor variable, 228
    rejection region, 229
    sampling distribution, 228
    true population slope parameter, 228,
        230–232
  standard deviation, 55, 131
  variance, 55, 129, 130
Power analysis, 115
  for a one-sample Poisson, 147–149
  for one-sample variance, 136–140
    MINITAB, 137, 138
  for two-sample Poisson rate, 199–201
  for two-sample variances, 193–195
Power of test, 115
Practical significance, 201
Predictor variable(s), 228
  categorical, 313–314
    characteristic of interest, 313
    example, 313
    MINITAB use, 315–321
  confidence intervals for mean response for
      specific value of, 232–233
    example, 232–233

population mean response value, 233
predictor value, 232–233
multicollinearity, 300
multiple regression analysis, 286
prediction intervals for response for specific
value of, 233–235
calculations, 233
example, 234
graph, 235
inferences, 234
variance inflation factors, 302
Probability distribution, 73–74
random variable, 58, 60
sample statistics, 61, 63
p-value, 107, 123, 397
confidence intervals, 109, 124
Kruskal–Wallis test, 406
linear regression model assumptions, 259
one-sample *t*-test, 107
population slope parameter, 230
statistical inference, 108, 124
Wilcoxon signed-rank test, 388, 393

Q

Quadratic trend analysis (MINITAB), 448
Qualitative data, 1
Quantitative data, 1

R

Randomized block design, 342
Random sample, 4, 13, 16
Random trend, time series analysis, 441
Random variable(s), *see also* Descriptive
statistics
continuous, 64
discrete variable, 58
mean, 61
probability distribution, 58, 61
representation, 58
summarized, 59
t-distribution, 94
Range, 52
Ratio variable, 8
Regression analysis, 9, *see also* Multiple
regression analysis
Regression inference
confidence intervals, calculation of, 226
MINITAB printout, 226
parameters, 227
standard error, 227
unknown population, 227

Regression line, 219
Regression sum of squares (SSR), 293
Rejection region, 102, 133, 187
Confidence Intervals, 121
correlation inference, 255–257
one-sample *t*-test, 106–14
population regression parameters, 229–230
statistical inference, 123, 181
two-way analysis of variance, 423, 423
Wilcoxon signed-rank test, 388
Residual(s), 217
equation of line of best fit, 221–224
error, mean square error due to, 292
histogram of
one-way ANOVAs, 361, 368
two-way analysis of variance, 428
values, multiple regression model
assumptions, 305
*i*th, 217
normal probability plot of, 306
plots, linear regression model assumptions,
259
standardized, 272–273
Right-tailed test, 103, 133, 176, 178
Root mean square error, 226
Round-off error, 216
Row factor, 417
Rows, of data set, 1
Ryan–Joiner test, 355
Kruskal–Wallis test, 410
MINITAB use, 266–268
normality assumption, 266–268
one-way ANOVAs, 361, 370
statistic, 305–306
two-way ANOVA, 434, 434, 440
Wilcoxon signed-rank test, 391, 392

S

Sample, 4
average, 48
mean, 48
size, 199
standard deviation, 53
statistic, 93
Sampling distributions, 61–64, 74, 190
area under the curve, 65
central limit theorem, 64
continuous random variable, 64
example, 61
graph, 63
nonstandard normal distribution, 69–73
normal distribution, 65–69

Sampling distributions (*cont.*)
   population parameter, 63
   probability distribution, 61, 62
   sample mean, 62
   standard normal distribution, 66–69
   standard normal table, 67
Scatter plots, 32–33
   Cartesian plane, 33
   data set, 33
   example, 32
   MINITAB use, 33
   predictor, 32
   response, 32
   simple linear regression model, 214–215
   variables, 32
Seasonal trend, time series analysis, 441
Selection bias, 6
Simple linear regression, 213–246, 247–283
   analysis, 214
   assessing linear regression model
      assumptions, 259
      MINITAB use, 259
      p-values, 258
      residual plots, 259
   cause-and-effect relationship between
      variables, 214
   coefficient of correlation, 250–253
      example, 252
      formula, 250
      linear relationship, 253
      linear relationship between variables, 250
      negative relationship, 250
      no relationship, 250
      positive relationship, 250
      sample standard deviation, 250
      scatter plot, 253
   coefficient of determination, 247–249
      example, 248
      MINITAB use, 249
      predictor variable, 247
      sample mean, 247
      SAT–GPA data set, 247
      scatter plot, 248
   confidence intervals for mean response for
      specific value of predictor variable,
      232–233
      example, 232
      population mean response value, 232
      predictor value, 232–233
   correlation inference, 250–253
      example, 255, 256
      negative linear correlation, 254
      null and alternative hypotheses, 255, 256

      population coefficient of correlation, 254
      positive linear correlation, 254
      rejection region, 255, 256, 257
      sampling distribution, 254
      test statistic, 255
      true population coefficient of correlation,
      255
   dependent variable, 214
   equation of line of best fit, finding, 221–224
      formulas, 221
      mean square error, 226
      population parameters, 221
      residuals, 221–223
      root mean square error, 226
      statistics, 221
      unknown population parameters, 222
   exercises, 242–246, 277–283
   histogram, residual values, 261
   independent variable, 214
   least squares line, 219
   line of best fit, 219
   mean square error, 226
   MINITAB use for correlation analysis,
      257–258
      dialog box, 258
      printout, 258
   MINITAB use for Ryan–Joiner test, 266–268
   MINITAB use for simple linear regression,
      224–226
      dialog box, 224
      fitted line plot, 225
      printout, 226
      regression dialog box, 225
      scatter plot, 224
   MINITAB use to calculate leverage values,
      269–272
      example, 271
      printout, 271–272
      regression analysis, 272
      storage dialog box, 270
      storage of leverage values, 270
   MINITAB use to calculate standardized
      residuals, 272–273
      printout, 274
      regression dialog box, 273
      storage of residuals, 274
   MINITAB use to create exploratory plots of
      residuals, 259–264
      area under standard normal curve,
      262–263
      dialog box, 260
      fitted line plot, 260
      histogram of residuals, 260, 261, 264

normal probability plot, 262
regression dialog box, 261
regression graphs box, 261
regression storage options, 259
residual data, 262
scatter plot, 260
MINITAB use to find coefficient of
determination, 249
error sum of squares, 249
example, 249
printout, 249
total sum of squares, 249
MINITAB use to find Cook's distance,
275–276
MINITAB use to find confidence and
prediction intervals, 235–242
confidence and prediction intervals, 241
data set, 241
dialog box, 236
example, 235–236
fitted line plot, 235, 237, 241
options box, 236
printout, 240
regression options tab, 237
scatter plot, 241
MINITAB use to test population slope
parameter, 230–232
assumptions, 231
confidence intervals, calculation of,
231–232
hypothesis test, 232
intercept of true population equation, 231
null hypothesis, 231
printout, 231
p-value, 231
standard error, 231
test statistic, 231
model assumptions, 220–221
errors, 220–221
population error component, 220–221
normality assumption, formal test of,
266–268
error component, 264
null and alternative hypotheses, 265
Ryan–Joiner test, 266–268
test statistic value, 266
violated assumption of normality, 268
normal probability plot, 262
normal score, 262
outliers, assessment of, 266–268
Cook's distance, 274–275
leverage values, 269–272
standardized residuals, 272–273

unusual observation, 268
outliers, dealing with, 276–277
example, 276
MINITAB printout, 277
unusual observation, 276
population regression parameters,
inferences about, 227–230
confidence intervals, calculation of, 227
example, 229
null and alternative hypotheses, 229
predictor variable, 228
rejection region, 229
sampling distribution, 228
true population slope parameter, 228,
230–232
prediction intervals for response for specific
value of predictor variable, 233–235
calculations, 233
example, 234
graph, 235
inferences, 234
predictor variable, 214
regression inference
confidence intervals, calculation of, 224
MINITAB printout, 226
parameters, 227
standard error, 227
unknown population, 227
regression line, 219
residual, 217
response variable, 214
root mean square error, 226
SAT–GPA data set, 213–214
simple linear regression model, 214–220
equation of line, 216
ith residual, 217
least squares line, 219
line of best fit, 219
line usefulness, 217
marginal plot with histograms, 214
regression line, 219
scatter plot, 214–215
unknown population linear equation,
216, 219–220
vertical distance, 217–218
simple regression analysis, 214
standard errors for estimated regression
parameters, 227
confidence intervals, calculation of, 228
standard error, 227
unknown population parameters, 227
standardized residuals, 272–273
unusual observations, 268

Slope-intercept form, 216
SSE, *see* Sum of the squares of the residual
        errors
SSR, *see* Regression sum of squares
SSTR, *see* Sum of squares for the treatment
Standard error, 63
    for estimated regression parameters, 227
        confidence intervals, calculation of, 227
        standard error, 227
        unknown population parameters, 227
Standardized residuals, 272–273
    MINITAB calculation, 273–274
Standard normal distribution, 66–69
Standard normal table, 67, 123
Statistical inference, 56–57
    confidence interval, 93–99
        calculation, 95–96
        degrees of freedom, 94
        example, 96–97
        MINITAB, 99–100
        point estimate, 93
        t-distribution, 94, 98
        theory for population mean, 94
        unknown population mean, 95
        use, 93
    confidence interval, difference between two
            means, 157–160
        calculation, 157
        degrees of freedom, 159
        example, 158
        hypothesis tests, 155
        MINITAB use to calculate, 160–162
        population mean lifetimes, 159
        unequal variances, 158
    confidence intervals, hypothesis tests for
            proportions and, 120–124
        distribution of sample proportion, 121
        example, 121
        population proportion, 120–121
        p-value, 123
        rejection region, 123
        standard normal tables, 123
        test statistic, 122, 126
    difference between two means, testing of,
            162–166
        degrees of freedom, 163
        descriptive statistics, 165
        hypothesis test, 162
        MINITAB, 166–167
        pooled standard deviation, 165
        rejection region, 166
        test statistic, 163, 165
    differences between two proportions, 178–182

        example, 179
        formula, 179
        p-value, 182
        rejection region, 181
        sampling distribution, 178
    exercises, 151–155
    hypothesis testing (one-sample t-test for
            population mean), 100–106
    interval plot, 167–170
    MINITAB use for one-sample count variable
        commands, 142
        dialog box, 143
        hypothesis test, 146–147
        printout, 144, 147
    MINITAB use for one-sample proportion,
            124–127
        commands, 125
        dialog box, 125, 128
        options box, 126, 128
        printout, 126, 128
    MINITAB use for one-sample t-test, 106–114
        dialog box, 107
        example, 109–113
        options box, 107
        printout, 108
        p-value, 108
        rejection region, 108, 111
        sample mean, 110
        sample standard deviation, 110
        unknown population, 110
        value of test statistic, 112
    MINITAB use for one-sample variance,
            134–136
        dialog box, 134, 136
        options box, 135
        printout, 135, 137
        summarized data, 134
    MINITAB use for power analysis for one-
            sample t-test, 116–120
        commands, 117
        dialog box, 117
        mean difference, 116
        options tab, 119
        printout, 118
        results, 120
        sample size, 116, 118–119
    MINITAB use for two-sample proportion
            confidence intervals and hypothesis
            tests, 182–184
        commands, 183
        dialog box, 182
        options box, 184, 185
        printout, 185, 186

MINITAB use to calculate confidence interval for difference between two means, 160–162
  commands, 160
  dialog box, 161
  options dialog box, 161
  printout, 162
MINITAB use to calculate confidence intervals for population mean, 99–100
  commands, 99
  dialog box, 100
  printout, 100
  summary data, 99
MINITAB use to create interval plot, 167–170
  commands, 160
  dialog box, 169
  example, 170
for one sample, 93–155
one-sample proportion, power analysis for, 127–129
  hypothesized proportion test, 128
  MINITAB use, 127
  printout, 128
one-sample *t*-test, 100–106
  alternative hypothesis, 101, 103, 111
  decision process, 102
  error types, 103
  graph, 102
  left-tailed test, 103
  level of significance, 102
  mean of random sample, 101
  MINITAB use, 106–114
  null hypothesis, 101, 103, 112
  rejection region, 102, 105–106
  right-tailed test, 103
  sampling error, 101
  significance level, 103
  two-tailed test, 103
  value of test statistic, 105–106, 111
one-sample *t*-test, power analysis for, 115–116
  conclusions, 115–116
  null hypothesis, 115
  power analysis, 115
  power of test, 115
  test statistic, 114
one-sample variance
  confidence interval for, 129–131
  hypothesis tests for, 132–133
paired confidence interval and *t*-test, 172–176
point estimate, 93
p-value, 107
two-count variables, 195–197

two-sample Poisson, 198–201
two-sample proportion, power analysis for, 184–187
  example, 181
  MINITAB printout, 184
  sample sizes, 184
two-sample *t*-test, use of MINITAB
  dialog box, 169, 171
  options box, 171
  for power analysis for, 170–172
  printout, 171
  smaller effect, 172
two variances, 184–191
  MINITAB, 191–193
  power analysis for, 193–195
types, 93
Statistical significance, 201
Statistics, 4
  definition of, 1
  descriptive, 3–4
  graphical methods, 2
  inferential, 4
Stem-and-leaf plots, 21–22
  example, 21
  leaf creation, 21, 23, 24
  MINITAB use, 22–24
  purpose, 21
  stem, 23
Stratified random sample, 13
Studies
  experimental, 5
  observational, 5, 6
  types of, 5–6
Summary tables and graphical displays, 2
Sum of squares for the treatment (SSTR), 346
Sum of the squares of the residual errors (SSE), 219

T

t-distribution, 77, 94, 98, 298
Test(s)
  Bartlett's test, 350
  distribution-free tests, 385
  F-test, 292, 296
  hypothesis test, 101, 157
  Kruskal–Wallis test, 400–405
    χ^2 distribution, 403
    example, 402
    MINITAB use, 405–410
    null and alternative hypotheses, 402
    ranking of data, 400
    rule of thumb, 405

Test(s) (*cont.*)
    test statistic, 403, 404
  left-tailed test, 103
  Levene's test, 352, 409
  Mann–Whitney test, 395–400
    example, 395
    null and alternative hypotheses, 395
    ranking of data, 395
    test statistic, 396
  one-sample *t*-test, 100–106
    alternative hypothesis, 101, 103
    decision process, 102
    error types, 103
    graph, 102
    left-tailed test, 103
    level of significance, 102
    mean of random sample, 101
    null hypothesis, 101, 103
    power analysis for, 115–116
    rejection region, 102, 105–106
    right-tailed test, 103
    sampling error, 101
    significance level, 103
    two-tailed test, 103
    value of test statistic, 105–106
  power of, definition of, 115
  right-tailed test, 103
  Ryan–Joiner test, 266–268, 355, 361
    Kruskal–Wallis test, 410
    normality assumption, 266–268
    one-way ANOVAs, 361, 362, 369
    two-way analysis of variance, 434, 434, 438
    Wilcoxon signed-rank test, 392, 393
  two-sample *t*-test, use of MINITAB for
    power analysis for, 170–172
    dialog box, 169, 171
    options box, 171
    printout, 172
    smaller effect, 172
  two-tailed test, 103
  Wilcoxon signed-rank test, 385–389
    assumption, 386
    example, 386
    MINITAB use, 389–395
    null and alternative hypotheses, 431
    population median, 386
    p-value, 388, 392
    ranking of observations, 386
    rejection region, 388
    sample size, 386
    symmetric distribution, 386
Time series analysis, basic, 440–449
  chronological order of data, 440

cyclical trend, 441
  example, 440
  linear forecast model, 444
  long-term trend, 441
  mean absolute deviation, 447
  mean absolute percentage error, 445
  mean squared deviation, 448
  MINITAB commands, 441
  MINITAB dialog box, 442, 443, 445, 448
  MINITAB output, 443
  MINITAB quadratic trend analysis, 448
  random trend, 441
  regression analysis, 443–444
  response variable, 444
  seasonal trend, 441
  time series plot, 440, 445
  trend analysis graph, 446
  trends, 441
Treatment group, 5
Trimmed mean, 90
t-test, 172–176, 395
  one-sample, 106–114
    dialog box, 107
    example, 109–113
    options box, 107
    printout, 107
    p-value, 107
    rejection region, 108, 111
    sample mean, 110
    sample standard deviation, 110
    unknown population, 110
    value of test statistic, 112
  two-sample, use of MINITAB for power
    analysis for, 170–172
    dialog box, 169, 171
    options box, 171
    printout, 171
    smaller effect, 172
Two-count variables, confidence intervals and
    hypothesis tests for, 195–197
Two-sample Poisson, 198–201
  MINITAB
    commands, 198
    dialog box, 199
    option box, 198
    printout, 200
   power analysis for, 199
Two-sample variances, power analysis for,
    193–195
Two-tailed hypothesis test, 103, 149–150
Two variances
  confidence intervals and hypothesis tests
    for, 184–191

example, 188
MINITAB use for testing two sample
    variances, 191–193
    dialog box, 192
    options box, 192
    printout, 199
Two-way analysis of variance, 417–424
    calculation of mean squares, 419
    column factor, 417, 420
    example, 417
    exercises, 449–452
    F-distribution, 418
    interaction between factors, 418
    MINITAB, 424–440
        box plot, 432, 438
        commands, 425, 429
        dialog box, 425, 427, 429
        example, 431–432
        four-way residual plots, 437
        histogram of residuals, 434
        interaction plot, 426, 436, 439
        main effects plot, 435, 439
        normal probability plot of residuals, 428
        printout, 432, 436
        residual plots, 434
        residual *versus* fitted values, 359
        Ryan–Joiner test, 434, 435, 438
    rejection region, 422, 423
    row factor, 417
    SSInteraction, 421
    test statistics, 418
Type I error, 103

U

Unknown population
    linear equation, 216, 219–220
    mean, confidence interval, 95
    one-sample *t*-test, 110
    parameters
        equation of line of best, 221
        standard errors for estimated regression
            parameters, 227
    standard deviation, 131
    variance, 189
Unusual observation, outliers, 268, 276
Upper and lower limits, box plots, 27–28

V

Variable(s), 9
    binary, 314
    categorical, 24

cause-and-effect relationship between, 214
continuous, 6, 24
control, 285
dependent, 32, 33, 214
discrete, 6
independent, 32, 33, 214
indicator, 314
interval, 8
mathematical properties, 9
nominal, 7
numeric, median of, 50
ordinal, 7
predictor, 32, 214, 228
    categorical, 313–314
    confidence intervals for mean response
        for specific value of, 231–232
    multicollinearity, 300
    multiple regression analysis, 286
    prediction intervals for response for
        specific value of, 233–235
    variance inflation factors, 302
random, 58–60
    continuous, 64
    discrete variable, 58
    mean, 61
    probability distribution, 57, 60
    representation, 58
    summarized, 59
    t-distribution, 94
ratio, 8
relationship between, 32
response, 32, 214
sample, linear trend between, 250
scales of, 7–9
straight line model, 214
types of, 6–7
Variables, graphing of, 15–46
    bar charts, 24
        discrete data, 24
        MINITAB use, 24–25
        variables, 24
    box plots, 27–31
        construction, 27, 28
        example, 28–30
        general form, 28
        median, 29
        MINITAB use, 31
        multiple, 31
        quartiles, 27
        upper and lower limits, 27–28
        whiskers, 27, 30
    frequency distribution, histogram drawn
        from, 15

Variables, graphing of (*cont.*)
  histograms, 15–17
    construction, 16
    frequency distribution, histogram drawn
      from, 15
    MINITAB use, 17–21
    purpose of drawing, 16
    software use, 17
  marginal plots, 33–35
  scatter plots, 32–33
    Cartesian plane, 33
    data set, 33
    example, 32–33
    MINITAB use, 33
    predictor, 32–33
    response, 32
    variables, 32
  stem-and-leaf plots, 21–22
    example, 21
    MINITAB use, 22–24
    purpose, 21
    stem, 23
Variance inflation factor (VIF), 302–303
  calculation, 302
  example, 302
  predictor variables, 302
  printout, 303

use of MINITAB to calculate, 303–304
value too high, 304
VIF, *see* Variance inflation factor

W

Weighted mean, 89
Wilcoxon signed-rank test, 385–389
  assumption, 386
  example, 386
  MINITAB use, 389–395
    commands, 390
    dialog box, 389
    example, 392
    histogram, 385, 391
    one-sample *t*-test, 393
    printout, 391, 393
    results, 394
    Ryan–Joiner test, 392
  null and alternative hypotheses, 431
  population median, 386
  p-value, 388, 392
  ranking of observations, 386
  rejection region, 388
  sample size, 386
  symmetric distribution, 386

Printed in the United States
by Baker & Taylor Publisher Services